T0345043

Biotechnology and Polymers

Biotechnology and Polymers

Edited by
Charles G. Gebelein
Youngstown State University
Youngstown, Ohio

Springer Science+Business Media, LLC

Library of Congress Cataloging-in-Publication Data

American Chemical Society Symposium on Polymers from Biotechnology
  (1990 : Boston, Mass.)
    Biotechnology and polymers / edited by Charles G. Gebelein.
       p.   cm.
    "Proceedings of an American Chemical Society Symposium on Polymers
  from Biotechnology, held April 22-23, 1990, in Boston,
  Massachusetts"--T.p. verso.
    Includes bibliographical references and index.
    ISBN 978-0-306-44049-6   ISBN 978-1-4615-3844-8 (eBook)
    DOI 10.1007/978-1-4615-3844-8
    1. Polymers--Biotechnology--Congresses.   I. Gebelein, Charles G.
  II. American Chemical Society.   III. Title.
  TP248.65.P62A42   1990
  668.9--dc20   91-28416
                                                              CIP

Proceedings of an American Chemical Society Symposium on
Polymers from Biotechnology, held April 22–23, 1990,
in Boston, Massachusetts

ISBN 978-0-306-44049-6

© 1991 Springer Science+Business Media New York
Originally published by Plenum Press, New York in 1991

PREFACE

The term biotechnology has emerged on the contemporary scene fairly recently, but the basic concept of utilizing natural materials, either directly or in modified versions, dates back to antiquity. If we search the ancient literature, such as the Bible, we find hundreds of examples wherein people employed, or modified, natural materials for a variety of important uses. As far back as the days of Noah we find pitch, a natural material, being used as a caulk. Clothing was made from animal skins and the products of several plants. Today, we would consider these things as important biotechnological developments.

Likewise, the human use of polymeric materials also has a long history. In fact, many of the original materials used by mankind were polymers derived from nature, such as wood, flax, cotton, wool and animal skins, which were used for shelter and clothing.

In recent years, however, the concept of biotechnology has taken on a new and renewed role in our society. This is due to a combination of factors, including an increased interest in environmental concerns and the desire to break free from the stranglehold that petrochemicals have placed on our society. If we can manufacture some of our polymers from renewable resources, then we can expect to prepare them for many more years into the future than we might if we could only depend on the petrochemical resources. Additionally, many naturally derived polymers are biodegradable, which could aid in alleviating some of our current landfill problems. In the area of drug administration, natural polymers might prove to be more effective, and biodegradable, drug carriers than the synthetic polymers now being used for this purpose.

This book considers several applications of biotechnology in the field of polymers. For convenience, we've subdivided this book into three general areas: Section I. NOVEL BIOTECHNOLOGY-DERIVED POLYMERS, Section II. POLYSACCHARIDE BASED SYSTEMS, and Section III. PROTEIN AND ENZYME BASED SYSTEMS. Admittedly there is much overlap between these subdivisions, but it makes the overall thrust of the book more apparent.

The first part of this book deals with a wide variety of biotechnology-derived polymers, and applications. The opening Chapter (Gebelein) sets forth the overall thesis of this book - that we can obtain hundreds of different types of polymers from biotechnology, a theme which is expanded and expounded in subsequent Chapters. The second Chapter (Levy & Salazar) describes how an artificial nucleic acid might be used to treat AIDS. Some further examples, and uses, of synthetic nucleic acid analogs are shown in the next Chapter (Takemoto et al.). Some natural, biodegradable, poly(esters) are the subject of the next two Chapters (Marchessault & Monasterios, and Dave et al.). The emphasis then shifts to some novel polymers which can be prepared from biotechnology: polyesterimides (Ghosh) and polyamides (Giannos et al.). The following pair of papers (Dirlikov et al. and Sperling et al.) considers various aspects of an epoxy monomer obtained from vernonia, and some other natural, functional triglycerides. The final paper in this Section considers tin-modified lignin (Carraher, et al.).

Section II is devoted exclusively to polysaccharide systems, but includes a fairly broad variety of polymers and applications. Kennedy et al. describe some industrially important polysaccharides, while Morris covers various bacterial-derived polysaccharides used in the food and agricultural industries. Carraher et al. describe some tin modified dextrans in the third Chapter of this Section. Some acidic, heparin-like, polysaccharides are described in the following Chapter (Linhardt, et al.); then Kobayashi reveals some cellulose derivatives which form liposomes and membranes. The final three papers of this Section deal with chitin and chitosan systems. Alkaline chitosan gels are presented by Hirano et al. Kikuchi and Kubota then describe some polyelectrolytes derived from chitosan derivatives. Finally, Seo and Iijima consider the sorption behavior of chitosan gels.

The third Section of this book concentrates on biotechnological polymers derived from, or based upon, proteins or enzymes. In the first paper (Rzepecki and Waite) cover basic research on bioadhesives, while the second paper (Masilamani et al.) describes an industrially produced bioadhesive stemming from this research. In the third Chapter of this Section, Urry et al. detail a synthetic, elastin-like, poly(pentapeptide) which functions as a reversible thermoplastic. The pair of papers by Yannas et al. and Li et al. treats nerve conduits and nerve regeneration using polymers based largely on collagen. Protein modifications designed to reduce antigenicity are described in the Chapter by Nitecki and Aldwin. Albert et al. then describe some semisynthetic enzymes prepared by the conformational modification of some proteins. Next, Wu and Hilvert consider a semisynthetic selanoenzyme, selenolsubtilisin. Finally, Hayashi and Ikada discuss spacer effects on the enzymatic activity of polymer-immobilized enzymes.

This book is derived, in part, from a Symposium held in Boston, April 22-23, 1990, sponsored by the Polymeric Materials: Science and Engineering Division of the American Chemical Society). Acknowledgment is made to the Donors of The Petroleum Research Fund, administered by the American Chemical Society, for partial support of the symposium from which this book is derived. Additional financial support was provided by the Polymeric Materials Division, Sci-Tec Symposium Associates, Inc., Allied-Signal, Inc., Rohm & Haas Company, and Anatrace, Inc. The Editor wishes to thank the individual authors for their fine contributions to this book. The book was typeset by CG Enterprises.

Charles G. Gebelein
Department of Chemistry
Youngstown State University
Youngstown, OH 44555

CONTENTS

II. POLYSACCHARIDE BASED SYSTEMS

III. PROTEIN AND ENZYME BASED SYSTEMS

NEW AND TRADITIONAL POLYMERS FROM BIOTECHNOLOGY

Charles G. Gebelein

Department of Chemistry
Youngstown State University
Youngstown, OH 44555

Biotechnology can be defined as the use, modification or mimicing of naturally-occurring materials. In this context, several types of biotechnology-derived polymers exist, although they're not always in mass production. This paper will summarize some pertinent information for certain biotechnology derived polymers, such as wool, cotton, silk, rubber casein and the cellulosics. In addition, some potential new polymers, and polymer applications, from biotechnology are briefly discussed.

INTRODUCTION

Traditionally, the earliest polymers used by mankind came from what we today call biotechnology. These polymers include cotton, wool, paper, rubber, leather, the cellulosic derivatives, and casein plastics or adhesives. We must, however, define our use of the term biotechnology because the term means many different things to various people. In our present context, we define biotechnology as the use, modification or mimicing of naturally-occurring materials. Although some seem to consider biotechnology as merely another aspect of biology, this newly emerging methodology is actually a new merger of chemistry and biology on a more grandiose scale than in biochemistry. As we have become more aware of the chemical basis and nature of biological systems, we have been able to adapt these biosystems in improved ways for our use and benefit. This new marriage is clearly evident in several recent biotechnology books.[1-3]

TRADITIONAL BIOTECHNOLOGICAL POLYMERS

A review of polymer history is beyond the scope of this paper and several such reviews are available.[4,5] Most early polymer applications were modifications of poly(peptides) or poly(saccharides), although some other classes of natural polymers found utility. For example, shellac is a natural poly(ester) derived from from the excretions of the insect *Laccifer lacca*. While these materials were not recognized as polymers until

Biotechnology and Polymers, Edited by C.G. Gebelein
Plenum Press, New York, 1991

comparatively recent times, these natural polymers were valuable resources.

The tanning process appears to be the first example of deliberate polymer modification to improve the properties, although the tanners had essentially no understanding of the chemistry involved. Likewise, the basic vulcanization process was developed without much understanding of the crosslinking reaction, or even the macromolecular nature of rubber.

Many additional modifications of natural polymers became the basis of large industries. These included the cellulosic systems, natural rubber and casein. Again, essentially all of this progress was made with negligible chemical or polymer theory or background. Thus, cellulose, Figure 1, was nitrated, acetylated, or esterified without recognizing that the raw material was macromolecular in nature. Nevertheless, these modifications became useful plastics and coatings for a wide variety of uses. Several cellulosics are still commercially important materials, and various modified cellulosics are made in tonnage quantities each year. These include cellulose acetate and the various forms of rayon. Sodium carboxymethylcellulose (CMC) is used in most synthetic laundry detergent formulations, as a soil redeposition preventative. Cellulosic derivatives, including CMC, various ethers, and other modifications, are widely used in cosmetic products.[6] In addition, cellulose itself can be converted into a very large number of low molecular weight materials via acid or enzyme hydrolysis, or into sugars via fermentation.[7] In a sense, wood, and its applications, could be considered as a polymer usage, although we seldom do.

In a similar manner, starch was used in the coatings industry long before the structure of amylose (Figure 2) was delineated. Starch still finds widespread usage in various coatings for paper and textiles, partly because it's considerably less expensive that the synthetic polymers. These uses are, of course, dwarfed by the consumption of starches in foods.

The basic thrust of the polymer industry began changing in the early 1900s, when chemists began to make commercial scale synthetic polymers. Although the first synthetic polymers were made in the late 1830s, they

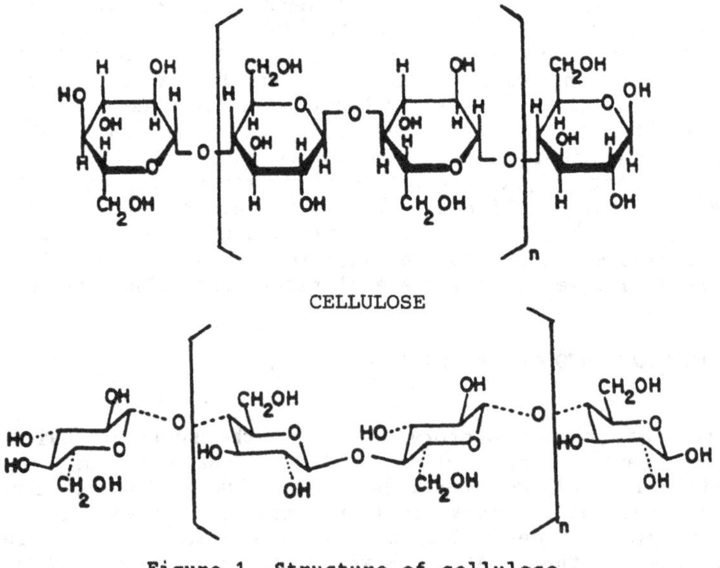

Figure 1. Structure of cellulose.

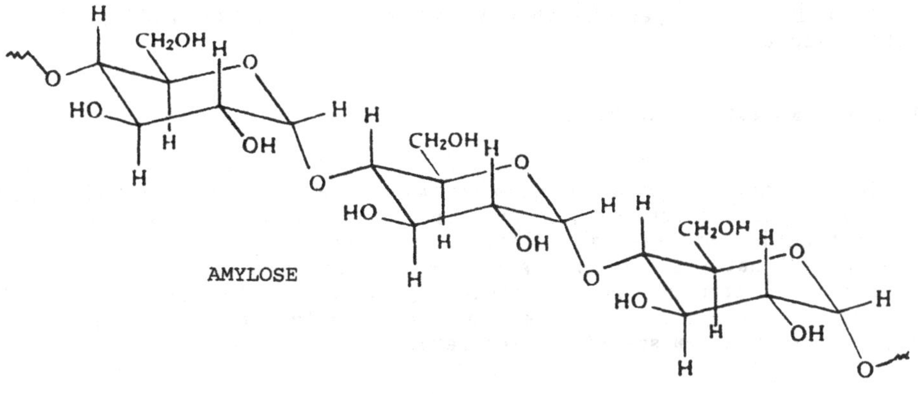

Figure 2. Structure of amylose.

remained laboratory curiosities, or annoyances. A few synthetic polymers did, however, achieve success in the coatings field where monomeric drying oils were oxidatively polymerized into paint films. In a sense, these were the first polymers derived from monomeric materials, rather than the phenolics, but the latter still remain as the first totally synthetic polymers made on a commercial scale.

The success of the synthetic polymers, coupled with the readily available raw materials derived from petroleum or coal, led to a decreased emphasis on natural polymers. In addition, the simpler nature of the synthetic polymer molecules made them easier to understand from a theoretical standpoint. As we learned later, many natural polymers are extraordinarily complex in both structure and synthesis. It was much easier to deduce structure-property relationships with the synthetic polymers. A third factor shaping this synthetic polymer revolution was the broad range of properties which could be achieved - something seemingly beyond the reach of the natural polymers, at that time.

Several new factors have infiltrated our considerations in recent years which could redirect our attention back toward the natural systems. These include (1) biodegradability and (2) long-range renewable resource considerations. The natural resources of petroleum and coal clearly are not unlimited and many additional chemical materials, such as medicines, depend on this supply, in addition to fuel utilization. While biological in origin, petroleum and coal are not readily renewable resources. Additionally, most petroleum based polymers either do not biodegrade at all, or show very limited degradation, posing potentially major disposal problems. Recycling can partially alleviate this difficulty, but it cannot solve these problems completely. The fact that most natural polymers biodegrade provides another impetus for the re-examination of these materials.

This does not mean we will see a mega-ton return to the old style polymers, such as casein plastics, cellulose nitrate and cellulose acetate. Many of these older polymers have severe deficits. For example, wool is eaten by moths and other insects; cotton shrinks and does not hold a crease, unless treated with another polymer; cellulose acetate is not solvent resistant, and cellulose nitrate is highly flammable. However, these older polymers come from renewable resources, which are also biodegradable, and this is a virtue in today's throw-away society. This alone should resurrect interest in natural polymers. Additionally, we have learned many vital things in the past century which will enable us to develop new and better polymers from biotechnology - polymers which

3

may have better properties than those we now prepare via synthetic methods alone.

NEW POLYMERS FROM BIOTECHNOLOGY

We are now discovering a expansive spectrum of new polymeric materials with properties which are superior to those achievable with the "old synthetic" systems, in addition to a renewed interest in the more conventional natural polymers. A full review of these new polymeric systems, derived from or based on natural polymers, is beyond the scope of this paper; some are illustrated in other chapters of this book. We'll only highlight a few specific cases here.

POLY(SACCHARIDES)

Various medical or pharmaceutical agents are routinely used to treat abnormal skin conditions, but this same objective has been a mainstay of the cosmetic industry for decades. Recently, some presumed inter-active cosmetics have appeared on the market, including derivatives of retinoic acid. Several polymers of the poly(saccharide) class have shown some emollient behavior, and the Japanese company Shiseido is marketing the sodium salt of hyaluronic acid, a natural poly(saccharide), Figure 3, as a skin treatment cosmetic.[8] Part of the cosmetic utility of hyaluronic acid probably arises from its hygroscopic nature, enabling it to impart moisture to the skin. Hyaluronic acid does possess good skin adhesive characteristics, and this would aid in its cosmetic utility.

Hyaluronic acid naturally occurs in the connective tissues of most vertebrate animals and can be isolated from such scrap material as rooster combs. Medically, hyaluronic acid has been used in the treatment of arthritis and in eye surgery. In both cases, the poly(saccharide) achieves its special functions because it produces highly viscous solutions. Hyaluronic acid has also been used as a polymeric matrix in controlled release applications, where it behaves, at least partially, as a hydrogel.[9,10]

Chitin, a poly(saccharide) closely related to cellulose and shown in Figure 4, is being studied by many research groups for a wide variety of biomedical, agricultural and cosmetic applications.[11-13] Chitin is found mainly in insect and crustacean shells. Most current research centers on the deacetylated chitin, which is called chitosan, Figure 5. Chitosan is now finding some new uses in the textile industry, waste water treatment and medicine.[14] While neither material is likely to be made *synthetically* on a commercial scale, both polymers are derived from formerly useless

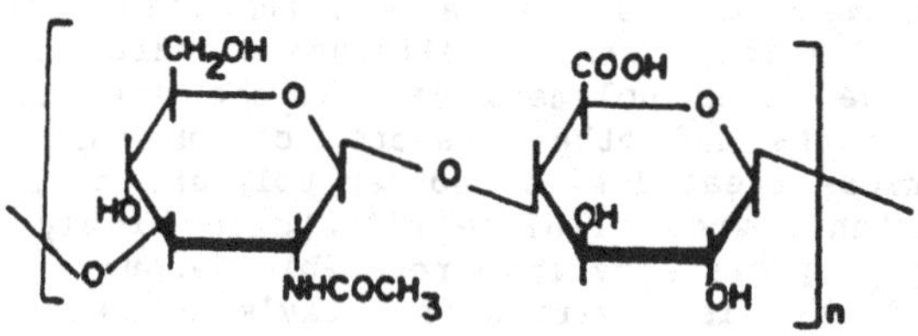

HYALURONIC ACID

Figure 3. Structure of hyaluronic acid.

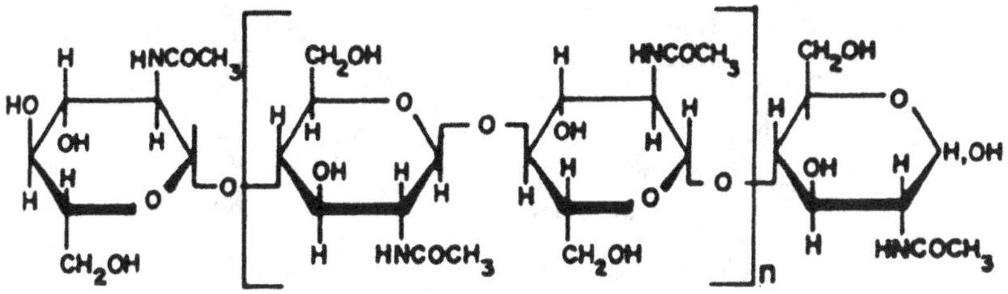

CHITIN

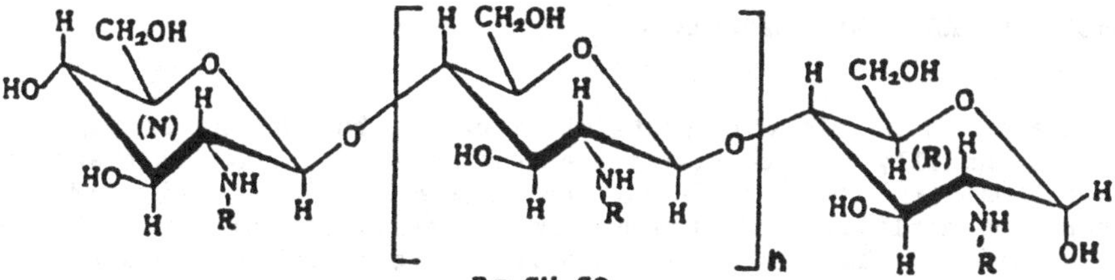

R= CH₃CO-

Figure 4. Structure of chitin.

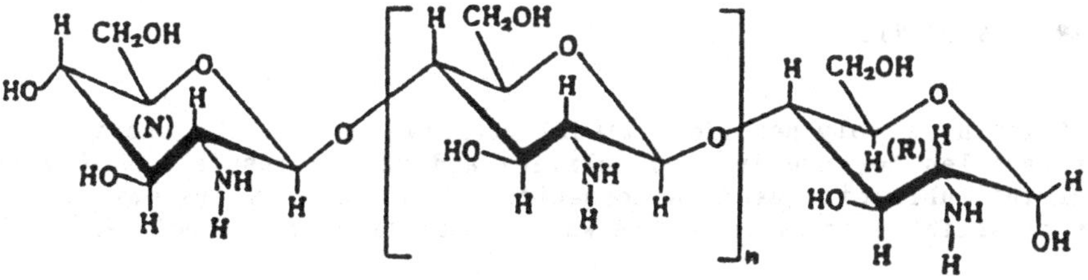

CHITOSAN

Figure 5. Structure of chitosan.

waste materials. Several chitin or chitosan derived materials are currently marketed in Japan.

Cosmetic uses of chitin and chitosan, and derivatives are comparatively new, but are growing in importance. These include the use of chitin derivatives in various hair and skin care formulations. Chitin derivatives also are being used for artificial skin replacements (wound dressings).[15,16]

Other poly(saccharides) find practical applications. For example, konjac and carrageenan have numerous cosmetic and medical uses. Konjac is used in controlled release applications and as a barrier film to protect wounds. Carrageenans are used in dentifrices, hand lotions, shaving creams, shampoos and controlled release systems.[17] Poly(saccharides) derived from fungi have been shown to possess anti-tumor properties,[18] and some recent research suggests curdlan sulfate may have antiviral activity, including potential anti-AIDS activity.[19] The topic of biologically active poly(saccharides) has been reviewed recently.[20]

Another poly(saccharide) which finds extensive biomedical usage is heparin, which has been extensively reviewed recently,[21] and continues as the subject of much research, including its use in blood substitutes.[22]

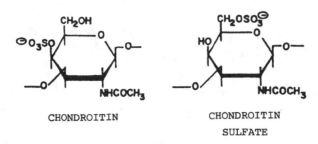

CHONDROITIN

CHONDROITIN
SULFATE

Figure 6. Structure of chrondroitin and chrondroitin sulfate.

COLLAGEN AND RELATED BIOPOLYMERS

Collagen is an old poly(peptide) material, but dozens of papers are published annually on potential biomedical uses, including skin and nerve regeneration[23-26] and artificial blood vessels.[27] The skin and nerve regeneration applications often utilize a matrix of collagen and chrondroitin sulfate, which is shown below in Figure 6. Collagen has also been used to immobilize percutaneous implants.[28] Collagen has been used in controlled release applications as well.[29] For many decades, collagen has been used in cosmetic surgery applications, such as treating the smallpox scars and correcting facial defects.

OTHER POLY(PEPTIDES)

Gelatin, a poly(peptide), finds extensive use in the preparation of microcapsules for use in drug delivery and the encapsulation of various water-insoluble oils used in cosmetics. These gelatin systems are normally crosslinked to overcome the water solubility of the polymer.

Poly(elastin), another poly(peptide), is also being examined in both the medical and controlled release areas.[30-32] Although this material is relatively new, and of synthetic origin, it could probably find utility in cosmetics. Other important poly(peptide) applications include their use as drug carriers,[33] and biodegradable drug delivery systems.[34]

Enzymes are well know poly(peptides) with very high catalytic activity. Several groups are synthesizing modified proteins with "artificial" catalytic activity.[35] More recently, catalytic antibodies, which are also poly(peptide) in nature, have been developed.[36] In addition, the concept of totally artificial enzymes has been pursued vigorously by several groups; this has been reviewed recently.[37] Research also continues on immobilized enzyme systems.[38]

Another special type of poly(peptide), the enkephalins and endorphins, are powerful analgesics. Many modifications have been made on these natural polymers and some may eventually be marketed as drugs.[39-41] Many other poly(peptides) have been studied as potential pharmaceutical agents, and this topic has been reviewed recently.[42]

PROTEIN ADHESIVES

In recent studies, J. H. Waite determined the structure of the polymer used by mussels to adhere to boat hulls.[43] This polymer has now been synthesized and appears likely to become an excellent adhesive with a biotechnology origin.[44,45] Additional work in this area is covered in later chapters of this book (Rzepecki & Waite; D. Masilamani, et al.).

NUCLEIC ACID ANALOGS

A considerable amount of research has been conducted on synthetic versions of the nucleic acids. These have included poly(alkylene phosphates) and related species,[46-49] and vinyl or (meth)acryloyl derivatives of the nucleic bases.[50-4] Some polymers of nucleic bases also act as immune system activators.[55]

Part of our research program has centered on nucleic acid analogs, in particular, using 5-fluorouracil and 6-methylthiopurine units in place of the usual nucleic bases. These compounds are known anti-tumor drugs. Accordingly, we have synthesized methacrylate analogs of these imitation nucleic bases and have studied their polymerization and the controlled release of the anti-tumor agents. We have found that these materials give the interesting, and potentially valuable, zero-order release of these anti-tumor drug agent.[56-8]

OTHER POLYMER SYSTEMS

Many other natural polymers are being studied for potential practical applications. One unusual system is melanin, the natural polymer which forms the basis of hair and skin coloration. This polymer is being explored as a potential sunscreening agent.[59]

Two recent reviews summarize the potential applications of several other natural and modified natural polymers.[60,61] The other papers in this book explore additional facets of this newly emerging domain of biotechnological polymers.

REFERENCES

1. John E. Smith, "*Biotechnology*," 2nd Edition, Edward Arnold, Baltimore, MD, 1988.
2. J. M. Walker & E. B. Ringold, Eds., "*Molecular Biology & Biotechnology*," 2nd Edition, Royal Society of Chemistry, London, 1988.
3. Mary L. Good, Ed., "*Biotechnology and Materials Science*," American Chemical Society, Washington, DC, 1988.
4. R. B. Seymour & C. E. Carraher, Jr., "*Polymer Chemistry*," 2nd Edition, Dekker, New York, 1988, Chapter 1.
5. R. B. Seymour, Ed., "*Pioneers in Polymer Science*," Kluwer Academic Publ., Boston, 1989.
6. J. R. Conklin, Proc. Poly. Mater. Sci. Eng., 63, 233-42 (1990).
7. J. F. Kennedy & E. H. M. Melo, Prit. Polymer J., 23, 193-8 (1990).
8. Literature from Shiseido Company.
9. N. E. Larsen, E. A. Leshchiner, E. G. Parent and E. A. Balazs, Proc. Poly. Mater. Sci. Eng., 63, 341-3 (1990).
10. E. A. Balazs, Proc. Poly. Mater. Sci. Eng., 63, 689-91 (1990).
11. S. Hirano, Y. Noishiki, J. Kinugawa, H. Higashijima & T. Hayashi in: "*Advances in Biomedical Polymers*," C. G. Gebelein, Ed., Plenum Press, New York, 1985, pp. 285-297.

12. A. Kato & T. Kondo in: "*Advances in Biomedical Polymers*," C. G. Gebelein, Ed., Plenum Press, New York, 1985, pp. 299-310.
13. S. Hirano, M. Hayashi, K. Murae, H. Tsuchida & T. Nishida in: "*Applied Bioactive Polymeric Materials*," C. G. Gebelein, C. E. Carraher, Jr. and V. R. Foster, Eds., Plenum Press, New York, 1988, pp. 45-59.
14. H. Struszczyk & O. Kivekäs, Brit. Polymer J., **23**, 261-5 (1990).
15. S. Hirano, H. Seino, Y. Akiyama and I. Nonaka, in: "*Progress in Biomedical Polymers*," C. G. Gebelein & R. L. Dunn, Eds., Plenum, New York, 1990, pp. 283-90.
16. S. Hirano, K. Hirochi, K. Hayashi, T. Mikami and H. Tachibana, Proc. Poly. Mater. Sci. Eng., **63**, 699-703 (1990).
17. R. J. Tye, Proc. Poly. Mater. Sci. Eng., **63**, 229-232 (1990).
18. K. Matsuzaki, I. Yamamoto, K. Enomoto, Y. Kaneko, T. Mimura & T. Shiio in: "*Applied Bioactive Polymeric Materials*", C. G. Gebelein, C. E. Carraher, Jr. and V. R. Foster, Eds., Plenum Press, New York, 1988, pp. 165-174.
19. I. Yamamoto, K. Takayama, T. Gonda, K. Matsuzaki, K. Hatanaka, T. Yoshida, T. Uryu, O. Yoshida, H. Nakashima, N. Yamamoto, Y. Kaneko & T. Mimura, Brit. Polymer J., **23**, 245-50(1990).
20. C. Schuerch, in: "*Bioactive Polymeric Systems*," C. G. Gebelein & C. E. Carraher, Jr., Eds., Plenum Press, New York, 1985, pp. 365-386.
21. R. J. Linhardt and D. Loganathan, in: "*Biomimetic Polymers*," C. G. Gebelein, Ed., Plenum, New York, 1990, pp. 135-173.
22. F. Prouchayret, F. Bonneaux, M. Leonard, D. Sacco & E. Dellacherie, Brit. Polymer J., **23**, 251-6 (1990).
23. I. V. Yannas, J. F. Burke, D. P. Orgill & E. M. Skrabut in: "*Polymeric Materials and Artificial Organs*," C. G. Gebelein, Ed., ACS Symposium Series #256, Washington, DC, 1984, pp. 191-197.
24. I. V. Yannas, D. P. Orgill, J. Silver, T. V. Norregaard, N. T. Zervas & W. C. Schoene, in: "*Advances in Biomedical Polymers*," C. G. Gebelein, Ed., Plenum Press, New York, 1985, pp. 1-15.
25. I. V. Yannas, E. Lee & M. D. Bentz, in: "*Applied Bioactive Polymeric Materials*," C. G. Gebelein, C. E. Carraher, Jr. and V. R. Foster, Eds., Plenum Press, New York, 1988, pp. 313-318.
26. A. Chang, I. V. Yannas, S. Perutz, H. Loree, R. R. Sethi, C. Krarup, T. V. Norregaard, N. T. Zervas and J. Silver, in: "*Progress in Biomedical Polymers*," C. G. Gebelein & R. L. Dunn, Eds., Plenum, New York, 1990, pp. 107-120.
27. S.-T. Li in: "*Advances in Biomedical Polymers*," C. G. Gebelein, Ed., Plenum Press, New York, 1985, pp. 171-183.
28. T. Okada and Y. Ikada, in: "*Progress in Biomedical Polymers*," C. G. Gebelein & R. L. Dunn, Eds., Plenum, New York, 1990, pp. 97-105.
29. A. Stemberger, M. Unkauf, D. E. Arnold and G. Blümel, Proc. Poly. Mater. Sci. Eng., **63**, 344-346 (1990).
30. D. W. Urry, R. D. Harris, H. Sugano, M. M. Long & K. U. Prasad in: "*Advances in Biomedical Polymers*," C. G. Gebelein, Ed., Plenum Press, New York, 1985, pp. 335-354.
31. D. W. Urry, Poly. Mater. Sci. Eng., **62**, 587-593 (1990).
32. D. W. Urry, J. Jaggard, R. D. Haris, D. K. Chang and K. U. Prasad, in: "*Progress in Biomedical Polymers*," C. G. Gebelein & R. L. Dunn, Eds., Plenum, New York, 1990, pp. 171-178.
33. J. M. Whiteley in: "*Bioactive Polymeric Systems*," C. G. Gebelein & C. E. Carraher, Jr., Eds., Plenum Press, New York, 1985, pp. 345-363.
34. R. V. Petersen in: "*Bioactive Polymeric Systems*," C. G. Gebelein & C. E. Carraher, Jr., Eds., Plenum Press, New York, 1985, pp. 151-177.
35. M. H. Keyes and D. E. Albert, in: "*Biomimetic Polymers*," C. G. Gebelein, Ed., Plenum, New York, 1990, pp. 115-133.
36. D. Hilvert, in: "*Biomimetic Polymers*," C. G. Gebelein, Ed., Plenum, New York, 1990, pp. 95-113.
37. Y. Imanishi, in: *Bioactive Polymeric Systems*," C. G. Gebelein & C. E.

Carraher, Jr., Eds., Plenum, New York, 1985, pp. 435-511.
38. M. K. Keyes & S. Sarawathi, in: *Bioactive Polymeric Systems*," C. G. Gebelein & C. E. Carraher, Jr., Eds., Plenum, New York, 1985, pp. 249-78.
39. N. R. Plotnikoff, R. E. Faith, A. J. Murgo & R. A. Good, Eds., *"Enkephalins and Endorphins, Stress and the Immune System*," Plenum Press, New York, 1986.
40. Gavric W. Pasternak, Ed., *"The Opiate Receptors*," Humana Press, Clifton, NJ, 1988.
41. S. H. Snyder, *"Brainstorming. The Science and Politics of Opiate Research*," Harvard University Press, Cambridge, MA, 1989.
42. J. Samanen, in: *Bioactive Polymeric Systems*," C. G. Gebelein & C. E. Carraher, Jr., Eds., Plenum, New York, 1985, pp. 279-344.
43. J. H. Waite, Int. J. Adhesion & Adhesives, 7, 9-14 (1987).
44. L. M. Rzepecki & J. H. Waite, Proc. Poly. Mater. Sci. Eng., 62, 571-4 (1990).
45. D. Masilamani, A. J. Salerno, H. R. Bhattacharjee, I. Goldberg, P. D. Unger & M. Oleksiuk, Proc. Poly. Mater. Sci. Eng., 62, 558-70 (1990).
46. S. Penczek and P. Klosinski, in: *"Progress in Biomedical Polymers*," C. G. Gebelein & R. L. Dunn, Eds., Plenum, New York, 1990, pp. 291-308.
47. S. Penczek and P. Klosinski, in: *"Biomimetic Polymers*," C. G. Gebelein, Ed., Plenum, New York, 1990, pp. 223-241.
48. S. Penczek and P. Klosinski, in: *"Biomimetic Polymers*," C. G. Gebelein, Ed., Plenum, New York, 1990, pp. 243-252.
49. S. Penczek, J. Baran, T. Biela, G. Lapienis, A. Nyk, P. Klosinski & B. Pretula, Brit. Polymer J., 23, 213-20 (1990).
50. K. Takemoto, in: *Bioactive Polymeric Systems*," C. G. Gebelein & C. E. Carraher, Jr., Eds., Plenum, New York, 1985, pp. 417-33.
51. K. Takemoto, J. Macromol. Sci., Rev. C5, 29-102 (1970).
52. K. Takemoto, in: *"Polymeric Drugs*," L. G. Donaruma & O. Vogl, Eds., Academic Press, New York, 1978, pp. 103-129.
53. Y. Inaki, T. Ishiwaka & K. Takemoto, Adv. Polym. Sci., 41, 1-51 (1981).
54. K. Takemoto, E. Mochizuki, T. Wada and Y. Inaki, in: *"Biomimetic Polymers*," C. G. Gebelein, Ed., Plenum, New York, 1990, pp. 253-267.
55. H. Levy & T. Quinn, in: *Bioactive Polymeric Systems*," C. G. Gebelein & C. E. Carraher, Jr., Eds., Plenum, New York, 1985, pp. 387-415.
56. C. G. Gebelein, T. Mirza & R. R. Hartsough in: *"Controlled Release Technology*", P. I. Lee & W. R. Good, Eds., ACS Symposium Series 348, 1987, pp. 120-126.
57. C. G. Gebelein, M. Chapman & T. Mirza in: *"Applied Bioactive Polymeric Materials*", C. G. Gebelein, C. E. Carraher, Jr. and V. R. Foster, Eds., Plenum Press, New York, 1988, pp. 151-163.
58. C. G. Gebelein, M. Davison, T. Gober & M. Chapman, Polym. Mater. Sci. Eng. Proc., 59, 798-802, (1988).
59. S. Nacht, Proc. Poly. Mater. Sci. Eng., 63, 600 (1990).
60. C. G. Gebelein, in: *Bioactive Polymeric Systems*," C. G. Gebelein & C. E. Carraher, Jr., Eds., Plenum, New York, 1985, pp. 1-15.
61. C. G. Gebelein, C. E. Carraher, Jr. & V. R. Foster in: *"Applied Bioactive Polymeric Materials*", C. G. Gebelein, C. E. Carraher, Jr. and V. R. Foster, Eds., Plenum Press, New York, 1988, 1-15.

USE OF THE ANTIVIRAL AND IMMUNE MODULATOR, POLY(ICLC), IN THE TREATMENT OF AIDS

Hilton Levy[a] and Andres Salazar[b]

(a) Office of the Scientific Director
NIAID
Bethesda, Md.
(b) Department of Neurology
Armed Service University
Bethesda, Md.

The interferon inducer and immune modulator, poly(inosinic)·poly(cytidylic) acid, complexed with poly(lysine) and carboxymethyl cellulose, poly(ICLC) was administered i.m., plus AZT, orally to patients with advanced AIDS. The purpose of this part of the study was to evaluate the safety of the treatment in such a series of patients, to test for toxic effects, and to see if there were any alterations in some of the subsets of lymphocytes. There did not appear to be any significant clinical or laboratory indications of toxicity. The disease was not worsened by the treatment. There were transient boosts in a number of immune parameters, such as NK cell activity, T4 and T8 cells, and the expression of DR antigens on both B and T lymphocytes. These data will be used in the next phase of the study to determine a rational dose and time schedule to maximize effectiveness.

INTRODUCTION

While the initial report of the existence of interferon in 1957 did not meet with universal acceptance,[1] there were a number of investigators who felt that the vaguely defined material potentially was a broad spectrum antiviral agent of interest in the treatment of human disease. The realization of this potential has been slow indeed, but meaningful clinical studies are now being carried out. The reasons for the lengthy start up time are several, but the most obvious derives from a consideration of the following quantitative aspects. Current studies often involve giving a patient $10^{6.5}$ units of interferon daily for several weeks. Until a few years ago this daily dose represented the crude interferon yield from about 5 liters of tissue culture fluid. Until the development of newer methods for large scale production of human interferon, there was no realistic way to prepare the amount of interferon needed for a clinical trial. The required amounts were even greater than indicated if one considers the losses encountered in concentrating and partially purifying the crude material.

Biotechnology and Polymers, Edited by C.G. Gebelein
Plenum Press, New York, 1991

Investigators were led to search for nonreplicating compounds that would cause the host to activate strongly its own interferon producing systems. A number of compounds were found (for review see 2), but they were either too toxic, or induced too small amounts of interferon to be of interest.

Two paths of investigation converged to lead to the development of effective chemicals for the induction of interferon. (1) A number of naturally occurring compounds that could induce IFN were found to contain double stranded RNA and (2) a group of investigators from Merck, Sharp, and Dohme published a series of papers in which they reported that a variety of natural and synthetic double stranded (d.s.) RNAs were effective interferon inducers in mice and rabbits.[3,4] Of these, the most effective inducer was a double stranded complex with one strand of poly-(riboinosinic acid) and one of poly(ribocytidylic acid) - poly(I)·poly(C). Poly(A)·poly(U) was much less effective as an interferon inducer. Poly(I)·poly(C) also proved to be a good adjuvant for antibody production but so did poly()A)·poly(U), suggesting a separation of the immune modulating function from that of interferon induction.[5] One of us (H. L.) showed that poly(I)·poly(C) was an effective agent against a number of spontaneous, virally induced or transplanted tumors in mice.[6] Table 1 summarizes some of these data. With the exception of the J 96132 reticulum cell sarcoma, some Ehrlich ascites tumors, and a few Walker carcinosarcomas all animals ultimately died.

Poly(I)·poly(C) induced little or no interferon in primates. In clinical studies with cancer patients, poly(I)·poly(C), even at relatively high doses, induced only very low levels of interferon, and showed no antitumor activity.

Toxicity was minimal. Work with poly(I)·poly(C) in humans was stopped, not because of toxicity, but because of ineffectiveness. It was shown that the loss of effectiveness in primates was attributable to a high concentration of hydrolytic enzyme activity in primates serum that

Table 1. Effect of poly(I·C) on animal tumors.*	
	% increase in median survival over control
J96132-Reticulum cell sarcoma (subcutaneous)	130
J96132-Reticulum cell sarcoma (ascities)	96
Carcinosarcoma Walker 256	100
Reticulum cell sacroma A-RCS	89
Ehrlich Ascities tumor	70
S 91 Melanoma	55
Fibrosarcoma	52
B 1237-lymphoma (ascities)	45
L 1210 Leukemia	42
Plasma cell YPC-1	39
B 1237-lymphoma (subcutaneous)	28
MT-1 tumor (subcutaneous)	26
Reticulum cell sarcoma	20
Leukemia P388	16
Leukemia 1964	12

Treatment, in most cases, was 150-200 µg/mouse, three times weekly, by intraperitoneal route.

Dose mg./m²	Mean Peak Serum Titers (IU/mL)[a] Rhesus Monkeys	Man
2.5	–	15 (0–25)[b]
7.5	–	200 (25–250)
12	1500 (600–2000)[b]	2000 (200–5000)
18	–	4500 (600–15,000) (toxic)
27	–	6000 (2000–10,000)
	(toxic)	
36	10,000 (3000–15,000)	–
240	100,000 (20,000–200,000) (lethal)	–

Table 2. Interferon induction with poly(ICLC).

(a) VSV-CPE assay; 8-hour sample post-first dose: > 3 trials at each dose.
(b) Numbers in parentheses = range.

inactivated poly(I)·poly(C). This activity was present in much lower levels in rodent serum.[6]

The addition of a complex of poly(lysine) and carboxymethylcellulose yields a derivative called poly(ICLC) which partially resists such hydrolysis, and which is an effective interferon inducer in monkeys, chimpanzees and man. In addition, poly(ICLC) is an immune modulating agent in mice and primates.[7] The ability of poly(ICLC) to induce interferon in monkeys and in man is summarized in Table 2.

In early studies, i.v. injection of 12 mg./m² was used, but this proved to be too toxic, and 6 mg./m¹² was used. Currently, intramuscular injection is being favored, because side effects are minimal. In addition to inducing the production of interferon in primates, poly(ICLC) also enhances a number of immune functions (see below).

ANTIVIRAL ACTIVITIES

Poly(ICLC) has been tested against a number of viruses in several species of animals. A few studies will be reviewed in slight detail, and then all the studies will be listed.

1. Rabies

Classical treatment of rabies consists of two phases. In the first, antibody to rabies is given to neutralize the infecting virus with passive immunity. In the second phase, vaccine is given to stimulate the host to develop his own active immunity to neutralize the virus at later stages. These two treatments tend to negate each other, and repeated injections of both are needed. In highly advanced industrial countries purified antibody and vaccine are available, but in those countries where rabies is a major human problem, the antibody and vaccine are much less purified and less potent, and large numbers of injections of both products are needed.

Table 3. Effects of poly(ICLC) treatment in post exposure prophylaxis of rabies in monkeys.	
Treatment	Dead/Treated
Poly(ICLC) + Vaccine 24 and 72 hours post infection	1/8
Vaccine 24, and 72 hours post infection	7/8
Controls (untreated)	8/8

As a result, perhaps as much as 50% of the patients in some countries do not complete the course of treatment. The use of poly(ICLC) to control the infecting virus, together with vaccine, has resulted in a drastically reduced treatment in an experimental monkey model (Table 3). One dose of the inducer followed by two doses of vaccine has resulted in control of lethal infection in monkeys.[6]

There are many problems involved in mounting a competent rabies trial in humans. For example, as will be mentioned later, poly(ICLC) is a potent immune adjuvant with some antigens. There was concern that the use of poly(ICLC) might increase the incidence of allergic encephalitis in patients receiving a vaccine made from suckling mouse brain. However, in tests in rats, this was found not to be the case.

2. CHRONIC HEPATITIS B

Chronic hepatitis B infection is a major worldwide problem for which there is no therapy. It is carried by millions of people in the world, particularly in the Orient. There is a model of this disease in chimpanzees which bears some resemblance to the human disease.

A study was begun to determine the effects of long-term therapy with poly(ICLC) on chronic hepatitis B viral infection. Poly(ICLC) was administered once daily for two six-day periods separated by one non-treatment day, and every other day thereafter for a total of seven weeks, to two infected chimps and one noninfected one.

During the period when chimps were being treated with poly(ICLC) they produced IFN (and presumably enhanced immune reactivity). During this time, evidence of the disease, as measured by the level of DNA polymerase in serum, declined to a vanishingly low level. When treatment was stopped after 16 weeks, the IFN levels disappeared and evidence of disease returned.[7] These data resemble those seen by Merigan in humans treated with IFN. It was only after treatment in man was continued for a year, some apparent cures were noted. It is not known what would be the result if chimpanzees would be treated for a year.

Table 4 summarizes the data for a variety of viral diseases in animals treated with poly(ICLC). In most instances there was some beneficial effect. However, in the case of Bolivian Hemorrhagic fever, the disease may have been worsened by treatment.[7]

Table 4. Virus diseases of animals that have been treated with poly(ICLC).

Disease	Animal	Results
Simian hemorrhagic fever	Monkey	Complete protection if given before virus, none if given after virus
Venezuelan equine encephalitis	Monkey	No animals with light virus challenge died; poly(ICLC) reduced viremia by 50%
Yellow fever	Monkey	75% protection up to 8 hr post challenge
Japanese encephalitis	Monkey	50% protection up to 24 hr post challenge
Tacaribe virus	Mouse	No effect by poly(ICLC)
Rabies	Monkey & Mouse	See text
Hepatitis	Chimpanzee	Virus controlled while on drug. Control ends when treatment stopped.
Bolivian hemorrhagic fever	Monkey	Possible worsening of disease
Tick-borne encephalitis	Monkey	Strong protection
Vaccinia	Monkey	Strong protection
Vaccinia skin lesions	(Rabbit topical treatment)	Spread of lesions stopped

IMMUNE MODULATORY ACTIONS OF POLY(ICLC)

During the early 1970s, there were reports indicating that interferon and interferon inducers had effects other than as viral inhibitors.[7,8] The enhancing effect of poly(I)·poly(C) on graft vs. host reaction in mice was reported by two groups.[9,10] However, with the demonstration that poly(I)·poly(C) is almost without interferon inducing capacity in primates, the possibility of immune modulating effects of poly(I)·poly(C) in primates was not investigated.

Poly(ICLC), which induces interferon in monkeys, chimpanzees, and humans, also modifies immune responses in primates as well as in mice. This review will consider separately effects on humoral antibody and on cell associated immune functions.

EFFECTS OF POLY(ICLC) ON HUMORAL ANTIBODY PRODUCTION

Three assay systems have been used in studying the effects of the interferon system on antibody production: (1) A totally *in vivo* system in which animals are immunized, and after a suitable time, the amount of circulating antibody is measured. (2) An *in vivo-in vitro* system, where animals are immunized, and after a short period of time, spleens are removed. The number of splenic lymphocytes capable of engaging in antibody production to the immunogen is determined by use of a plaque-forming technique in tissue culture (Jerne's plaque technique). (3) A totally *in vitro* system, the Mishelle-Dutton technique, which is done completely in tissue culture.

The realization that interferon inducers can modify antibody production developed early, with studies on synthetic polyneucleotides playing a pivotal role.[11,12] It was shown that natural nucleic acids were active in augmenting antibody formation to sheep red blood cells (SRBC) in mice, as were double-stranded synthetic polynucleotides. When single stranded synthetic polynucleotides were used, no augmentation was obtained.

In connection with the question of whether the immune modulation brought about by the inducers is attributable solely to the interferon induced, it is important to note that insofar as antibody response to sheep red blood cells in mice is concerned, poly(A)·poly(U) is as good as poly(I)·poly(C), even though poly(A)·poly(U) is a poor interferon inducer, while poly(I)·poly(C) is a very good interferon inducer. These studies were all done in mice.

Poly(ICLC) modifies antibody response by monkeys to several, but not all antigens. For example, monovalent influenza virus subunit vaccine designated A/swine x-53, prepared from A.NJ/76 (New Jersey; swine) strain of virus is only moderately to weakly effective when given as a single dose to young people and young monkeys. When the vaccine was given to monkeys simultaneously with one dose of poly(ICLC), hemagglutination inhibition (HAI) antibody titers in the serum were detected earlier and rose to levels several times higher than in monkeys receiving just vaccine.[13] The adjuvant activity of poly(ICLC) was particularly pronounced in young monkeys, where as little as 10 µg of drug/kg of body weight was effective. This level of poly(ICLC does not induce detectable levels of serum interferon, and no fever was produced.

Analogous results were obtained in monkeys using inactivated Venezuelan Equine Encephalomyelitis (VEE) virus vaccine.[14] Some of the data is shown in Figure 1. It can be seen that there was an increase of about 40-fold when one compares levels attained after primary administration of vaccine along with poly(ICLC) to that attained with vaccine alone, and perhaps 200-fold after a secondary immunization. There was no alteration in the progression of IgM and IgG development. At the peak of antibody levels, most of the antibody was IgG. Poly(L-lysine) complexed to carboxymethylcellulose (CMC) without poly(I)·poly(C) had no adjuvant action.

A polysaccharide vaccine made from Hemophilus influenza gives a poor response in very young children, where the disease threat is maximum. The vaccine is also poor in young monkeys. The data that is presented in Table 5 are normalized values obtained by radioimmunoassays carried out by Dr. Porter Anderson. The value of 100 was assigned, in each case, to the amount of radioactivity found prior to immunization. The vaccine alone caused a small boost in antibody production, but when given with poly(ICLC) there was a more pronounced boost (H. B. Levy, E. Stephen, and D. Harrington, unpublished observations). Poly(A)·poly(U) complexed to CMC and poly(L-lysine) was not as effective as poly(ICLC).

In three experiments there was no effect by poly(ICLC) at these dose levels on platelets, total and differential leukocytes, packed cell volume, erythrocyte sedimentation rate, serum urea nitrogen, total and

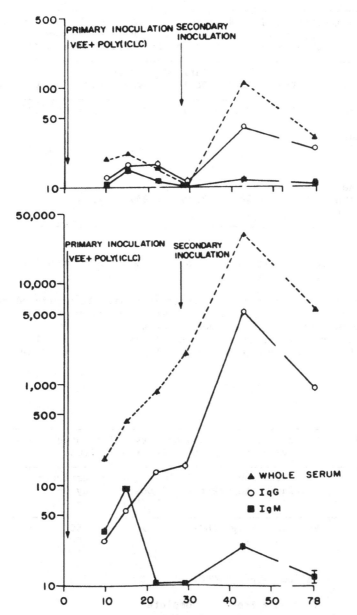

Figure 1. Effect of poly(ICLC) on antibody production by
monkeys given inactivated Venezuelan equine
encephalomyelitis (VEE) vaccine.

direct bilirubin, glutamic oxalacetic transaminase, lactic dehydrogenase,
alkaline phosphatase, and creatinine.

In an experiment using a different type of endpoint, it was shown
that poly(ICLC), given with a vaccine versus Rift Valley fever (RVF)
virus, increased the survival of mice subsequently challenged with live
RVF virus. In Table 6 are shown the increased survival rates of mice

Table 5. Effect of low doses of poly(ICLC) on antibody production by rhesus monkeys in response to a polysaccharide vaccine for *Hemophilus influenza*.

Treatment	Antibody Levels						
	Pre	Day7	Day14	Day20	Day28	Day34	Day42
Vaccine	100	590	348	286	225	187	113
Vaccine + poly(ICLC) 0.3 mg/kg	100	5643	5040	3340	2063	2162	780
Vaccine + poly(ICLC) 0.3 mg/kg	100	6589	3904	1884	1132	839	721

given poly(ICLC) and vaccine, as compared with those found in mice immunized with vaccine alone.

The last of these adjuvant effects that will be mentioned deals with observations made by Klein, et al.,[15] using an envelope antigen made from herpes virus. In these experiments, both antibody levels and survival after subsequent challenge with whole virus were measured. Both aspects of immunity were enhanced when the envelope antigen was administered with poly(ICLC).

Table 6. Effects of adjuvants on survival of mice immunized with vaccine to rift valley fever virus.

Treatment			% Survivors
Vaccine	Adjuvant	Dose (Ug/kg)	Day 35 (N = 16)
Vaccine + poly(ICLC)		20	50
		100	50
		200	13
Saline + poly(ICLC)		200	6
Vaccine + Freund's complete adjuvant			6
Saline + Freund's complete adjuvant			6
Vaccine + Freund's complete adjuvant			6
Saline + Freund's complete adjuvant			0
Vaccine controls			19
Saline controls			0

It should be emphasized that poly(ICLC) is not a universal adjuvant. When used with serum albumin or pheumococcal polysaccharide III, an inhibition of production of circulating antibody was seen (H. Levy and P. Baker, unpublished observation). With rabies virus or interferon protein as antigens, no effect was noted.[6] Alterations of dosage of poly(ICLC), or of the temporal relationship between antigen and inducer administration, could conceivably give different results.

Anderson has carried out experiments in rats that shed some light on physiological changes associated with administration of antigen plus poly(ICLC) . (A. D. Anderson, H. B. Levy, and D. Harrington, unpublished observations). Antibody response by rats to VEE vaccine was augmented by poly(ICLC) as it is in monkeys. The weight of the lymph nodes draining the area where the poly(ICLC) was injected, increased as the dose of the drug increased. With 40 μg/rat there was an increase of 100%. The number of small lymphocytes in the high venules of the paracortex of these nodes also increased with the dose of poly(ICLC) within 24-48 h after injection. Furthermore, in individual rats, the number of small lymphocytes in the lymph node paracortex at 2 days is related to the antibody level in that animal at 28 days.

In summary it was observed that: (1) injection of poly(ICLC) increases the size of the node; (2) this increase in size is associated with an increase in small lymphocytes, which are cells capable of producing antibody; and (3) the increase in number of these lymphocytes is related to the increase in antibody production.

EFFECTS ON CELL ASSOCIATED IMMUNE REACTIONS

1. Graft-Versus-Host Reaction

Interferon, given to mice, inhibited the graft-versus reaction.[9a] Contrary results were seen with interferon inducers. In a graft-versus-host test, Cantor, et al.,[9] showed that the interferon inducer poly(I)-poly(C) could enhance the activity of the donor spleen cells used in the test by as much as 3-fold. Poly(A)·poly(U) exerted a similar but lesser effect. In this study interferon gave equivocal results. Comparably, Turner, et al.,[10] showed that poly(I)·poly(C) increased rather than decreased allograft rejection.

2. Delayed-Type Hypersensitivity

Interferon has a multiphasic effect on delay-type hypersensitivity (DTH). When mice that have been sensitized to SRBC or Newcastle disease virus (NDV) are challenged in the footpad with the antigen there is a marked swelling of the footpad. If 10^6 units of interferon are administered to the mice 24 hours before sensitization there can be a complete suppression of sensitization and no footpad swelling develops. However, if interferon is given a few hours after the sensitizing antigen, there is an enhancement of swelling when the mice are challenged with antigen after 4 days. On the other hand, if the mice are sensitized normally, and the interferon is given a few hours before the footpad challenge, a complete suppression of swelling is found.[16,17] The essential effector cell for DTH is the T-cell.

It is postulated that the inhibition of sensitization, seen when interferon is given before the antigen, is associated with inhibition of blast formation.[18] It has also been suggested that the augmentation seen when the antigen is given may be given later may be attributable to the inhibition of development of suppresser T-cells.

In marked contrast to the above, poly(ICLC), given to already sensitized mice, strongly enhanced DTH to SRBC, as measured by foot pad swelling. This was true whether the double-stranded RNA was given 3, or 2, or 1 day before the antigen or even along with the antigen.[19] These differences between the effects of interferon and the interferon inducer may be related to the effects on lymphocyte inhibition or stimulation discussed in the next section.

EFFECT ON HEMATOPOIETIC COLONY-FORMING CELLS

Mouse granulocyte-macrophage precursor cells can grow into colonies in soft agar provided that a glycoprotein colony-stimulating factor (CSF) is present. Single cells, arising from pluripotential stem cells, can give rise to such colonies. Such cells are called colony-forming cells (CFC). Interferon *in vivo* inhibits the number of CFC found in mice, and the interferon *in vitro* exerts an inhibitory effect on colony growth. The addition of sufficient CSF can prevent the inhibition by interferon.[20] Poly(I)·poly(C) *in vivo* in mice causes a short-lived depression in CFC in the spleen. However, after a few hours, even though interferon was present in serum, there was an increased number of CFC in the spleen.

Six days after injection of 50 - 100 µg poly(I)·poly(C) there is an 8-fold increase in the number of CFC in the spleen. Serum CSF levels were depressed for 6 hours after injection of poly(I)·poly(C), but by 48 hours there was an increase of 4-fold over controls. *In vitro*, poly(I)·poly(C) also increased the number of CFC in spleen cell cultures, both because the double-stranded RNA increases the ability of cells to respond to CSF, and because poly(I)·poly(C) leads to the release of CSF from leukocytes in culture.[19]

Effects on radiation damage are possibly related to the effects of interferon and its inducers on colony formation.[21,22] A number of compounds that can induce interferon, including good inducers like poly-(ICLC) and very poor inducers like the amino thiols, are able to exert significant protection against death by X-irradiation in mice. The degree of protection bore no relationship to the amount of interferon induced. Since the inducers undoubtedly induce, *in vivo*, a complex group of lymphokines, it has been suggested that it may not be the interferon that is the protective agent, but one or more of the lymphokines. With poly(ICLC) maximum protection was observed when the drug was administered 48 - 72 hours before irradiation. The irradiation was given, therefore, when little or no circulating interferon was present. When irradiation was done at times of maximum induced serum interferon, no protection was seen. As part of that study, mice that had received poly(ICLC) before irradiation were sacrificed 8 days after irradiation. The number of colony-forming cells in their spleens were found to be up to 10 times as many as in untreated X-irradiated mice.

Interferon, on the other hand, gives some increase in survival time in mice against lethal X-irradiation when given one dose 2 - 4 hours after irradiation. In this instance there was no increase in number of colony-forming cells, as there was with poly(ICLC), but there was an increase in the activity of natural killer cells, and possibly in other immune functions. Presumably the increase on survival is attributable to increased resistance to infection.[23]

EFFECTS ON MACROPHAGES

Macrophages have long been implicated in host defense against tumors.[24] In order to be maximally effective, they must be converted from low cytotoxicity to high cytotoxicity, through activation by specific or nonspecific factors.

For example, infection with a variety of bacteria will increase the effectiveness of the tumoricidal action of macrophages *in vivo*. A soluble lymphokine, produced by lymphocytes appears to be needed for this type of action. This poorly defined factor is called macrophage activating factor (MAF). In contrast certain interferon inducers, such as endotoxins and poly(I)·poly(C) can transform macrophages without the need for lymphocytes.[24] Direct activation of macrophages by pyran copolymer and other polyanionic inducers without lymphocyte-derived MAF was also demonstrated.[26-28] Since these inducers stimulate macrophages to produce interferon, and interferon activates macrophages, it is reasonable to think that interferon is the effective activator in these instances.

Further, if antibody to interferon is present along with the inducer, activation does not take place.[24] Single-stranded RNA was not effective.[29] It has been suggested that macrophage activation accounts for a major part of the antitumor action of pyran copolymer. Schultz, et al.,[30] and Stringfellow[31] have shown that prostagladin E can reverse the activation of macrophage by interferon and interferon inducers. Since tumors can produce prostagladin E,[32] the tumors may partially protect themselves against the macrophage action.

Peritoneal macrophages from c57BL/6N mice were incubated *in vitro* with poly(ICLC) at varying doses. B16-B/6 cells labeled with I^{131}-IUDR served as target cells. Levels of drug from 0.001 to 0.1 µg/mL activated macrophages, levels below 0.001 µg/mL had no effect, while levels of 1 µg did not enhance or were inhibitory.[33]

1. Macrophage Activation *In Vivo*

Poly(ICLC) was injected i.p. into mice, and 24 hours later peritoneal macrophages were obtained by lavage. Doses of poly(ICLC) at 0.5 or 0.05 mg/kg activated the macrophages while both higher and lower doses did not.[33]

2. Allogeneic Mixed Lymphocyte Reaction

C3H/HeN(H-2d) cells were used to stimulate c57B1/6N(h-2b) spleen cells, in the presence of varying doses of poly(ICLC). After 4 days of cocultivation, the mixed cultures were pulse labeled with tritiated thymidine for 24 hours and the incorporated radioactivity determined. All doses from 0.01 µg/mL to 5 µg/mL of poly(ICLC) stimulated the MLR, but 10 µg/mL caused inhibition.[33]

CELL MEDIATED CYTOTOXICITY STIMULATED IN AN ALLOGENIC MIXED LYMPHOCYTE TUMOR REACTION (MLTR-CMC) ASSAY

Splenic lymphocytes from C3H/HeN(H-2K) mice were incubated with irradiated p815(H-2d) tumor cells, with or without poly(ICLC). After 5 days of incubation, the effector lymphocytes were washed, and tested for specific cytotoxicity against p815 cells, using a 51Cr release assay. Poly-(ICLC) *in vitro* did not induce the production of specific cytotoxic lymphocytes.[33]

1. Induction of Tumor Specific Cytotoxic T Lymphocytes *In Vivo*

Mice were immunized intradermally with irradiated collagenese-dissociated UV 2237 tumor cells, with or without poly(ICLC). Ten days later spleen cells were exposed to target cells, a 75Se methionine labeled clone of UV 2237, and the amount of released radioactivity released in the cultures of cells from animals that had received poly(ICLC) was measured.[33] Spleen cells from mice immunized with tumor cells show an increase in specific cytoxic activity as compared to spleen cells from mice inoculated with salt solution. Poly(ICLC) further enhanced the development of specific cytotoxic activity.[33]

EFFECTS ON NATURAL KILLER CELL ACTIVITY *IN VITRO*

Natural killer (NK) cells are a subpopulation of lymphocytes that can lyse a wide variety of tumor cells and some normal cells. They are classified as null cells, being neither B or T cells. The assay usually is an *in vitro* one in which target cells, labeled with 51Cr are exposed to the NK cell population, and the amount of radioactive 51Cr released is measured. NK cell activity is highest in young animals (mice 5-8 weeks of age), and is present in very low levels, if at all, in older animals. However, when older rats were exposed to C. parvum or BCG, there was an increase in the cytotoxicity of the lymphocyte population.

Similarly, pyran copolymer, in its original heterogeneous molecular weight form, was an active interferon inducer and augmentor of NK cell activity, but when certain defined molecular weight fractions of pyran copolymer, MVE-1 and MVE-2, were tested, loss of interferon-inducing capacity was observed along with loss of NK cell activation capacity.[34]

A wide range of poly(ICLC) levels augmented NK cell activity when cells were incubated with the drug *in vitro*. Doses of 0.001 μg/mL to 10 μg/mL increased NK activity 2-3 fold.[33]

1. NK Cell Activation *In Vivo* - Mice

Poly(ICLC) 0.5 mg/kg, in mice, significantly enhanced NK cell activity for up to 7 days when given i.v., i.p., and even longer when given subcutaneously or into the footpad. When given i.v., poly(ICLC) augmented NK cell activity in spleen, peritoneal cavity and lung.[19,33] In mice poly(ICLC) differed from other activators of NK cells in that repeated injections of the drug over at least a two week time period continued to be active. Other modulators, such as pyran or interferon, were effective after the first dose but the cells became unresponsive with subsequent doses.

The most striking effect of the inducers on NK cell activity is the augmentation discussed above. However, it should be borne in mind that several inducers show a late repressing effect of NK cells. The augmentation is seen within a few hours, while the inhibition is not seen until 5-7 days after administration of the inducer.[35,36]

Prostaglandins of the E series appear to have an antagonistic effect on NK activity. Animals bearing a variety of tumors show depressed NK cell activity. These tumors produce prostaglandin E.[39] Injection of prostaglandin synthetase inhibitors, such as aspirin or indomethacin, to tumor-bearing mice restored NK cell activity.

EFFECTS OF POLY(ICLC) ON NK ACTIVITY IN HUMANS

Since it is now considered likely that NK cell activity plays a role in host defenses against virus diseases and tumors, a number of laboratories have investigated ways in which such activity could be increased in patients.

Einhorn reported that in four of five patients with osteosarcoma, receiving one injection of 3×10^6 U of interferon there was an initial, very short-lived decrease in cytotoxicity of peripheral leukocytes versus Chang cells, followed by an increased activity lasting about 24 hours. Patients receiving 3×10^6 U of interferon three times weekly over 6 months maintained elevated NK cell activity during the 6 months of observation. Other investigators have found similar results.[37,38] However, a recent series of experiments by the National Cancer Institute did not show enhancement.[40] Rather it was found the interferon caused only a very short term increase in NK cell activity, followed by a decrease in such activity on further treatment.

The data dealing with the effects of poly(ICLC) *in vivo* in humans is also inconsistent. Krown, et al.,[41] reported that 5 of 11 patients showed a sustained increase in NK activity, 1 showed a decrease, and 5 showed no change. Herberman and colleagues, as a part of a phase I study of poly-(ICLC), found that of 7 patients studied, four had an increase in NK cell activity, 2 had a decrease, and one showed no change.[42] In a phase I immunology study using an intramuscular route of administration,[43] it was found that 5/12 patients receiving 1 mg/m² of poly(ICLC) showed a moderate but sustained elevation of NK cell activity. It does appear that low doses of poly(ICLC), particularly given i.m., may cause elevation, while higher doses were less effective. However, patients receiving either 1 or 4 mg/m² by the i.v. route did not show an elevation. At this writing it is not clear what factors may play a role in determining why sometimes there appears to be elevation and sometimes not.

In this survey of cell-associated immune responses, it was found that while there was often a dose-related relationship, the range of drug that enhanced the response was limited and positive effects usually occurred at relatively low doses. Higher doses of poly(ICLC) were often inhibitory. Thus, the usual approach to cancer chemotheraphy, using the maximum tolerated dose, would not be an effective way to utilize the immune enhancing potentiality of poly(ICLC).

PRELIMINARY STUDIES WITH AIDS

Because of the antiviral properties and the immune enhancing activities of poly(ICLC) it was decided to test its possible effect in patients with AIDS. Patients with advanced AIDS (Walter Reed stage 6) at Walter Reed Army Hospital and at Presbyterian Hospital in Puerto Rico

Table 7. Poly(ICLC) and zidovudine in AIDS. Laboratory values.

Patient	BUN Base	BUN 2mos	BUN 6mos	SGOT Base	SGOT 2mos	SGOT 9mos	HGB Base	HGB 4mos	HGB 9mos
PR-01	14	11	14	36	39	–	11	11.2	10.5
PR-02	–	18	29	78	49	41	14	14.1	12.3
PR-03	08	11	11	–	22	33	13.1	13.4	12.8
PR-04	12	12	10	64	21	31	12.4	13.2	12.8
PR-05	15	13	–	16	24	–	13.6	13.4	–
PR-06	12	10	15	24	42	50	11.4	10.1	8.1
PR-07	14	11	11	355	470	195	13	12.3	12.2
PR-08	20	18	20	27	38	21	13.5	12	11.5
04-056	13	20	09	31	38	39	9.4	9.7	7.7

were treated with AZT (0-1200 mg. daily) and poly(ICLC) (2 mg per patient weekly, i.m.).

Lymphocyte cell subset enumeration was done by the use of suitable fluorescent labeled monoclonal antibodies and a fluorescence activated cell sorter. Cytolytic activity of NK cells was tested against K562 cells, using a standard 51Cr release assay.

Some of the results obtained in the first phase of the open-ended pilot study with AIDS patients are given in Tables 3 to 5, and Figures 1 to 4. No adverse laboratory changes were found in these patients. Table 7 illustrates this for BUN, SGOT, and hemoglobin.

Table 7 shows that the treatment regimen is not associated with a decrease in hemoglobin, as is frequently seen with AZT alone. Poly(ICLC) has been shown to increase the production of colony stimulating factors. Table 8 shows data summarizing the effects of the treatment on white blood cells and lymphocyte subsets. There were no reproducible changes seen in this type of study. However, see Figures 1 to 5.

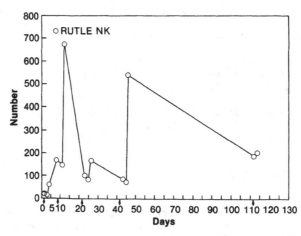

Figure 2. Effect of poly(ICLC) on NK cell numbers in an AIDS patient.

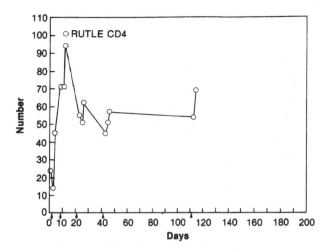

Figure 3. Effect of poly(ICLC) on CD4 number in an AIDS
patient.

Table 8. Poly(ICLC) and zidovudine in AIDS treatment white
blood cells and subsets.

Patient	Base	4mos	9mos	Base	4mos	9mos	Base	4mos	9mos
PR-01	2000	3600	2600	20	290	180	80	880	530
PR-02	4200	5000	3100	100	200	60	1000	1360	550
PR-03	2800	2300	2600	10	20	10	61	500	340
PR-04	3400	2500	3100	270	370	-	790	546	-
PR-05	2600	1100	-	40	24	-	690	603	-
PR-06	2555	4000	3700	200	94	-	440	756	-
PR-07	3500	3100	3700	250	280	130	720	580	860
PR-08	5600	5500	4600	130	120	100	1125	1310	1240
04-056	2300	2700	2500	003	006	008	216	218	455

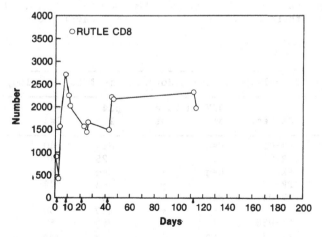

Figure 4. Effect of poly(ICLC) on CD8 number in an AIDS
patient.

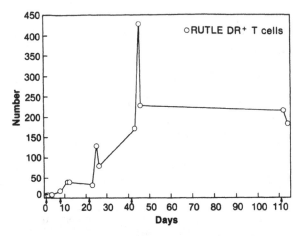

Figure 5. Effect of poly(ICLC) on expression of DR antigen on
T cells in an AIDS patient.

Tests for the presence of HIV and for the p24 antigen were performed at two times during this first part of the study. Table 9 shows that at the first testing time a number of the patients tested negative for virus, but that at the second period all were positive. Patients with this advanced stage of disease would be expected all to have virus present.

In the second phase of this study, a more detailed analysis of some of the subsets of lymphocytes have been undertaken. Specifically, enumeration of CD4 cells, CD8 cells, NK cells, total T cells and expression of DR antigens on B and T cells were performed. A few samples will be given in Figures 2-7.

Figure 2 shows that every time an injection was given, there is a transient elevation of NK cell activity.

Similar results are seen with CD4 cells (Figure 3), CD8 cells (Figure 4), and DR expression on T cells (Figure 5), and DR expression on B cells (Figure 6) and CD3 cells (Figure 7). The data shows that after almost every injection there was a short lived (1 to 2 days) boost in the immune

| | HIV Culture | | p24 Antigen | |
Patient	4mos	7mos	1mo	7mo
PR-01	Neg	+	Neg	440
PR-02	+	+	25	89
PR-03	Neg	++	Neg	80
PR-04	+	+	Neg	78
PR-05	Neg	–	Neg	–
PR-06	+	+	>200	430
04-056	+	+	Neg	57
PR-07	Neg	+	–	346
PR-08	Neg	++	–	78

Table 9. Poly(ICLC) and zidovudine in AIDS. Virology.

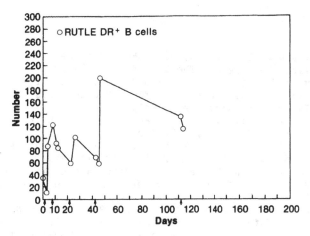

Figure 6. Effect of poly(ICLC) on expression of DR antigen on
T cells in an AIDS patient.

parameter being measured, followed by a decline to pretreatment values.
It might be reasonable to alter treatment schedules to try to maintain
that elevated level.

CONCLUSION

In a preliminary exploration of the possibility of using a combination of azidothymidine (AZT) plus poly(ICLC) in the treatment of patients
with AIDS, it was found that the combination of drugs can be given safely, and that the disease is not worsened by the treatment, and that there
was a suggestion of benefit. After each treatment there was a transient
enhancement in the number or activity of NK, CD4, CD8, and total T cells.
It is planned to use this information to try to prolong the length of
time that these enhancements occur.

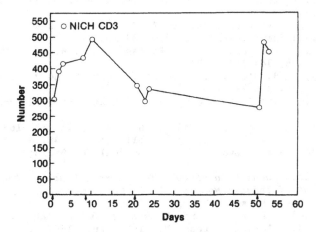

Figure 7. Effect of poly(ICLC) on CD3 cells number in an AIDS
patient.

REFERENCES

1. A. Isaacs, and J. Lindenmann, Proc. Roy. Soc. Ser. B., **147**, 258 (1957). "Virus Interferon, The Interferon."
2. H. B. Levy, and T. Quinn, in: *"Bioactive Polymeric Systems,"* C. G. Gebelein and C. E. Carraher, Jr., Eds., Plenum Press, New York, 1985, p. 387. "Interferon Induction by Polymers."
3. A. K. Field, A. A. Tytell, G. P. Lampson, and M. R. Hilleman, Proc. Natl. Acad. Sci., U.S., **58**, 1004 (1967).
4. A. K. Field, G. P. Lampson, A. A. Tytell, M. M. Nemes, and M. R. Hilleman, Proc. Natl. Sci., U.S., **58**, 2101 (1967).
5. W. Braun, M. Ishizuka, Y. Yajima, D. Webb, and R. Winchurch, in: *"Biological Effects of Polynucleotides,"* R. F. Beers and W. Braun, Eds., Springer, 1971.
6. G. M. Baer, J. H. Shaddock, S. A. Moore, P. A. Yager, S. Baron, and H. B. Levy, J. Infect. Dis., **136**, 286, (1977). "Successful Prophylaxis Against Rabies in Mice and Rhesus Monkeys: The Interferon System and Vaccine."
7. H. B. Levy, and F. L. Riley, in: *"Polymers in Medicine,"* E. Chiellini and P. Guisti, Eds., Plenum Pub., NY,, 1984, p. 45.
8. S. Baron, F. Dianzani, and G. J. Johnson, Texas Rep. Biol. Med., **41**, 1-12 (1982).
9. H. R. Cantor, R. Asofsky, and H. B. Levy, J. Immun., **4**, 1035 (1970).
9a. A. S. Hirsch, D. A. Ellis, P. W. Black, A. P. Monaco, and H. L. Wood, Transplantation, **17**, 234-236, (1974).
10. W. S. Turner, P. Chan, and M. A. Chirigos, Proc. Soc. Exp. Biol. Med., **133**, 334-338 (1970).
11. W. Braun, and W. Firschein, Bacteriol, Rev., **31**, 83-94 (1967).
12. W. Braun, and M. Nakano, Science, **157**, 819-821 (1967).
13. E. L. Stephen, D. E. Hilmas, J. A. Mangiafico, and H. B. Levy, Science, **197**, 1289-1290 (1970).
14. D. C. Harrington, C. L. Crabbs, D. E. Hilmas, J. R. Brown, C. A. Higbee, F. E. Cole, and H. B. Levy, Infect. Immun., **24**, 160-166 (1979).
15. R. J. Klein, E. Klein-Buimovici, H. Moser, R. Moucha, and J. Hilfenhaus, Arch. Virol. **68**, 63-80 (1981).
16. E. DeMaeyer, and J. DeMaeyer-Guingnard, Ann. N.Y. Acad. Sci., **350**, 1-11 (1980).
17. Ibid Tex. Rep. Biol. and Med., **41**, 420-426 (1982).
18. P. Lindahl, P. Leary, and I. Gresser, Proc. Natl. Acad. Sci., U.S., **69**, 721 (1972).
19. M. A. Chirigos, V. Papademetriou, A. Bartocci, E. Read, and H. B. Levy, Int. J. Immunopharmac, 3, 329-337 (1981).
20. T. A. McNeil, W. B. Blaze, D. E. Hilmas, and H. B. Levy, Tex. Rep. Biol. Med., **35**, 343-349 (1977).
21. E. Lvovsky, W. B. Blaze, D. E. Hilmas, and H. B. Levy, Tex. Rep. Biol. Med., **35**, 228-293 (1977).
22. J. R. Ortaldo, and J. L. McCoy, Radiation Research, **81**, 262-266 (1980).
23. M. Levy, and E. F. Wheelock, Adv. Cancer Res., **20**, 131-163 (1974).
24. R. M. Schultz, and M. A. Chirigos, Cell. Immun., **48**, 403-409 (1979).
25. P. Alexander, and R. Evans, Nature (London) New Biology, **222**, 444-452 (1971).
26. R. Evans, in: *"Modulation of Host Immune Resistance in the Prevention or Treatment of Induced Neoplasms."* A. Chirigos, Ed., Fogerty Int. Ctr., U.S. Govt. Press, 1976, pp. 295-301.
27. R. Evans, and P. Alexander, in: *"Immunobiology of the Macrophage,"* D. S. Nelson, Ed., Academic Press, NY, 1976, pp. 535-576.
28. A. M. Kaplan, P. S. Morahan, and W. Regelson, J. Natl. Cancer Inst., **52**, 1919-1921 (1974).

29. R. M. Schultz, J. D. Papamatheakis, and M. A. Chirigos, Cell. Immun., **29**, 403-409 (1977).
30. R. M. Schultz, N. A. Pavlidis, W. A. Stylos, and M. A. Chirigos, Science, **202**, 320-321 (1978).
31. D. A. Stringfellow, in: "*Interferon and Interferon Inducers*," D. A. Stringfellow, Ed., Dekker, NY, 1980, pp. 145-165.
32. R. Snyderman, and M. C. Pike, Science, **192**, 370-372 (1976).
33. J. A. Talmadge, J. Adams, H. Phillips, M. Collins, B. Lenz, M. Snyder, and M. Chirigos, Cancer Res., **45**, 105a (1985).
34. J. Y. Djeu, J. A. Heinbaugh, H. T. Holden, and R. B. Herberman, J. Immun., **122**, 175-181 (1979).
35. E. Ajo, I. Haller, A. Kimura, and H. Wigzell, Int. J. Cancer, **21**, 444-452 (1978).
36. A. Santoni, P. Puceceti, C. Riccardi, R. B. Herberman, and E. Bonmasser, Int. J. Cancer, **24**, 656-663 (1979).
37. M. Moore, and M. R. Potter, Br. J. Cancer, **41**, 378-387 (1980).
38. G. Riethmiller, G. R. Pape, M. R. Hadam, and J. G. Saal, in: "*Natural Cell Mediated Immunity Against Tumors*," R. B. Herberman, Ed., Academic Press, NY, 1980.
39. D. A. Stringfellow, in: "*Interferon and Interferon Inducers*," D. A. Stringfellow, Ed., 1980, pp. 145-165.
40. A. Maluish, E. R. Leavitt, S. A. Sherman, R. K. Oldham, and R. B. Herberman, J. Biol. Resp. Modifiers, **2**, 470-481, 1983.
41. S. E. krown, D. Kerr, W. E. Stewart, III., S. Pollod, C. Rundles, Y. Hisrschant, C. Perisley, H. B. Levy, and H. Otegen, in: "*Augmenting Agents in Cancer Therapy*," E. M. Hersch, M. Chirigos, and M. Mastrangelow, Eds., Raven Press, NY, 1981, pp. 175-176.
42. R. B. Herberman, M. J. Brunda, G. B. Cannon, J. Y. Djeu, M. E. Nunn-Hargrove, J. R. Jett, J. R. Ortaldo, C. Reynolds, C. Riccardi, and A. Santoni, *ibid.*, 253-266.

SYNTHESIS AND INTERACTION OF WATER SOLUBLE NUCLEIC ACID ANALOGS

Kiichi Takemoto, Takehiko Wada, Eiko Mochizuki and Yoshiaki Inaki

Department of Applied Fine Chemistry
Faculty of Engineering
Osaka University, Suita, Osaka 565, Japan

Water soluble nucleic acid analogs were prepared by grafting uracil, thymine, 5-fluorouracil, hypoxanthine, cytosine, and adenine derivatives containing a hydroxy group onto linear poly(ethyleneimine). These polymers contained about 90 unit% of the nucleic acid bases, and were easily soluble in water at neutral pH resign. The analogs were found to form complementary polymer complexes with each other and with polynucleotides. The interaction study was made on these polymers, and with polynucleotides; poly(uridylic acid) poly(U), poly(adenylic acid) poly(A), poly(inosinic acid) poly(I) and poly(cytidylic acid) poly(C) in water. They were found to form polymer complexes with polynucleotide by specific base-base interaction in aqueous solution.

INTRODUCTION

Recently, polymer chemists have been exploring synthetic methods for preparing both natural and synthetic macromolecular analogs for biological use. Evaluation and modification of these materials by the chemist has led to the preparation of an array of polymeric agents with various application for biological systems.[1] The most essential function of nucleic acids is based on the formation of specific base-base pairing through hydrogen bounding between purine and pyrimidine bases. Some synthetic polynucleotides are known to show biological activities. A synthetic double stranded polynucleotide complex of poly(inosinic acid), poly(I), and poly(cytidylic acid), poly(C), (poly(I):poly(C) complex) is effective as an interferon inducer,[2-6] while it has a high level of toxicity.[7] Very recently, it was revealed that the 1:1 complex of poly(uridylic acid), poly(U) and poly(adenylic acid), poly(A) (poly(U):poly(A) complex) was also effective as an interferon associated enzyme inducer, which showed a low level of toxicity,[3] and polymer complexes of polynucleotides with the synthetic nucleic acid analogs are expected to have enhanced bioactivity.[8]

The chemistry of synthetic nucleic acid analogs has recently received much attention, and a number of synthetic polymers containing nucleic acid bases have been prepared and their properties have been studied.[9-13] These synthetic nucleic acid analogs were almost insoluble in water at natural pH values, because most of them consisted of a hydrophobic polymer backbone with pendent nucleic acid bases. The interactions between the synthetic nucleic acid analogs, and with polynucleotides were studied. However, these studies were limited to organic solvents to water-organic mixed solvents. It is, therefore, very important to prepare the water soluble synthetic nucleic acid analogs for biomedical applications.

In the present study, water soluble poly(ethyleneimine) derivatives containing uracil, thymine, 5-fluorouracil, hypoxanthine, cytosine, and adenine were prepared and their interactions with polynucleotides, such as poly(A), poly(U), poly(C), poly(I) in natural aqueous solution were investigated. As 5-fluorouracil is known as a famous anticancer agent, it seems to be interest to study the interaction of the polymer containing 5-fluorouracil unit with polynucleotides. The bioactivity of the polymer complex of nucleic acid analogs with polynucleotides will be reported later.[14]

EXPERIMENTAL

1. Materials

Water soluble poly(ethyleneimine) derivatives containing nucleic acid bases were prepared by the method described in previous papers (Scheme 1).[15,16,17]

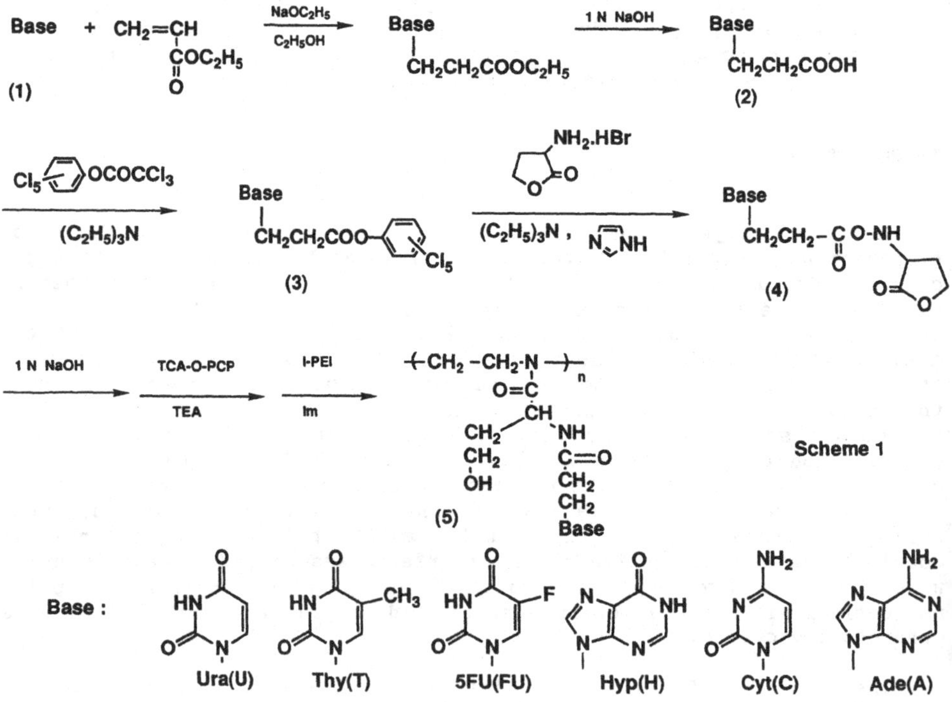

Scheme 1

2. Interaction Between Synthetic Polymers and Polynucleotides

Interaction of the synthetic polymers and polynucleotides were estimated from the hypochromicity values in UV spectra.[18] The UV spectra were measured with a JASCO UV-660 spectrometer equipped with a temperature controller at 20°C. Polynucleotides were obtained from Yamasa Shoyu Co. Ltd. Water soluble poly(ethyleneimine) derivatives containing nucleic acid bases and polynucleotides were dissolved in Kolthoff buffer (pH 7.0, 0.1M KH_2PO_4 −0.05M $Na_2B_4O_7 \cdot 10H_2O$). These solutions were stocked for 2 days at 20°C, and then mixed to give a polymer mixture of 10^{-4}M total concentration of nucleic acid base units in aqueous solution.

RESULTS AND DISCUSSION

1. Preparation of the Nucleic Acid Analogs

Natural and synthetic polynucleotides are known to form polymer complexes by specific base-base interactions between nucleic acid bases. The synthetic nucleic acid analogs such as poly(methacryamide), poly-(ethyleneimine) and poly(L-lysine) derivatives containing nucleic acid bases were also found to form polymer complexes with polynucleotides by specific base-base interactions.[12] Since the solubilities of these nucleic acid analogs in water were low, the specific interactions should be studied in organic solvents or water-organic mixed solvents, such as dimethyl sulfoxide, ethylene glucol, and water-propylene glycol.[12,19,20]

On the other hand, complex formation of polynucleotides and its application to biomedical field have been studied in aqueous solution, because organic solvent causes denaturation of polynucleotides. Therefore, it is necessary to prepare a water soluble polymer complex of nucleic acid analogs with polynucleotides for the biomedical uses.[1] The water soluble poly(ethyleneimine) derivatives of nucleic acid bases open a way to study the interaction with polynucleotides in neutral aqueous solution.

Various kinds of synthetic nucleic acid analogs have been prepared by the polymerization of the corresponding monomers or by grafting of the nucleic acid base derivatives on the polymers.[9-13] The most convenient method of preparing the nucleic acid analog is the grafting of the nucleic acid base derivatives onto the functional polymers, such as poly-(acrylate), poly(lysine), and poly(ethyleneimine) derivatives containing nucleic acid bases.[13] In order to prepare water soluble polymers containing high content of nucleic acid bases, hydrophilic units should be introduced to the side chains. Overberger and Inaki prepared poly(ethylene-imine) derivatives containing nucleic acid bases and in this case amino acids were used as the spacer.[21] One of these derivatives was the poly-(ethyleneimine) having nucleic acid base and serine as a spacer, which was soluble in water. For the preparation of the serine derivative, however, protection of the functional groups was necessary.

In the present study, (±)-α-amino-γ-butyrolactone hydrobromide was used as a starting compound. This lactone has a stable lactone ring with the hydroxyl and carbonyl groups protected, and gave a homoserine unit by a ring opening reaction. By using this lactone, ethyleneimine polymers containing both nucleic acid base and homoserine units were successfully prepared in high yield. Scheme 1 shows the preparation of the water soluble nucleic acid analogs containing various nucleic acid bases. The butyrolactone derivatives of nucleic acid bases were obtained as stable solid in high yield.

2. Homoserine Derivatives of Nucleic Acid Bases

Starting from nucleic acid bases, the carboxyethyl derivatives of the nucleic acid bases were prepared by a Michael type addition reaction of ethyl acrylate followed by hydrolysis. The carboxyethyl derivatives of nucleic acid bases were reacted with (±)-α-amino-γ-butyrolactone hydrobromide to give γ-butyrolactone derivatives of nucleic acid bases. For the coupling reactions, pentachlorophenyl ester derivatives were used with imidazole as a catalyst.[22]

3. Grafting Onto Poly(ethyleneimine)

The grafting of nucleic acid base derivatives with a hydroxyl group onto poly(ethyleneimine) polymer backbone was also carried out by the activated ester method. Since the reactivity of the γ-lactone is low, the direct reaction of the lactone derivative with poly(ethyleneimine) was not effective. Therefore, the lactone derivatives were at first hydrolyzed to the 3-hydroxybutyric acid derivatives, followed by condensation with poly(ethyleneimine) using the activated ester method. The grafting reaction was carried out in N,N-dimethylformamide, where a small amount of 4-pyrrolidinopyridine was added as an effective catalyst. Nucleic acid base contents of the polymers were determined by UV spectroscope of hydrolyzed samples. A quantitative calculations were made by using the corresponding carboxyethyl derivatives as standards. The nucleic acid base units (unit mol%) on the polymer are tabulated in Table 1.

4. Interaction of Synthetic Polymers with Polynucleotides

The formation and stoichiometry of the polymer complex between nucleic acid analogs are effected by several factors. One of the factors is the property of the polymer backbone: its flexibility, steric regularity, electric charge, branching, and molecular weight. These factors can reflect the compatibility, penetration ability of the polymer and the stability of the polymer complex. In general, polynucleotides have flexible, sterically regular and negatively charged polymer backbones. Whereas the synthetic analogs are probably less flexible, sterically inhomogeneous and neutral. Other important factors for the specific polymer complex include temperature and solvent conditions for the interaction. Further-

Table 1. Composition of samples.			
Nucleic acid base	Polymer	Content	Yield
Uracil	PEI-Hse-Ura	94%	78%
Thymine	PEI-Hse-Thy	97%	78%
5-Fluorouracil	PEI-Hse-5FU	91%	68%
Hypoxanthine	PEI-Hse-Hyp	92%	68%
Adenine	PEI-Hse-Ade	92%	80%
Cytosine	PEI-Hse-Cyt	86%	78%

more, the pH and concentration of salts are important for the complex formation in an aqueous solution.[23,24] In the present study, the formation of the polymer complexes of the poly(ethyleneimine) derivatives were investigated. Further study made also for the poly(ethyleneimine) derivatives-polynucleotides systems.

5. Interaction of PEI-Hse-Ura and PEI-Hse-Ade

The interactions between water soluble poly(ethyleneimine) derivatives containing uracil (PEI-Hse-Ura, 5U) and those containing adenine (PEI-Hse-Ade, 5A) were studied at pH 7.0 with continuous variation techniques (Job plot). As shown in Figure 1b, the stoichiometry of the complex based on nucleic acid base units was determined to be 1:1 (uracil: adenine) from the base ratio at the maximum hypochromicity value (15.2%). Hypochromicity has been widely used to indicate the interaction of nucleic acid base derivatives. The hypochromicity value obtained was higher than that for the PEI-Ade: PEI-Thy system which has no spacer groups.[22] The hypochromicity values for the PEI-Hse-Ura and PEI-Hse-Ade system (1:1 base unit) were determined at pH 2.2 and 5.5. At pH 2.2 the hypochromicity was negligible even after 3 days. On the other hand, the hypochromicity value at pH 5.5 was 9.1%, which was compared to the value at pH 7.0. Adenine base has a pKa value at 4.15, and exists in a protonated form at pH 2.2. The protonated adenine base can not form a complex with the complimentary thymine base. This may be the reason for the negligible hypochromicity value at pH 2.2 and the high hypochromicity value at pH 5.5.

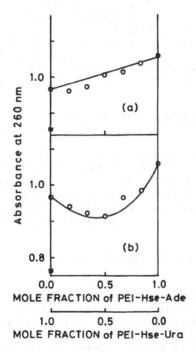

Figure 1. Continuous variation curve of PEI-Hse-Ura:PEI-Hse-Ade system. Absorbance at 260 nm after (a) 3 hours and (b) 3 days in 0.05 M Kolthoff buffer solution (pH 7.0) at 20°C. [PEI-Hse-Ura] = 1.1 x 10⁻⁴ mol/L, and [PEI-Hse-Ade] = 8.3 x 10⁻⁵ mol/L.

From these facts, the polymer complex between PEI-Hse-Ura and PEI-Hse-Ade at pH 7.0 was concluded to be formed by the complimentary nucleic acid base interaction.

For the system in question, time dependence of hypochromicity value was observed as shown in Figure 1. The hypochromicity was not observed in 3 hours after mixing of the polymer solutions (Figure 1a). The absorbance decreased (the hypochromicity increased) and then became constant after 3 days (Figure 1b). The conformational change of the synthetic polymers should be thus important for the formation of a stable complex.

The time dependency in Figure 1 suggests that the self-association of the nucleic acid bases in the poly(ethyleneimine) derivatives dissociated slowly by accompanying change in conformation, and formed the intermolecular polymer complex by the interaction between adenine and thymine bases. Similar results were reported on the polymer complex formation between poly(methacrylate) derivatives of uracil and adenine, where the self association of the uracil bases in the polymer inhibited the polymer complex formation.[22,25,26] Therefore, the conformational change of the synthetic polymers seems to be important for the formation of a stable polymer complex.

6. Interactions of PEI-Hse-Cyt and PEI-Hse-Hyp

Figure 2 shows the mixing curve between Pei-Hse-Cyt(5c) and PEI-Hse-Hyp(5h) in the Kolthoff buffer solution (pH 7.0). The figure shows the highest hypochromicity value at a base unit ratio of about 1:1 (hypoxanthine:cytosine), suggesting the formation of a stable 1:1 polymer complex due to complimentary nucleic acid base interactions. The hypochromicity value (21%) was higher than that of the PEI-Ade:PEI-Thy system in the literature.[24] As shown in Figure 2, time dependence of the hypochromicity value was observed. Therefore the conformational change of synthetic polymers may be important for formation of a stable polymer complex.

7. Interactions of PEI-Hse-Hyp with Poly(C)

Interactions between PEI-Hse-Hyp and poly(C) which contains the complimentary nucleic acid base can be clearly observed in aqueous solution at pH 7.0, as shown in Figure 3. The overall stoichiometry of the complex based on the nucleic acid base units was approximately 2:1 (hypoxanthine: cytosine) under the conditions used. The maximum hypochromicity value (26%) was smaller than that of the poly(I):poly(C) (33%) system.

8. Interactions of PEI-Hse-Cyt and Poly(I)

Figure 4 shows the mixing curve between PEI-Hse-Cyt and poly(I) after 3 h and 3 days at pH 7.0 in the Kolthoff buffer aqueous solution. The stoichiometry of the complex was 2:1 (hypoxanthine:cytosine) and the hypochromicity value was 7%. The stoichiometry of the PEI-Hse-Cyt:poly(I) complex was similar to that of PEI-Hse-Hyp:poly(C) system.

The hypochromicity value was small as compared with that of the PEI-Hse-Hyp:poly(C) system. Polynucleotides containing purine base form very stable structures by self association in aqueous solution. On the other hand, the structures of polynucleotides containing pyrimidine bases

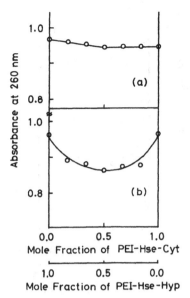

Figure 2. Continuous variation curve of PEI-Hse-Hyp:PEI-Hse-Cyt system. Absorbance at 260 nm after (a) 3 hours and (b) 3 days in 0.05 M Kolthoff buffer solution (pH 7.0) at 20°C [PEI-Hse-Hyp] = 9.5 x 10^{-5} mol/L, and [PEI-Hse-Cyt] = 9.6 x 10^{-5} mol/L.

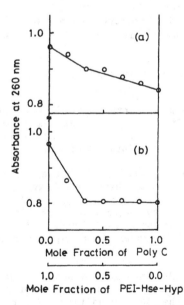

Figure 3. Continuous variation curve of PEI-Hse-Hyp:poly(C) system. Absorbance at 260 nm after (a) 3 hours and (b) 3 days in 0.05 M Kolthoff buffer solution (pH 7.0) at 20°C [PEI-Hse-Hyp] = 9.5 x 10^{-5} mol/L, and [poly(C)] = 1.2 x 10^{-4} mol/L.

37

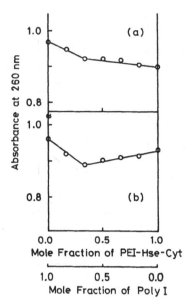

Figure 4. Continuous variation curve of PEI-Hse-Cyt:poly(I)
system. Absorbance at 260 nm after (a) 3 hours and
(b) 3 days in 0.05 M Kolthoff buffer solution (pH
7.0) at 20°C [poly(I)] = 1.2 x 10⁻⁴ mol/L, and
[PEI-Hse-Cyt] = 9.6 x 10⁻⁵ mol/L.

are not stable in aqueous solution. Poly(I) forms stable structures by
self-association in aqueous solution. Therefore, the intermolecular in-
actions of poly(I) with PEI-Hse-Cyt become small because of enhanced
intramolecular interactions of hypoxanthine bases in poly(I).[28,29]

9. Interaction of PEI-Hse-Ura with Poly(A)

Figure 5 shows the mixing curves for PEI-Hse-Ura with poly(A) at pH
7.0. From the curve in Figure 5b, the maximum hypochromicity value obtained was 54.4%. The overall stoichiometry of the complex based on
nucleic acid base units was 1:1 (uracil:adenine).

As a control experiment, interaction between poly(A) and poly(U) was
studied under the conditions used here. As shown in Figure 6, the complex
formation was observed immediately after mixing of the polymer solutions,
and for maximum hypochromicity value was 40.3%. The hypochromicity value
for the PEI-Hse-Ura:poly(A) system was higher than the poly(U):poly(A)
system, PEI-Hse-Ade:poly(U) system, and any other nucleic acid analog-
polynucleotide system.

To determine whether the interaction between the bases is due to the
specific base-base interaction, the interaction of PEI-Hse-Ura with poly-
(U) was measured under the same condition. Figure 7 shows the mixing
curves for PEI-Hse-Ura:poly(U) at pH 7.0. The hypochromicity, however,
could be observed neither after 3 hours (Figure 7a) nor after 3 days
(Figure 7b).

This result indicates that the uracil bases of PEI-Hse-Ura do not

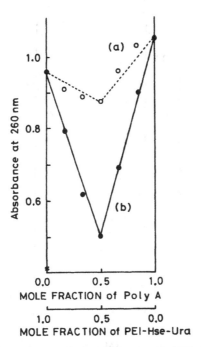

Figure 5. Continuous variation curve of PEI-Hse-Ura:poly(A)
system. Absorbance at 260 nm after (a) 3 hours and
(b) 3 days in 0.05 M Kolthoff buffer solution (pH
7.0) at 20°C [PEI-Hse-Ura] = 1.1 x 10⁻⁴ mol/L, and
[poly(A)] = 1.2 x 10⁻⁴ mol/L.

interact with the uracil bases nor the phosphate units of poly(A). Based
on these facts, the formation of the PEI-Hse-Ura:poly(A) complex in
Figure 5b is concluded to be caused by the complimentary base-base interaction between adenine and uracil.

10. Interaction of PEI-Hse-5FU with Poly(A)

Figure 8 shows the mixing curves of PEI-Hse-5FU (5FU) with poly(A) at
pH 7.0. The maximum hypochromicity value was obtained for the base unit
ratio of 2:1 (5-fluorouracil:adenine). The maximum hypochromicity value
of the PEI-Hse-5FU:poly(A) system was 40.3%, which was smaller than the
PEI-Hse-Ura:poly(A) system (54.4%), equal to the poly(U):poly(A) system
(40.3%), and higher than the PEI-Hse-Thy:poly(A) system (38.5%). These
facts suggest that the incorporation of the 5-fluorouracil base results
in a decrease in the stability of the polymer complex.

A remarkable effect of 5-fluorouracil was observed for the stoichiometry of the polymer complex. The stoichiometry of the polymer complex
between PEI-Hse-5FU and poly(A) was 2:1 (5-fluorouracil:adenine, Figure
8b), which is the same as the poly(A):poly(U) polymer complex. On the
other hand, the stoichiometry of the PEI-Hse-Ura:poly(A) polymer complex
was 1:1 (Figure 5b). The reason for the formation of the 2:1 complex of
poly(A) with poly(U) is due to the tendency of poly(A) to form a single
stranded structure. The 2:1 polymer complex of PEI-Hse-5FU:poly(A),
therefore, might be formed by the same way as the poly(A):poly(U)
complex.

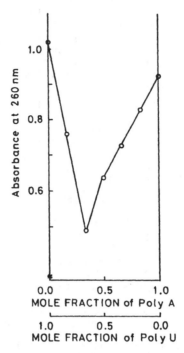

Figure 6. Continuous variation curve of poly(U):poly(A) system. Absorbance at 260 nm after (a) 3 hours and (b) 3 days in 0.05 M Kolthoff buffer solution (pH 7.0) at 20°C. [poly(U)] = 1.0 x 10⁻⁴ mol/L, and [poly-(A)] = 9.9 x 10⁻⁵ mol/L.

11. Interaction of the PEI-Hse-Ade and Poly(U)

Figure 9 shows the mixing curve between PEI-Hse-Ade (5a) and poly(U), for 3 hours (a) and 3 days (b) after mixing the polymer solution. In this case, the formation of the polymer complex was observed even after 3 hours, as shown in Figure 9a. The stoichiometry of the complex was 1:1 (thymine:adenine) and the maximum hypochromicity value was 49.6% at this base ratio. This value was a little higher as compared with the values obtained for PEI-Hse-Thy:PEI-Hse-Ade and poly(A):poly(U) systems. The formation of the polymer complex, however, was negligible at pH 2.2, where the adenine base exists in a protonated form. From these facts, it should be concluded that formation of the polymer complex between PEI-Hse-Ade and poly(U) was caused by the specific interaction between adenine and uracil, and the self association of PEI-Hse-Ade in aqueous solution was negligible.

12. Interaction of PEI-Hse-Ura with Poly(C)

It is well known that the complimentary base of uracil is not cytosine but adenine. However, as seen in Figure 10b, the hypochromic effect was also observed for the interaction of PEI-Hse-Ura and poly(C), for 3 days after mixing the polymers. The overall stoichiometry of the system was 1:1 (uracil:cytosine). The maximum hypochromicity value (20.1%),

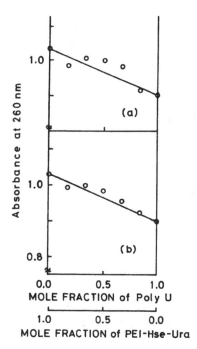

Figure 7. Continuous variation curve of PEI-Hse-Ura:poly(U) system. Absorbance at 260 nm after (a) 3 hours and (b) 3 days in 0.05 M Kolthoff buffer solution (pH 7.0) at 20°C. [PEI-Hse-Ura] = 1.3×10^{-4} mol/L, and [poly(U)] = 1.1×10^{-4} mol/L.

however, was smaller than that of the PEI-Hse-Ura:poly(A) system (Figure 5b).

As a control experiment, the interaction between poly(U) and poly(C) was also studied under the conditions used. The complex formation was observed immediately after mixing the polymer solutions, and a similar curve was obtained even after 3 days. The maximum hypochromicity (5.5%) was observed at base unit ratio of 1:1 (uracil:cytosine). The maximum hypochromicity value of the poly(U):poly(C) system, however, was smaller than that for the PEI-Hse-Ura:poly(C) system. Therefore, the interaction between PEI-Hse-Ura and poly(C) was concluded to be caused by the interaction between uracil and cytosine bases.

13. Selectivity of the Base-Base Interaction

The maximum hypochromicity values and stoichiometries of the polymer complexes are summarized in Table 2. In this table, the value of selectivity (S) was calculated by Equation 1, where $H_{(A)}$, and $H_{(C)}$ are the maximum hypochromicity values of the uracil derivatives with poly(A) and with poly(C), respectively. The selectivity value means the selectivity of the uracil derivatives to the complementary poly(A) based on poly(C). PEI-Hse-Ura has the highest hypochromicity value for poly(A), but the selectivity value (s:2.7) is low. On the other hand, PEI-Hse-Thy has the lowest hypochromicity value for poly(A), but the selectivity value is

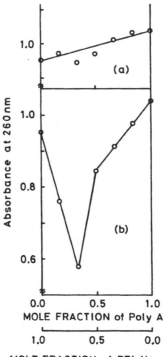

Figure 8. Continuous variation curve of PEI-Hse-5FU:poly(A)
system. Absorbance at 260 nm after (a) 3 hours and
(b) 3 days in 0.05 M Kolthoff buffer solution (pH
7.0) at 20°C. [PEI-Hse-5FU] = 9.6×10^{-5} mol/L, and
[poly(A)] = 1.2×10^{-4} mol/L.

high. PEI-Hse-5FU was found to be highly selective, and the hypochromicity value is the same with that of PEI-Hse-Ura.

$$S = H_{(A)}/H_{(C)}$$ (Equation 1)

Table 2. The maximum hypochromicity values. Stoichiometry and
the values of selectivity (S) for nucleic acid
analogs - polynucleotides systems.

	Poly(U)	PEI-Hse-Ura	PEI-Hse-5FU	PEI-Hse-Thy
Poly(A)	40.3%	54.4%	40.3%	38.4%
	2:1	1:1	2:1	2:1
Poly(C)	5.5%	20.1%	4.0%	3.0%
	1:1	1:1	1:1	1:1
S	7.3	2.7	10.1	12.8

(a) The maximum hypochromicity values are above the stoichiometry (uracil derivative:adenine or cytosine) ratios.

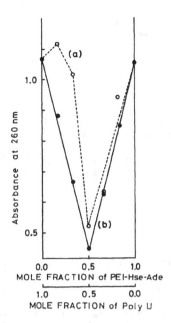

Figure 9. Continuous variation curve of PEI-Hse-Ade:poly(U) system. Absorbance at 260 nm after (a) 3 hours and (b) 3 days in 0.05 M Kolthoff buffer solution (pH 7.0) at 20°C. [Poly(U)] = 1.1 x 10^{-4} mol/L, and [PEI-Hse-Ade] = 8.3 x 10^{-5} mol/L.

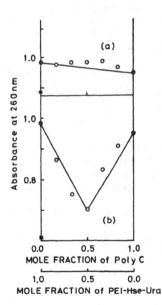

Figure 10. Continuous variation curve of PEI-Hse-Ura:poly(C) system. Absorbance at 260 nm after (a) 3 hours and (b) 3 days in 0.05 M Kolthoff buffer solution (pH 7.0) at 20°C. [PEI-Hse-Ura] = 1.2 x 10^{-4} mol/L, and [poly(C)] = 1.4 x 10^{-4} mol/L.

The data in Table 2, therefore, should indicate that the higher the hypochromicity values the lower the selectivity for the interaction with the complementary base.

CONCLUSION

Water soluble poly(ethyleneimine) derivatives containing both nucleic acid bases and homoserine units were prepared. The poly(ethyleneine) derivatives were found to form the 1:1 polymer complex by complementary base pairing in aqueous solution. These polymers also formed the polymer complexes with polynucleotides by specific interaction between complementary nucleic acid bases.

REFERENCES

1. R. M. Ottenbrite, "*Anionic Polymeric Drugs*," L. G. Donaruma, R. M. Ottenbrite, O. Vogl, Eds., John Wiley & Sons, Inc., p. 1, (1980).
2. J. H. Park, M. A. Gallin, A. Billiau, and S. Baron, Arch. Ophthal., 81, 840 (1969).
3. H. B. Levy, J. Bioactive and Compatible Polymers, 1, 348 (1986).
4. A. K. Field, A. A. Tytell, G. P. Lampson, and M. R. Hillman, Proc. Nat. Acad. Sci. U.S.A., 58, 1004 (1967).
5. M. M. Nemes, A. A. Tytell, G. P. Lampson, A. K. Field, and M. R. Hillman, Proc. Soc. Exp. Biol. Med., 132, 776 (1969).
6. V. DeVita, G. Canellos, P. Carbone, H. Levy, and H. Graenick, Proc. Am. Assoc. Cancer Res., 11, 21 (1970).
7. M. Absher and W. R. Stinebring, Nature, 223, 715 (1969).
8. T. Sato, K. Kojima, T. Ihda, J. Sunamoto, and R. M. Ottenbrite, J. Bioactive and Compatible Polymers, 1, 448 (1986).
9. K. Takemoto and Y. Inaki, Adv. Polym. Sci., 41, 1 (1981).
10. J. Pitha, Adv. Polym. Sci., 50, 1 (1983).
11. A. G. Ludwick, K. S. Robinson, and J. McCloud, Jr., J. Polym. Sci. Polym. Symp., 74, 55 (186).
12. Y. Inaki and K. Takemoto, "*Current Topics in Polymer Science*," Volume I, R. M. Ottenbrite, L. A. Utracki, and S. Inoue, Eds., Hanser Publisher, p. 79, (1987).
13. K. Takemoto and Y. Inaki, "*Functional Monomers and Polymers*," K. Takemoto, Y. Inaki, and R. M. Ottembrite, Eds., Marcel Dekker, Inc., p. 149, (1987).
14. J. Sunamoto and K. Takemoto, to be published.
15. T. Wada, Y. Inaki, and K. Takemoto, Polym. J., 21, 11 (1989).
16. T. Wada, Y. Inaki, and K. Takemoto, Polym. J., 20, 1059 (1988).
17. T. Wada, Y. Inaki, and K. Takemoto, J. Bioactive and Compatible Polymers, 4, 25 (1989).
18. G. J. Thomas, Jr., and Y. Kyogoku, J. Am. Chem. Soc., 89, 4170 (1967).
19. J. Pitha, P. M. Pitha, and E. Stuart, Biochemistry, 10, 4595 (1971); Biopolymers, 9, 965 (1970).
20. C. G. Overberger, Y. Inaki, and Y. Nambu, J. Polym. Sci. Polym. Chem. Ed., 17, 1759 (1979).
21. C. G. Overberger and Y, Inaki, J. Polym. Sci. Polym. Chem. Ed., 17, 1739 (1979).
22. Y. Inaki, Y. Sakuma, Y. Suda and K. Takemoto, J. Polym. Sci. Polym. Chem. Ed., 20, 1917 (1982).
23. D. Thiele, W. Guschlbauer, and A. Faver, Biochim. Biophys. Acta, 272, 22 (1972).

24. K. Fujioka, Y. Baba, A. Kagemoto, and R. Fujishiro, Polym. J., 12, 843 (1980).
25. S. Fang, Y. Inaki, and K. Takemoto, J. Polym. Sci. Polym. Chem. Ed., 22, 2455 (1984).
26. Y. Inaki, S. Sugita, T. Takahara, and K. Takemoto, J. Polym. Sci. Polym. Chem. Ed., 24, 3201 (1986).
27. P. O. P. Ts'o, Ed., "*Basic Principles of Nucleic Acid Chemistry,*" Vols. 1 and 2, Academic Press, New York (1974).
28. S. Higuchi, Biopolymers, 23, 493 (1984).
29. S. Higuchi, Biopolymers, 23, 831 (1984).

INFLUENCE OF COPOLYMER STRUCTURE ON PROPERTIES OF

POLY-β-HYDROXYALKANOATES

R. H. Marchessault and C. J. Monasterios

McGill University
Chemistry Department
3420 University St.
Montreal, Canada H3A 2A7

Poly-β-hydroxyalkanoates are a family of natural polyesters which are often in the form of copolymers with homologous alkane sidechains. Commercial PHA is a microbial, random copolymer of butyrate and valerate repeating units. The product is prepared by isolating the nascent granules which are in the cytoplasmic fluid of the cell. The range of properties covered by the various copolymers that have been isolated and studied can be demonstrated by plotting versus the "average number of sidechain carbons." In this fashion, one notes glass transition temperatures from 10 to -40°C as one progresses from C_1 to C_6 sidechain. Similarly, melting point goes from 184 to about 50°C, and density from 1.24 to 1.04 g/cc. As the sidechain length increases, the thermoplastic material of relatively high crystallinity (~60%) changes to a thermal elastomer of ~25% crystallinity. As the sidechain lengthens, the base of the unit cell gets increasingly rectangular and the molecule develops a distinct herringbone or comb-like conformation. The organic synthesis of PHA copolyesters using alumoxane catalysts and racemic lactones yields crystalline products which mimic the properties of the natural polyesters.

INTRODUCTION AND STRUCTURES

A family of natural polymers which are thermoplastic and water stable, but nevertheless biodegrade, are industrial biotechnology mimics of polyolefins.[1] These molecules, functionally akin to starch are the food reserve of the bacterial world.[2] They are simple polyesters with properties like polypropylene, but environmentally benign because in soil or water they are substrates for microorganisms.

$$[-O-\overset{R}{\underset{|}{C}H}-CH_2-\overset{O}{\overset{\|}{C}}-] \qquad R = CH_3, C_2H_5, CH_5H_{11}, C_6H_{13}, etc.$$

Biotechnology and Polymers, Edited by C.G. Gebelein
Plenum Press, New York, 1991

While the first member of the series in the above formula, called poly-β-hydroxybutyrate (PHB), is ubiquitous, it is now well established that certain bacterial species, on given alkane substrates, provide a whole family of the copolyesters of the above type.[3] This occurs by incorporating monomers with the same number of carbons as in the alkane substrate or differing by ± 2 carbons as a result of the classical β-oxidation physiological cycle. The presence of copolyesters of the PHA type in nature demonstrates adaptability to substrate variation.

Similarly, for commercial exploitation of PHB, it soon became apparent that a family of materials differing in properties was required to satisfy the variety of markets that were bound to develop. This was achieved by Imperial Chemical Industries (ICI) by a classical fermentation route using a pure carbohydrate substrate to which varying amounts of propionic acid was added to produce copolyesters containing only butyrate (HB) and valerate (HV) comonomers.[4]

$$
\begin{array}{cc}
\underset{\text{HB}}{-[-\overset{\displaystyle\overset{CH_3}{|}}{CH}-CH_2-\overset{\displaystyle\overset{O}{\|}}{C}-O-]_x} \text{——} & \underset{\text{HV}}{[-\overset{\displaystyle\overset{\overset{CH_3}{|}}{\underset{|}{CH_2}}}{CH}-CH_2-\overset{\displaystyle\overset{O}{\|}}{C}-O-]_y-}
\end{array}
$$

BIOSYNTHESIS

The biosynthesis of PHB/V would appear to be like an emulsion polymerization with synthetase enzymes forming a micellar structure around each granule and perhaps providing some control on the size of the individual granules.[5] The isolation process used by ICI resembles wood pulping technology in that the granules are made to persist while the cell wall, proteins and nucleic acids, etc. are washed away by a variety of mechanical and enzymatic processes.[4,6] In the isolated state the spray dried powder of PHA is a colloid similar to the original PHA granules, as shown in Figure 1. The average particle size is about 0.5 μm and one frequently sees that two granules have coalesced. This supports the concept that the nascent state is paracrystalline with a mobility which favors the required enzymatic transformations, which would be less true if the nascent state were genuinely crystalline as was originally postulated by one of us.[7]

PROPERTIES

The range of properties which are now available in the PHB/V family from ICI is demonstrated in Figure 2 where the glass transition of PHA copolyester as a function of the sidechain length is plotted for naturally occurring materials.[3,8,9] For sidechains ranging from C_1 (PHB) to C_6, the range of T_g is from about 10°C to -40°C. The latter is typical of an elastomer and, indeed, when the sidechain is in the range of C_3 to C_6, the materials are truly thermal elastomers with greatly decreased crystallinity compared to PHB and the PHB/V copolyesters which are the thermoplastics.[9,10,11]

The crystalline properties of PHB/V materials have been well described, especially the high crystallinity of PHB and its tendency to form

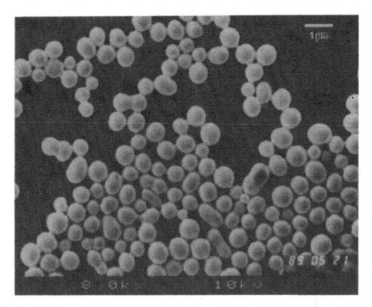

Figure 1. Scanning electron micrograph of PHB particles prepared by Imperial Chemical Industries using enzyme purification methods that preserve the native granules. (1cm = 1μm)

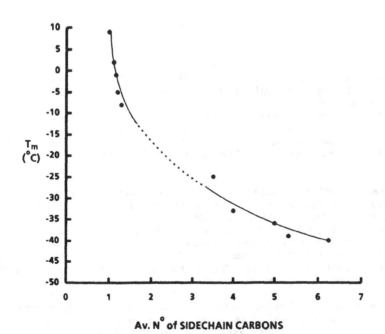

Av. N° of SIDECHAIN CARBONS

Figure 2. Glass transition temperature (T_g) of PHA as a function of average sidechain length.

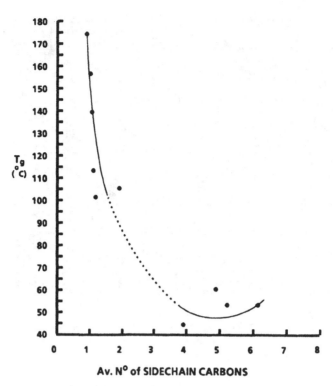

Figure 3. Melting point (T_m) of PHA versus average number of sidechain carbon atoms.

large spherulites.[8] This necessitates the inclusion of nucleating agents for commercial plastics applications, to control the spherulite size and accelerate the rate of crystallization, which is distressingly slow in PHB/V with valerate contents in the 10-30% range.[12] Nevertheless, these PHB/V materials eventually develop a level of crystallinity comparable to PHB itself because the system is isodimorphic.[10] Figure 3 shows the melting point (T_m) variation as a function of sidechain length for natural PHA.

In the case of the longer sidechain materials the degree of crystallinity that develops at room temperature after melting is 25% or less (see Figure 4). This lower degree of crystallinity is due to the high content of mobile alkyl component which extends from the backbone. The relatively high mobility of the sidechain carbons versus the low mobility of the backbone carbons was demonstrated by direct comparison of the ^{13}C spin lattice relaxation times (T_1) of the sidechain carbons compared to backbone carbons in samples with a C_5 sidechain content of 85%.[13]

As the sidechain length increases, the baseplane dimensions of the PHA becomes more rectangular as shown in Figure 5. At the same time the density of the material decreases in keeping with differences in inter-molecular packing forces (see Figure 6). Nevertheless, the dipolar inter-actions of the backbone atoms are retained since the 2_1 helix symmetry persists although the fiber repeat ($\underline{c}$-axis of unit cell) itself drops

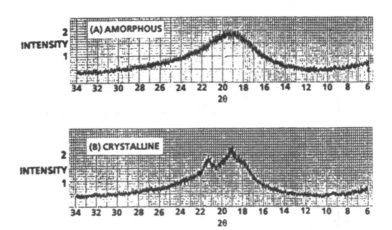

Figure 4. X-ray diffraction of PHA(C₆).

regularly from 5.96 A to 4.55 A as the sidechain increases its length. The latter extends from the helical backbone in a planar zig-zag conformation forming a planar molecule which is a herringbone or a two-sided comb depending on the dihedral angle at the C_β-C_g bond.[13]

ORGANIC SYSTHESIS

The organic synthesis of PHA copolyesters is a natural extension of the pioneering work by Shelton, Agostini and Lando using alumoxane catalysts for a ring opening lactone polymerization.[14] The crystalline properties of the synthetic PHB product of a racemic monomer feed suggested a stereoblock structure. The recent work of Bloembergen, et al., in making synthetic PHB/V whose properties mimic those of the natural polyester opens a new chapter in the field.[15] The synthetic polymers are both steric and chemical copolymers. Hence the physical properties of PHA are strongly dependent on the tacticity index. However, the high tacticity analogs of PHB/V are similar to the bacterial polyesters in most respects.

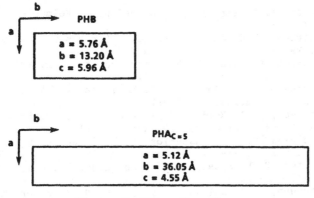

Figure 5. PHA unit cells.

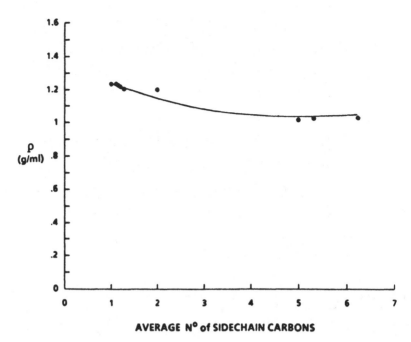

Figure 6. Density ($\gamma/\langle$) of PHA as a function of average
sidechain length.

REFERENCES

1. P. A. Holmes, Phys. Technol.,**16**, 32 (1985).
2. E. A. Dawes, "*Microbial Energetics,*" Blackie, Glasgow (1986).
3. R. A. Gross, C. DeMello, R. W. Lenz, H. Brandl and R. C. Fuller, Macromolecules, **22**, 1106 (1989).
4. P. A. Holmes, L. F. Wright and S. H. Collins, European Patent Application 0 052 459 (1981); P. A. Holmes, L. F. Wright, and S. H. Collins, European Patent Application 0 069 497 (1981).
5. D. Ellar, D. Lundgren, K. Okamura and R. H. Marchessault, J. Mol. Biol., **35**, 489 (1968).
6. ICI, Agricultural Division, Billingham, U. K., private Communication.
7. Jeremy K.M. Sanders and Glenn N. Barnard, FEBS Letters, **231**, 16 (1988).
8. P. J. Barham, A. Keller, E. L. Otun and P. A. Holmes, J. Mater. Sci., **19**, 2781 (1984).
9. S. Bloembergen, Ph.D. Thesis, Chemistry Department, University of Waterloo, Waterloo, Ontario, Canada (1987).
10. T. L. Bluhm, G. K. Hamer, R. H. Marchessault, C. A. Fyfe and R. P. Veregin, Macromolecules, **19**, 2680 (1986).
11. R. H. Marchessault, S. Coulombe, H. Morikawa, K. Okamura and J. F. Revol, Can. J. Chem., **59**, 38 (1981).
12. S. Bloembergen, D. A. Holden, G. K. Hamer, T. L. Bluhm and R. H. Marchessault, Macromolecules, **19**, 2865 (1986).
13. R. H. Marchessault, C. Monasterios, F. Morin and P. R. Sundararajan, J. Biol. Macromolecules, in press.
14. J. R. Shelton, D. E. Agostini and J. B. Lando, J. Polym. Sci., Chem. Ed., **9**, 2789 (1971).
15. S. Bloembergen, D. A. Holden, T. L. Bluhm, G. K. Hamer and R. H. Marchessault, Macromolecules, **22**, 1663 (1989).

BIODEGRADATION OF BLENDS CONTAINING POLY(3-HYDROXYBUTYRATE-CO-VALERATE)

P. Dave,[a] R. A. Gross,[b] C. Brucato,[b] S. Wong,[b] and
S. P. McCarthy[a]

University of Lowell
Departments of Plastics Engineering[a] and Chemistry[b]
Lowell, MA 01854

Polymer blends containing poly(3-hydroxybutyrate-co-16%-3-hydroxyvalerate, (P(HB-coOHV), were investigated by incorporating poly(ε-caprolactone), (PCL), poly(styrene-co-35%-acrylonitrile), (SAN), or poly(styrene), (PS), as the second component. Blend concentrations of 20/80, 50/50 and 80/20 weight percents were solvent cast from chloroform and analyzed for miscibility using DSC. Biodegradation studies were conducted using an exposure to the extracellular depolymerase from *Penicillium funiculosum* at 28°C for 48 and 96 hours. Weight loss and characterization of the exposed samples using GPC, SEM and NMR were used to determine degree of biodegradation. Blends of P(HB-co-HV) with PCL or PS were found to be immiscible at all concentrations investigated. Blends of P(HB-co-HV) with SAN were found to be miscible in all concentrations investigated. Biodegradation results for the miscible blend showed a lower weight loss as compared to the immiscible blends investigated. Weight loss increased with increasing P(HB-co-HV) content for all blends investigated.

INTRODUCTION

The development of polymers and blends that exhibit a high susceptibility to microbial attack is of great importance as a viable approach for relieving environmental concerns caused by the current use of non-degradable disposable plastics. There is also a growing interest within the medical community for specific applications of polymers where degradation within the body is important. Indeed, the spectrum of biological properties that would be ideally required for applications ranging from sutures, skin staples, fascia clips, bone plates, and other surgical fixation devices is much wider than can be achieved with existing materials. A number of studies have been published which indicate the utility of P(HB-co-HV) for sustained release dosage applications.[1,2]

Poly(3-hydroxyalkanoates), PHAs, are a class of polyesters which occur naturally in a wide variety of bacterial microorganisms throughout nature.[3-5] The organism *Alcaligenes eutrophus* has been used to produce copolymers of 3-hydroxybutyrate (HB) and 3-hydroxyvalerate (HV), where the composition of the polymer ranged from 100% HB to 95% HV.[6-10] At low

Biotechnology and Polymers, Edited by C.G. Gebelein
Plenum Press, New York, 1991

HV content, P(HB-co-HV) is hard and brittle, resembling unplasticized polyvinylchloride or polystyrene.[6,7] As one increases the HV content of P(HB-co-HV) from 0 to 20 mole%, the extension to break (%) increases from 6% to 100%.[11] As a logical extension of the above, a thorough investigation of blends which contain bacterial polyesters as a degradable component should provide diverse new materials where the bacterial polyester component will degrade to non-toxic, readily metabolized compounds. Many bacteria, when presented with P(HB), were shown to be capable of using this substance as an exogenous source of carbon by excreting extracellular enzymes that depolymerize it.[12-14] Deasy et. al.[14] carried out the isolation of 16 bacterial strains from soil which were capable of growing with P(HB) as the sole source of carbon. Fukui and coworkers isolated from activated sewage sludge a bacterial stain which shows exceptionally high activity in degrading P(HB).[12,13]

Blends of polyethylene and starch have gained increasing scientific interest in recent years. Griffin[15] has reported the use of silicones and isocyanates as compatabilizers for the polyethylene/starch blends. Biodegradability of these blends was reported as proceeding initially by a degradation of starch particles leaving a cellular structure which is more readily attacked by oxidation, hydrolysis, direct enzyme action or combination of these processes.[15] Maddever and Chapman have reported a decrease in the molecular weight of the polyethylene in a 7% starch blend after 12 weeks in a compost.[16] This blend also contained an auto-oxidant which is proposed to cause degradation through the generation of peroxides. It was not concluded by the investigators as to whether the decrease in the polyethylene molecular weight was the result of microbial attack or peroxide generation. Otey and Westoff [17] have done similar work with blends of starch and an ethylene acrylic acid copolymer. Soil burial tests conducted on these materials have shown significant degradation for the high starch concentrations within weeks of exposure.

Blends of PHB and poly(ethylene oxide) have been shown to be miscible from work reported by Avella and Martuscelli.[18] Degradability studies were not conducted on these blends although the individual components have been shown to be biodegradable.[7-19] Yasin et. al. have shown an increase in the hydrolytic degradation of blends of P(HB-co-HV) with polysaccarides over that of pure P(HB-co-HV).[20] SEM photomicrographs showing the residual porous spongelike matrix left behind after the polysaccaride dissolves out indicates an immiscible blend where the polysaccaride phase is removed preferentially.[20] Work reported by Holmes[21] suggests that blends of P(HB) that contain 25% or more chlorine or nitrile groups may show miscibility.

In the present study, the dependence of biodegradation rates on the morphology of the phase domains and miscibility of blend components has been investigated. The miscibility of P(HB-co-16%HV) with other commercial polymers was investigated using differential scanning calorimetry (DSC). Degradation rates were measured using an extracellular depolymerase. Exposure to simulated landfill and marine environments as well as an aerobic composting facility is ongoing. Further characterization of blends before and after degradation exposures utilized GPC, SEM, NMR, tensile strength, and weight loss data.

EXPERIMENTAL

1. Materials

    Poly(3-hydroxybutyrate-co-3-hydroxyvalerate), P(HB-co-HV), with 0% to

22% hydroxyvalerate content, was supplied by ICI, PLC, Agricultural Division, Billingham, UK, under the tradename BIOPOL.. Poly-caprolactone (PCL) was supplied by Union Carbide Corp., NJ. General purpose crystal polystyrene (PS) was supplied by Polysar, Inc., MA. Poly(L-lactide) (PLA) was supplied by Boehringer Ingelheim KG, W. Germany. Poly(styrene-co-acrylonitrile) (SAN) was supplied by Monsanto Chemical Co., MA. Poly-(acrylonitrile-co-butadiene-co-styrene) (ABS) was supplied by Monsanto Chemical Co., MA. Polyvinylchloride (PVC) was supplied by Exxon Chemical Co., Canada.

2. Methods

2A. Blend Miscibility

Blends of P(HB-co-16%HV) with the PS, ABS, PLA, PCL and SAN polymers were prepared by dissolving the polymeric binary mixture in a solution of chloroform (5% wt./v polymer) at blend ratios 20/80, 50/50, and 80/20, respectively. Before preparing the blends, P(HB-co-16%HV) was purified. The polymer was dissolved in chloroform (5% solution) and precipitated in cold methanol (methanol:chloroform, 10:1) with constant stirring. The precipitate was filtered, washed with cold ether/acetone, and finally dried under vacuum for 24 h at room temperature. Films were prepared onto glass slides by allowing the solvent to evaporate at 25°C for 12 h. All test films were dried for 12-18 h at 50°C under vacuum, cut to dimension of approximately 1 x 1 cm, weighed accurately (using Mettler AE 260), and the film thickness measured (average film thickness of approximately 0.090 mm).

Blends of P(HB-co-16%HV) with compounded PVC (2.0 phr Ba-Cd stabilizer and 0.20 phr stearic acid) were prepared by melt blending on a Haake Torque Rheometer equipped with a mixing chamber at blend ratios of 20/80, 50/50, and 80/20, respectively. Both the cast films and the processed melt blends were analyzed for miscibility using DSC.

2Aa Differential Scanning Calorimetry, (DSC).

DSC was performed using a Perkin-Elmer DSC-2C instrument. The amount of sample used from 2 to 8 mg and the heating rate was at 20°C/min. Calibration of the instrument was carried out using high purity indium. The samples were initially heated from room temperature to 175°C. The peaking melting temperature (T_m) was determined from the melting endotherm. After 1 min. at 175°C the samples were rapidly quenched to - 23°C, and then reheated once again to 175°C. The glass transition temperatures (T_g) of the blend components and the blends were obtained from this second heating run.

2B. Enzymatic Degradation of Films

Biodegradation or biological degradation of polymers consists of those processes which result in chain cleavage from an attack on the material by living organisms, eg. bacteria, fungi, insects, and rodents. Degradation by the extracellular depolymerase will be referred to as enzymatic degradation.

Penicillium funiculosum (ATCC 9644), obtained from National Regional Research Labs, Peoria, Ill., was maintained in slant tubes containing a basal mineral salts media in noble agar with 0.1% PHB. A slant tube of the culture was used to inoculate a Bio-Flo fermenter containing 200 mLs of a basal mineral salts broth (pH 6.5) with 0.1% PHB as the sole carbon source. The fermentation was maintained at 28°C. Once the culture had reached its stationary phase, the culture was harvested and centrifuged at 10,000 x g for ten minutes. The supernatant, which contained the extracellular depolymerase, was used to carry out the degradation studies.

Three identical sets of films were used, where each film sample was analyzed in triplicates. The first two sets were incubated in the presence of enzymes for 2 days and 4 days, respectively. The third set of films were incubated for 4 days in a phosphate buffer solution (28°C, pH 6.0). The polymer films used were chemically sterilized by first rinsing them in a 1% o-phenylphenol, o-benzyl-p chlorophenol solution. The films were then rinsed two times in sterile water. The sterile films were placed in sterile 10 mL vials and 5 mLs of the supernatant containing the extracellular depolymerase was added to each vial under aseptic conditions. The vials were then incubated at 28°C for 2 and 4 days, respectively. The material which remained in the film form after the above exposures was removed from the vials, dipped in sterile water, vacuum dried overnight, weighed, dried a second time for 12 to 18 h at 50°C under vacuum and reweighed. The second weight values were used herein to determine the weight-loss from polymer films. The polymeric material which was floating in solution after the incubation period was collected from each vial and is presently under investigation. Further characterization of blends before and after degradation exposures is in progress using SEM and GPC.

RESULTS AND DISCUSSION

1. DSC Results

The DSC results obtained for the component polymers and blends are summarized in Table 1. Miscibility was determined by observing a temperature shift in the T_g or T_m. The blends investigated for P(HB-co-16%HV) with PCL showed that the glass transition temperature (T_g) of the bacterial polyester remains at 5°C and the absence of any shift in the melting temperature (T_m) of PCL or P(HB-co-16%HV). A similar behavior was observed for the T_g's with blends of the bacterial polyester with PS and also for the blends with PLA. These results indicate immiscibility for blends of P(HB-co-16%HV) and PCL, PS, or PLA.

Blends of P(HB-co-16%HV)/PVC in ratios of 20/80, 50/50, and 80/20, were investigated. The T_g of PVC decreases from 85°C for 100% PVC, to 60°C for 80% PVC blend, and to 18°C for the 50% PVC blend. The T_m of the bacterial polyester is also depressed in the blends. These DSC results indicate miscibility for the P(HB-co-16%HV)/PVC blends.

An intermediate behavior was observed for binary blends of P(HB-co-16%HV) with SAN or ABS. The T_g of SAN (108°C) decreases to 50°C fir tge 80% of SAN blend, and to 47°C for the 50% SAN blend. Blends of P(HB-co-16%HV)/ABS exhibited a single T_g observed at 63°C, 59°C, and 49°C for the 20/80, 50/50, and 80/20 blends (see Figure 1). These results indicate the formation of a phase which has an intermediate T_g but for which the T_g does not vary proportionately with blend composition. These blends will be regarded as partially miscible.

Melt blending studies are currently underway where each composition will be mixed and injection molded into a strip 0.020 inches in thickness. These blends will be analyzed for miscibility, domain size distribution, crystallinity, mechanical properties, and degradability.

Table 1. DSC results for component polymers and blends.		
COMPONENT		
POLYMERS	T_g (°C)	T_m (°C)
P(HB-co-16%HV)	3	135
PCL	–	60
PS	107	–
PLA	60	172
SAN	108	–
ABS	108	–
PVC	85	–
BLENDS (A/B)	T_g (°C) of A	T_g (°C) of B
(HB-co-16%HV)/PCL		
(20/80)	5	–
(50/50)	5	–
(80/20)	5	–
P(HB-co-16%HV)/PS		
(20/80)	4	92
(50/50)	4	95
(80/20)	4	89
P(HB-co-16%HV)/PLA		
(20/80)	2	62
(50/50)	2	63
(80/20)	2	63
P(HB-co-16%HV)/PVC		
(20/80)	60	–
(50/50)	18	–
(80/20)	–	–
P(HB-co-16%HV)/SAN		
(20/80)	50	–
(50/50)	47	–
(80/20)	–	–
P(HB-co-16%HV)/ABS		
(20/80)	63	–
(50/50)	59	–
(80/20)	49	–

2. Degradation Study

The results obtained by the weight-loss in film samples due to the enzyme activity are summarized in Figures 2-5. Weight-loss data for P(HB-co-16%HV)/PS are shown in Figure 2. Blends of PS with P(HB-co-16%HV) (in ratios of 20/80, 50/50, and 80/20) display a weight-loss of 14.6%, 20.4%, and 48.7%, for the day 2 test samples. Greater weight-loss was seen for the day 4 samples. Since blends of P(HB-co-16%HV) with PS are immiscible it may be that the increased degradation rate of the 80% P(HB-co-16%HV) blends relative to films containing only P(HB-co-16%HV) is due to the

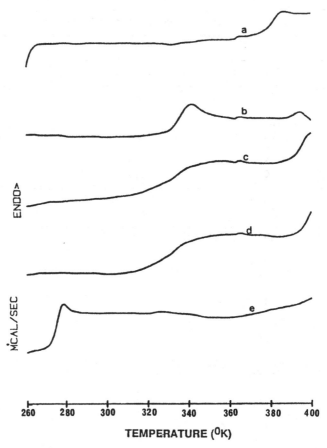

Figure 1. Thermogram profiles of (a) plain ABS; (b) 80% ABS blend; (c) 50% ABS blend; (d) 20% blend; (e) plain P(HB-co-16%HV).

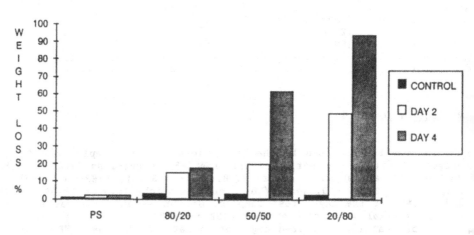

Figure 2. Biodegradation of PS/P(HB-co-16%HV) blends.

degradation of the bacterial polyester which surrounds domains of the dispersed PS phase. The domains of PS may, then, be released from the film surface into the medium.

Results of weight-loss for blends of P(HB-co-16%HV)/PCL are shown in Figure 3. These blends also show significant degradation. Weight-loss for the 20/80, 50/50, and 80/20 blends of P(HB-co-16%HV)/PCL are 3.2%, 25.7%, and 45.9% for the day 2 test samples. The low weight-loss (3.2%) in the 20% P(HB-co-16%HV) blend relative to the P(HB-co-16%HV)PS 20/80 blend (14.6%) may be due to the differences in sample morphology (crystallinity and phase domain size). Degree of crystallinity as a function of blend composition and degradation is currently being measured by X-ray analysis to investigate the role of crystallinity.

The lack of degradation in the PCL is surprising since degradation of PCL by *Penicillium Funiculosum* has been reported by Cook et. al. (22) and preliminary trials here have shown that the *Penicillium Funiculosum* is able to degrade the PCL. A hypothesis is that the enzyme which is produced when the microorganism is degrading P(HB-co-HV) is different from that which is produced when degrading PCL. Since the extracellular depolymerase supernatant was obtained from a feeding of pure P(HB) the enzyme which was produced would degrade P(HB-co-HV) rather than PCL. Ongoing research to conduct the replicate trials using supernatant produced from *Penicillium Funiculosum* grown on PCL is being pursued to validate this hypothesis.

Results of weight-loss for blends of P(HB-co-16%HV)/SAN are shown in Figure 4. These blends in general show a significantly lower weight-loss in comparison to our previous observations. Blends of P(HB-co-16%HV)/SAN in ratios of 20/80, 50/50, and 80/20 indicate a weight-loss after a 2 day incubation period of 8.5%, 19.9%, and 23.1%, respectively. Since degradation under the conditions investigated would be anticipated to occur by an initial enzyme catalyzed hydrolysis in the amorphous state, this result suggests that the intimate contact between P(HB-co-16%HV) and SAN chains in the amorphous phase caused a decreased rate of degradation relative to similar blend rations of immiscible P(HB-co-16%HV)/PS blends. An AB block copolymer of P(HB-co-PS) has been prepared as a compatibilizing agent for the blends of PS/P(HB-co-HV) to further investigate this phenomenon.

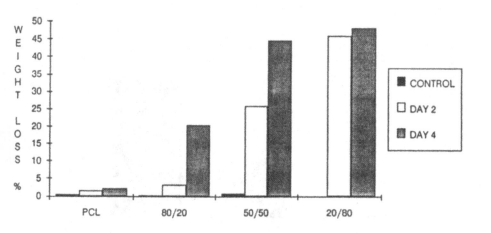

Figure 3. Biodegradation of PCL/P(HB-co-16%HV) blends.

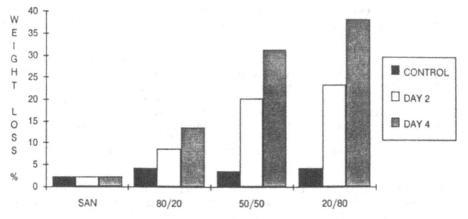

Figure 4. Biodegradation of SAN/P(HB-co-16%HV) blends.

All the test samples incubated in buffer solution in the absence of depolymerase of enzyme(s) showed negligible weight-loss (within experimental-error). Pure P(HB-co-16%HV) incubated in an enzymatic PHA depolymerase medium showed (Figure 5) a weight-loss of 37.4% for day 2 samples and 88% for day 4 samples.

The blend compositions are currently being exposed to accelerated landfill, simulated marine, and aerobic composting environments to measure the degradability of these blends in more realistic biodegradation tests.

CONCLUSIONS

Blends of the bacterial copolyester with PS, PCL, and PLA were completely immiscible as determined by the DSC. This conclusion was based on a negligible effect of blending on the glass transition for each component.

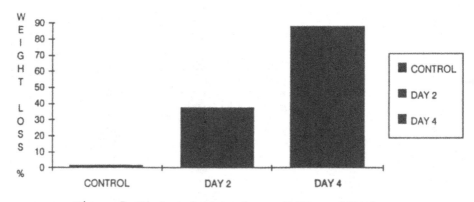

Figure 5. Biodegradation of pure P(HB-co-16%HV).

The DSC results indicate complete miscibility of P(HB-co-16%HV) with PVC, and partial miscibility with SAN and ABS, at all ratios. This was concluded from the existence of a single T_g at intermediate temperatures as well as a depression in the melting point of P(HB-co-16%HV). It appears, as suggested by Holmes [21], that polyvinyls which contain adequate quantities of the polar repeating units vinyl chloride and acrylonitrile form miscible blends.

It appears from this work that miscible blends which contain a degradable and non-degradable component biodegrades by enzymatic attack at a relatively slower rate when compared to similar blend systems which are immiscible.

ACKNOWLEDGEMENTS

This research was conducted under the Polymer Degradation Research Consortium at the University of Lowell. The authors are grateful to the corporate and governmental members of PDRC for their financial support.

REFERENCES

1. K. Juni, M. Nakano, and M. Kubota, J. Controlled Release, **4** (1), 25 (1986).
2. W. Koenig, H. R. Seidel, and J. K. Sandow, Eur. Pat. Appl. EP-133988 A2, 1985.
3. R. H. Findlay and D. C. White, Appl. Environ. Microbial., **45**, 71 (1983).
4. R. J. Capon, R. W. Dunlop, E. L. Ghisalberti, and P. R. Jeffries, Phytochemistry, **22**, 1181 (1983).
5. L. L. Wallen and W. K. Rohweeder, Environ. Sci. Technol., **8**, 576 (1974).
6. E. R. Howells, Chem. Ind., 508 (1982).
7. P. A. Holmes, Phys. Technol., **16**, 32 (1985).
8. R. Leaversuch, Modern Plastics, Aug., 1987, 52.
9. T. L. Bluhm, G. K. Hamer, R. H. Marchessault, C. A. Fyfe, and R. P. Veregin, Macromolecules, **19**, 2871 (1986).
10. Y. Doi, M. Kunioka, and A. Tamaki, Polym. Preprints, Am. Chem. Soc., Div. Polym. Sci., **29** (1), 588 (1988).
11. P. L. Ragg, Symp. in Biotechnology, Div. of Polym. Chem., June, 1988.
12. T. Tanio, T. Fukui, Y. Shirakura, T. Saito, K. Tomita, and S. Masamune, Eur. Biochem., **124**, 71 (1982).
13. Y. Shirakura, T. Fukui, T. Saito, Y. Okamoto, T. Narikawa, K. Koide, K. Tomita, T. Takemasa, and S. Masamune, Biochemica et Biophysica Acta, **880**, 46 (1986).
14. M. Regina Brophy and P. B. Deasy, Int. J. Pharm., **29**, (2-3), 223 (1986).
15. G. J. L. Griffin, *"Biodegradable Fillers in Thermoplastics,"* ACS Advances in Chemistry Series, 1975, p. 134.
16. W. J. Maddever and G. M. Chapman, in: *"Proc. Symp. Degradable Plastics,"* SPI, 1987, p. 41.
17. F. H. Otey, R. P. Westoff, and W. M. Doane, Ind. Eng. Chem. Prod. Res. Dev., **19**, 592 (1980).
18. M. Avella and E. Martuscelli, Polymer, **29**, 1731 (1988).
19. F. Kawai, CRC Crit. Rev. in Biotechnol., **6**, 273 (1987).
20. M. Yasin, S. J. Holland, A. M. Jolly, and B. J. Tighe, Biomaterials, **10**, 400 (1989).
21. P. A. Holmes, Eur. Pat. Appl. EP-81305188.5 (1981).
22. W. J. Cook, J. A. Cameron, J. P. Bell, and S. J. Huang, J. Polym. Sci., Polym. Letters, **19**, 159 (1981).

THE USE OF BIOTECHNOLOGY DERIVED MONOMERS IN THE SYNTHESIS OF NEW

POLYMERS: DEVELOPMENT OF POLYESTERIMIDES

Malay Ghosh*

Department of Chemistry and Chemical Engineering
Stevens Institute of Technology
Hoboken, NJ 07030

The applications of polyesterimides as heat resistant material has many advantages over the conventionally used polymers. This paper will summarize our research on the synthesis and characterization of two new polyesterimides from N-(4-carboxyphenyl) trimellitimide [7] and two other monomers, 4,4'-diacetoxybiphenyl [2] and cis-1,2-diacetoxy-3,5-cyclohexadiene [4] respectively. The physical, physicochemical and thermal properties of the polymers have been investigated. The monomers [2] and [4] have been prepared via enzymatic reactions from commercially available inexpensive starting materials followed by acetylation. This shows, in general, the potential of biotechnology on the development of new polymers, and heat resistant polymers in particular.

INTRODUCTION

The search for more efficient heat resistant, high performance polymers has provided the incentives for continuing research involving extensive study of these polymers.[1] Excellent thermooxidative stability over a prolonged period of time is considered to be the major requirement for a high temperature polymer. In addition, ease of processibility is certainly necessary for practical application. The polymers are also expected to be chemically stable under harsh conditions. Numerous classes of synthetic polymers have been developed and polyimides are considered to be outstanding from the viewpoint of thermal stability.[2,3] However, the insoluble nature of polyimides in readily available polar organic solvents and their narrow processing temperature window restricted their applications to a large extent. Several approaches are developed by the polymer scientists to overcome these shortcomings of polyimides.[2] One unique way is the synthesis of copolyimides. This approach has been found to be highly successful. Copolyimides, namely polyamideimides and polyesterimides possess excellent thermal stability along with acceptable solubility, mechanical strength and environmental stability.[4-9]

*Present Address: Schering-Plough Corp.,
2000 Galloping Hill Road, Kenilworth, NJ 07033

The major amount of starting materials needed for the synthesis of polyimides and polyesterimides come from petroleum base chemicals. In the past years, biotechnology, has shown great promise to do subtle chemical manipulation on an existing compound which previously was difficult to carry out by conventional methodologies.[10-12] These discoveries stimulated many efforts toward the applications of enzymes to carry out chemical transformations. The enzyme catalytic reactions are extremely selective, high yielding and easy to carry out in large scale. Thus they are highly practical from the view point of process chemistry. We have used the advantage of enzymatic oxidation reaction to prepare monomers for the synthesis of target polyesterimides. 4,4'-Dihydroxybiphenyl and cis-3,5-cyclohexane-1,2-diol were prepared from commercially available phenol [1] and benzene [3] respectively.[19-20] The synthesis, characterization and thermal stability data of the polymers are given in this paper.

EXPERIMENTAL

1. Materials

Phenol, trimellitic acid anhydride [TMA], 4-aminobenzoic acid [PABA], thionyl chloride, DMF, DMAc, DMSO and NMP were purchased from Aldrich Chemical. Phenol, thionyl chloride and acetic anhydride were distilled prior to use. [TMA] and [PABA] were crystallized from appropriate solvents. All the solvents were purified according to the method described in the literature.[21] *Pseudomonas putudia* and horseradish peroxidase and were obtained from Sigma Chemical. Water was doubly distilled. Oxygen was technical grade.

2. Characterization of Monomers and Polymers

Infrared spectra were recorded on a Perkin Elmer model 1310 spectrometer using nujol mull. ^{1}H-NMR and ^{13}C-NMR spectra were done on a Bruker 200 AP instrument using CDCl$_3$ and/or DMSO-d$_6$ as solvent and TMS as internal standard. Viscosity measurements were carried out in 0.5% (g/mL) solution in DMAc at 27°C. Thermal analysis of the polymers were done on a DuPont 9900 thermal analyzer. The chemical stability of the prepared polymers were checked on keeping them in concentrated sulfuric acid and in 5% aqueous sodium hydroxide solution. Elemental analysis was performed in Schwarzkopf Microanalytical Laboratory.

3. Synthesis of Monomers

4,4'-Diacetoxybiphenyl [2] was prepared via the reaction of acetic anhydride with 4,4'-dihydroxybiphenyl which in turn was obtained by enzymatic oxidation of phenol [1] by horseradish peroxidase.[19] Stereospecific bis-hydroxylation of benzene [3] by *Pseudomonas putida*[20] and subsequent acetylation gave cis 1,2-diacetoxy-3,5-cyclohexadine [4]. Condensation of [TMA] and 4-aminobenzoic acid [6] in DMAc at 150°C gave N-(4-carboxyphenyl) trimellitimide [7] in excellent yield.[22]

4. Synthesis of Polymers

The polyesterimides were synthesized by thermal polytransesterification reaction of imidodiacid [7] with [2] and/or [4] at elevated temperature under nitrogen atmosphere. A typical experimental procedure is given below.

A mixture of (40.5 g, 0.15 m) of 4,4'-diacetoxybiphenyl [3] and (46.65 g, 0.15 m) of imidodiacid [7] were taken in a cylindrical glass reactor and was stirred vigorously. The temperature of the reaction mixture was maintained at 280°C for 1 h while the dry nitrogen was purged in the reactor. The temperature of the reactor was then raised to 340°C and was stirred for another 2 h. At this point most of the acetic acid was removed by distillation. The reaction mixture was cooled, diluted with 2 L of ice cold water to precipitate out the polymer. The brown precipitate was filtered, washed with cold water and dried. Purification of the polymer [8] was done by reprecipitation from m-Cresol and water. The yield of the purified polymer is 82%. In the case of monomer [4], the temperature of the reaction mixture was initially kept at 250°C and then raised to 300°C for 1 h.

RESULTS AND DISCUSSION

The polyesterimides have been conveniently synthesized by classical polytransesterification reaction. The synthetic route utilized to develop monomers [2], [4] and [7] is shown in scheme 1 whereas scheme 2 illustrates the polymerization reaction.

The yield and physical characteristics of the synthesized polymers are listed in Table 1.

The characteristics IR bands of polymer [8] are observed at 1785 and 1725 cm^{-1} which are due to the presence of the imide functional group in the repeat unit of the polymer. A sharp band at 1735 cm^{-1} is assigned to the ester functionality. Similarly polymer [9] exhibits peaks at 1775, 1730, 1715, 1640 and 1630 cm^{-1}. The last two bands are due to the presence of olefinic system in polymer [10]. The signals in the ^1H-NMR spectra of the polymers [8] and [9] ranges from 5.3 to 8.1 ppm. The characteristic A_2B_2 pattern was observed for polymer [8] at 7.05 and 8.1

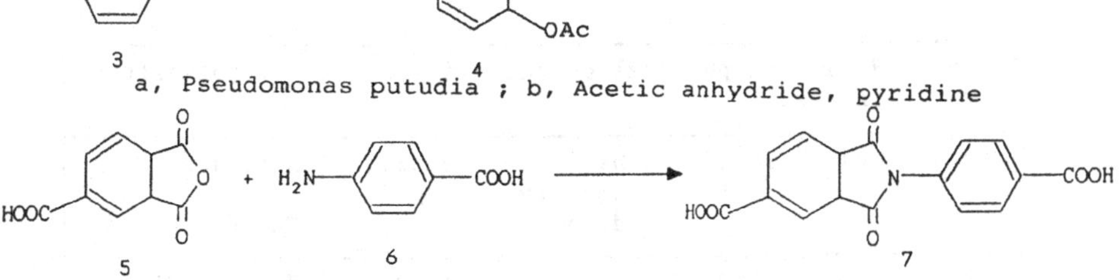
HO
a, b
AcO
OAc
1
2
a, Horseradish peroxidase, Oxygen
b, Acetic anhydride, pyridine
a, b
OAc
OAc
3
4
a, Pseudomonas putudia ; b, Acetic anhydride, pyridine
O
O
HOOC
O
+ H2N
COOH
O
N
HOOC
O
COOH
5
6
7

Scheme 1. Synthesis of monomers by enzymatic oxidation.

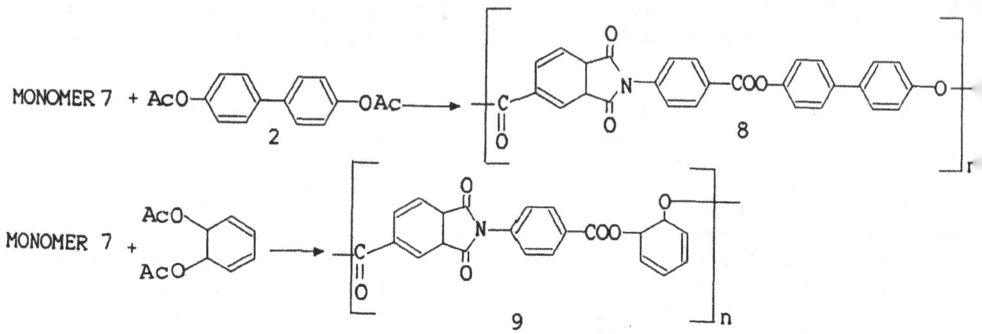

Scheme 2. Synthetic route to prepare polyesterimides.

ppm indicating the presence of 1,4-disubstituted phenyl system. The other aromatic protons of [8] gave complex multiplets in the range of 7.2 to 7.4 ppm. Proton noise decoupled ^{13}C-NMR shows peak from 120 to 190 ppm. Four small signals ranging from 170 to 190 ppm indicates the presence of four carbonyl systems in the repeat unit of the molecule. This is consistent with our earlier observation with polyesterimide systems.

The inherent viscosity of the synthesized polymers are given in Table 1. The low inherent viscosity corresponds to low molecular weight. Comparable values of inherent viscosity of other polyesterimide at same concentration have been reported.[2] The solubility behavior of the polymers are listed in Table 2. The polymers are found to be insoluble in most polar organic solvents and soluble in DMAc, m-Cresol, formic acid and sulfuric acid, etc.

The thermal properties of the polyesterimides have been evaluated by thermogravimetric analysis and differential scanning calorimetry. Preliminary DSC studies of the polymers at 10°C/min ramp failed to provide T_g of the compounds. However, we have confirmed the T_g of polymer [8] at 253°C by using 40°C/min temperature ramp. In contrast, no detectable T_g was observed for polymer [9] after several experiments. TGA analysis indicates that an initial weight loss of 2-5% takes place at around 100°C, which probably is due to loss of water or entrapped solvent. The annealed sample does not show this weight loss. The polymer [8] exhibits excellent thermal integrity until nearly 400°C, at which temperature a 20% weight loss takes place. The degradation of both the polymers has been found to be a single step process and starts around 425°C. The maximum weight loss takes place in the temperature range of 450-600°C. The isothermal aging of the polymers were evaluated at 275°C in air and in nitrogen. The data are presented in table 3. It can be said that the polymers could be used at 250 to 270°C without significant decomposition.

Table 1. Yield and physical characteristics of the polyesterimides.

Polymer	Color	Yield (%)	Inherent Viscosity (dL/g)	Density (g/cm³)	Nitrogen (%)	
					Calcd.	Found
8	Brown	82	1.04	1.33	72.88	72.74
9	White	87	1.91	1.45	68.57	68.39

Table 2. Solubility behavior of the polymers.		
Solvent	Polymer 8	Polymer 9
Benzene	–	–
Acetone	–	–
Chlorofoam	–	±
Cyclohexane	–	–
n-Butanol	–	–
Ethyl acetate	–	–
1,4-Dioxane	–	–
Tetrahydrofuran	–	–
Ethyl methyl ketone	–	±
N,N-Dimethylacetamide	+	+
N,N-Dimethylformamide	+	+
Dimethylsulfoxide	±	+
N-Methyl-2-pyrrolidone	+	+
Ethanol	–	–
HMPA	+	+
m-Cresol	+	+
Formic acid	+	+
Sulfuric acid	+	+

Solubility keys:

+ soluble
– insoluble
± partly soluble

An investigation on the chemical stability of the polymers showed that they hydrolyze quickly in the presence of both acid (sulfuric acid) and base (5% aqueous sodium hydroxide) at room temperature. However, completion of the hydrolysis requires more than 5 h. The polymers are found to be more vulnerable to acids. This is probably due to the faster hydrolysis rate of ester group of polymer by acids.

The moisture absorption property of the polymers [8] and [9] have also been investigated gravimetrically at 27°C under 70% relative humidity. The polymers were dried before the measurement. It has been found that both of them exhibit low moisture absorption property (0.1 and 0.3% by weight for compound 8 and 9 respectively).

Table 3. Isothermal aging data of polymers heated at 275°C in air and in nitrogen.

Polymer	Atmosphere	Weight loss in % after (hrs)						
		1	4	8	16	32	64	128
8	Air	0	0	0.2	1.2	1.9	2.7	4.6
9	Air	0	0.3	0.7	1.6	2.8	3.9	5.9
8	Nitrogen	0	0	0.1	0.5	1.9	2.4	4.1
9	Nitrogen	0	0	0.6	1.1	2.0	3.9	5.1

ACKNOWLEDGMENT

Partial support of this work was obtained from New Jersey Commission for Science and Technology and is gratefully acknowledged. The author would like to thank Drs. S. Viti and C. Bottom for their help and encouragement.

REFERENCES

1. P.E. Cassidy, *"Thermally Stable Polymers: Synthesis and Properties,"* Marcel Dekker, New York, 1980.
2. K. L. Mittal, Ed., *"Polyimides: Synthesis, Characterization and Applications,"* Vol. 1 and 2, Plenum Press, New York, 1984.
3. P. Hergenrother and T. L. St. Clair, *"Proceedings of Second International Conference on Polyimides,"* Ellenville, New York, 1985.
4. M. Ghosh and S. Maiti, J. Macromol. Sci., A22, 1463 (1985).
5. S. S. Ray, A. K. Kundu, M. Ghosh and S. Maiti, J. Polym. Sci., Part A, **24**, 603 (1986).
6. M. Ghosh, Angew. Makromol. Chem., **172**, 165 (1989).
7. M. Ghosh, J. Macromol. Sci., Part A, in press.
8. M. H. Yi, S. G. Lee, K-Y. Choi and J. C. Jung, J. Polym. Sci., Part A, **26**, 1507 (1988).
9. S. S. Ray, A. K. Kundu, M. Ghosh and S. Maiti, Eur. Polym. J., **21**(2), 131 (1985).
10. P. H. Abelson, *"Biotechnology & Biological Frontiers,"* Amer. Assn. Adv. Sci., Washington, DC, 1984.
11. *"Commercial Biotechnology: An International Analysis,"* Office of Technology Assessment, U.S. Congress, OTA-BA-132, Washington, DC, 1984.
12. *"Impacts of Applied Genetics: Microorganisms, Plants and Animals,"* Office of Technology Assessment, U.S. Congress, OTA-HA-132, Washington, DC, 1981.
13. T. Bryan Jones, Tetrahedron, **42**, 3351 (1986).
14. G. M. Whiteside and C-H. Wong, Angew. Chem., Int. Ed., **24**, 617 (1985).
15. A. M. Klibanov, Science, **219**, 722 (1983).
16. G. M. Whiteside and C-H. Wong, Aldrichimica Acta, **16**, 27 (1983).
17. G. M. Whiteside in: *"Enzymes in Organic Synthesis,"* Ciba Found. Symp. 111, Pittman, London, 1985, p. 76.
18. H. W. Blanch and A. M. Klibanov, *"Enzyme Engineering,"* Ann. N. Y. Acad. Sci., Vol. 542, New York, 1988.
19. R. D. Schwartz and D. B. Hutchinson, Enzy. Microb. Tech., 3, 361 1981).
20. D. J. Ballard, A. Courtis, I. M. Shirley and S. C. Taylor, J. Chem. Soc., Chem. Commun., 954 (1983).
21. D. D. Perrin, W. F. F. Armareo and D. R. Perrin, *"Purification of Laboratory Chemicals,"* Pergamon Press, London, 1982.
22. S. Maiti and A. Ray, J. Appl. Polym. Sci., **27**, 4345 (1982).

THE BIOSYNTHESIS OF UNUSUAL POLYAMIDES CONTAINING GLUTAMIC ACID

Steven A. Giannos,[a] Devang Shah,[a] Richard A. Gross,[a] David Kaplan,[b] and Jean M. Mayer[b]

(a) University of Lowell
Department of Chemistry
Lowell, MA 01854

(b) Science and Advanced Technology Directorate
U. S. Army Natick Research
Development and Engineering Center
Natick, MA 01760-5020

A study of the effect of culture time and the availability of nutrients on the production and molecular weight of bacterially derived γ-polyglutamic acid (γ-PGA) has been carried out. The introduction of additional nutrients after a 3 day culture period improved polymer production (in γ-PGA g/L) and gave polymer of higher molecular weight relative to control samples. Modification of the product polyamides by synthesis of the corresponding ethyl and propyl esters has been carried out. Characterization of these products by NMR spectroscopy and thermal analysis is reported.

INTRODUCTION

The biosynthesis by specific micro-organisms of polyamides consisting of glutamic acid repeating units that are linked between the α-amino and γ-carboxylic acid functionalities, respectively (γ-PGA), have been known for some time. Reviews by Housewright[1] and Troy [2] on this interesting biopolymer have been published.

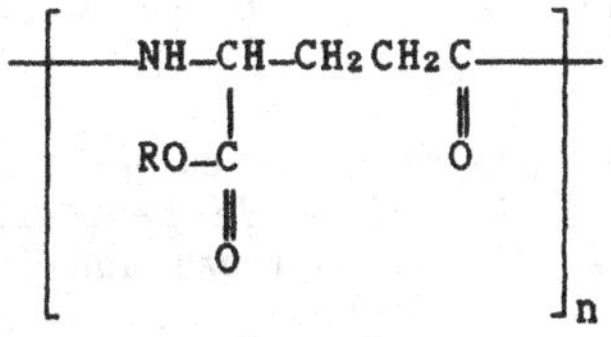
NH–CH–CH2CH2C
RO–C
O
O
n

R = H when produced by bacteria

D or L

Chemical proof that γ-PGA produced by certain *Bacillus sp.* contains repeating units which are joined by γ-linkages was presented by Troy.[3]

Biotechnology and Polymers, Edited by C.G. Gebelein
Plenum Press, New York, 1991

This was demonstrated by reacting the bacterial product with diazomethane, followed by lithium borohydride reduction and hydrolysis of the modified polymer to give exclusively γ-amino-δ-hydroxyvaleric acid.[3]

The γ-PGA is a water soluble polymer which for *Bacillus Anthracis* is present in the capsule of the bacterium serving as a structural component of the cell membrane.[1] In contrast, for certain strains of *Bacillus subtilis*, most if not all of the γ-PGA formed diffuses freely into the growth medium.[4]

Thorne and Leonard claimed that the stereochemistry of γ-PGA produced by *B. subtilis* (ATCC 9945a), which was later reclassified as *B. licheniformis* (ATCC 9945a), could be controlled by alternating the concentration of manganese [II] ions in the fermentation.[5] Using this mechanism, the products obtained had between 40 and 80% D-glutamic acid repeating units.[5] These authors also described the isolation of γ-PGA stereoisomers which consisted of predominantly, if not entirely, D-glutamic acid, L-glutamic acid, and 50% of each isomer respectively. Therefore, it appears that individual chains of γ-PGA produced by *B. licheniformis* are composed of a single glutamic acid stereoisomer. In addition, Thorne and Leonard described the precipitation of soluble aqueous solutions of D-γ-PGA and L-γ-PGA when mixed.[5] This suggests that a stereocomplex between poly-D- and poly-L-γ-PGA chains is obtained under these conditions. It is interesting to note that investigations carried out by Troy and co-workers with the identical strain of *B. Licheniformis* gave γ-PGA which contains at least 90% of the D-isomer.[2]

Fermentation conditions and physical properties of the γ-PGA produced by a *Bacillus subtilis sp.* were investigated by Omata and co-workers.[6,7] These authors report γ-PGA production from glucose of up to 18 g/L of crude product.[6] After purification of this crude material by dialysis and formation of the corresponding Na-γ-PGA derivative, a molecular weight average of 1,160,000 g/mol was obtained from sedimentation constants.[7]

The mucin from "Natto" contains a polysaccharide (levanform fructan) as well as large quantities of γ-PGA. The γ-PGA fraction may have variable contents of D- and L-glutamate stereoisomers.[8] "Natto", which is produced from soybean by *Bacillus subtilis* (natto), is one of the most traditional fermentation foods in Japan.[8] This suggests that this biopolymer may be safely consumed by oral ingestion.

The degradation of γ-PGA produced by bacterial fermentation has been studied by incubation of the polymer in various tissue extracts. Bovarnick and co-workers found that γ-PGA was degraded by liver, kidney, spleen, and brain extracts with the production of free glutamic acid.[9] Torii reported that enzymes found in dog liver extracts were active in forming free L-glutamic acid from blends of natural origin D- and L-γ-PGA.[10] Interestingly, these enzymes did not liberate free L- or D-glutamic acid from a natural origin D-γ-PGA sample.[10] This work clearly showed the dependence of the *in vitro* degradation rate of γ-PGA on polymer stereochemistry.

We are currently exploring the biosynthesis of novel polyamides from renewable resources using the bacterium *Bacillus licheniformis* (ATCC 9945a). The use of these polymers for specific biomedical and environmentally degradable polymer applications is currently under investigation. In this paper, we report the effect of culture time and the availability of nutrients on the production and molecular weight of γ-PGA. Modification of γ-PGA to form water insoluble derivatives will also be described. Finally, characterization of these products by NMR spectroscopy and thermal analysis is reported.

EXPERIMENTAL

1. Organism

Bacillus licheniformis (ATCC 9945a) was obtained from Dr. L. K. Naka-mura, of the U. S. Department of Agriculture, Peoria, IL. This freeze dried sample was suspended using 0.5 mL Tryptic Soy Broth (Difco Labs). This solution was transferred to a larger volume of Tryptic Soy Broth (100 mL) and incubated on a rotary-shaker (250 RPM, 37°C) for 18 h. From this inoculum four stock culture plates were prepared using Tryptic Soy Broth supplemented with 1.5% agar. These culture plates were then incubated at 37°C for 18 h.

2. Organism Growth and Polymer Production

Sterile Tryptic Soy Broth (100 mL) in a 250 mL Erlenmeyer flask was inoculated using an isolated mucoid colony from a stock culture plate. This culture was incubated on a rotary shaker (250 RPM, 37°C) for 18 h.[3] Aliquots of this inoculum (1.5 mL) were introduced into 500 mL Erlenmeyer flasks each containing 100 mL of sterile medium E[11] at pH 7.4. For a time course study, multiple culture flasks were prepared taking 100 mL aliquots of medium E from a larger volume. The cultures, inoculated in an identical manner, were incubated on a rotary shaker (250 RPM, 37°C). These cultures were then interrupted and analyzed at different time points.

3. Second Feeding of Cultures

A solution (solution A) which contained 5 times the initial concentration of the medium components (with the exception of the phosphate salts) was prepared and filter sterilized. As a control, solution B (double distilled water) was also filter sterilized. Solution A and solution B (20 mL portions) were each added to five culture flasks that had been incubated for 3 d bringing the culture volumes to 120 mL (see above). This gave five culture flasks which had been given additional nutrients after a 3 d growth period along with five control cultures.

4. Polymer Isolation

At specific time intervals, culture flasks were removed from the incubator shaker and their contents were poured into a blender containing 50 g of ice. The interior walls of the culture flasks were rinsed multiple times with distilled water and transferred to the blender. The resulting mixture was then blended for 2 min and centrifuged (16,000 RPM, 6°C) for 10 min to separate the cells. The cells were washed three times with distilled water warmed to 40°C and subsequently dried (40°C, 5 mmHg, 24 h) to obtain the cellular dry weight. The supernatant which contained the γ-PGA was poured into three volumes of 95% ethanol which had been previously cooled to -20°C. The resultant precipitate was isolated by twining around a glass rod and subsequently lyophilized to remove any remaining solvent.

5. Polymer Purification

A 3.5% solution of crude γ-PGA was dialyzed against 3, 5, 10, 5, and 3% solutions, respectively, of aqueous sodium chloride and then precipitated into 4 volumes of cold 95% ethanol as described above to give Na-γ-PGA. An aqueous solution of Na-γ-PGA was then acidified to pH 2 at 5°C and precipitated into 3 volumes of n-propanol to obtain the corresponding free acid derivative, H-γ-PGA.

6. Instrumental Procedures

The solution of ^{1}H NMR measurements were recorded at 200 MHz on a Varian XL-200 NMR spectrometer. Solution ^{13}C NMR measurements were recorded at 67.5 MHz on a Bruker WP-270 SY NMR spectrometer. Differential scanning calorimetry (DSC) was performed using the Perkin-Elmer Model DSC-2c.

The γ-PGA weight and number average molecular weights (M_w and M_n, respectively) along with the dispersity (M_w/M_n) were measured using a Waters Model 150-C ALC/GPC Gel Permeation Chromatography (GPC) System. Separations were achieved using one TSK-60 along with two TSK-50 Bio-Gel (Bio-Rad Laboratories, Richmond, CA) columns in series. A calibration curve was generated using a series of polysaccharide (linear β-D-glucan) standards ranging in molecular weight between 12,000 and 853,000 g/mol (Polymer Laboratories Std., Church Stretton, UK). An aqueous solution of sodium acetate (0.1 M) and sodium azide (0.05% wt/v) at pH 6.0 was used as the eluant at a flow rate of 1.0 mL/min. Sample concentrations of 0.1% wt/v and injection volumes of 300 µL were used. All samples were solubilized over a three day period (pH = 6.0, 25°C) before performing the analysis. Calculations were carried out neglecting contributions from low molecular weight sample impurities (less than 5,000 g/mol).

7. Synthesis of n-Alkyl-γ-PGA Derivatives

The procedure used was a modification of that described previously by Bocchi et. al.,[12] Due to the poor solubility of Na-γ-PGA in dimethylsulfoxide (DMSO), the free acid form (H-γ-PGA) was dissolved in DMSO and to this solution was added a five molar excess of either n-ethyl or n-propyl bromide. The reaction was allowed to continue for 48 h at room temperature after which the product was precipitated into a 6% sodium chloride aqueous solution. The precipitate was further purified by soxhlet extraction with acetone, ether, and then dried in vacuo.

RESULTS

A study was carried out to probe the question of how changes in concentrations of the carbon sources in the culture medium and increased culture time might alter polymer production and polymer molecular weight. The carbon sources used in this work were L-glutamic acid, citric acid, and glycerol (see reference 11 for the composition of the medium).

Investigation of changes in cellular dry weight and γ-PGA found in the medium during a 7-day culture period indicated that the stationary phase of growth was reached after approximately 3 days, and maximum

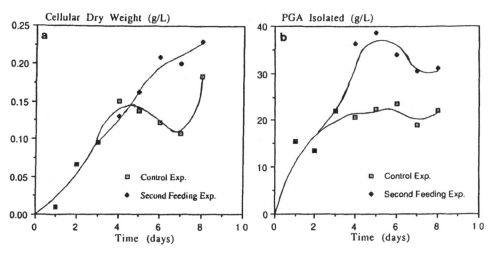

Figure 1. Results from two time course studies showing changes in: (a) cellular dry weight, and (b) crude γ-PGA isolated. The second feeding experiment (◆) was carried out with the addition of media nutrients after a 3 day culture time.

yields of crude γ-PGA were obtained between days 3 and 5. Therefore, it was decided that a second feeding of nutrients after a 3 d culture time would be appropriate for this preliminary study.

In Figure 1, results from two time course studies which compare cellular growth and γ-PGA production are shown. The cultures were grown with and without the addition of nutrients (a second feeding) after a 3 d culture period (see Experimental for details).

It is apparent that both cellular growth and γ-PGA production increase dramatically for cultures which have been fed a second time. At day 5, the yield of crude γ-PGA increased from 23 g/L to 39 g/L when nutrients were added after day 3. It is of interest to note that produc-

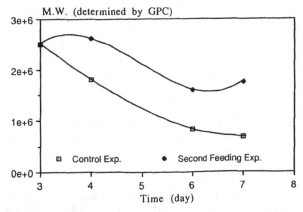

Figure 2. Molecular weight results for two time course studies.

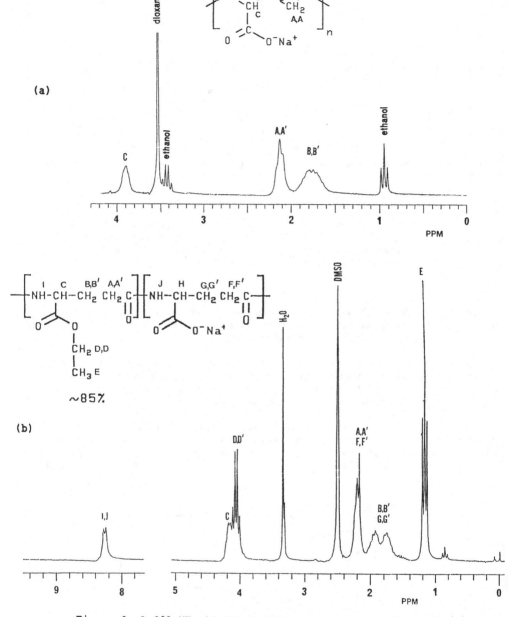

Figure 3. A 200 MHz ¹H NMR spectra, recorded at 25°C, of (a)
Na-γ-PGA in D₂O, and (b) the ethyl-γ-PGA derivative
in d₆-DMSO.

tion of 39 g/L of crude γ-PGA (not dialyzed) using the method above resulted in the highest yield (in γ-PGA g/L) thus far reported that we are
aware of (as compared to 18 g/L of crude γ-PGA).[6] It appears that the
crude γ-PGA sample obtained at day 5 from a two times fed culture has a
purity of approximately 70%, where most impurities were low molecular

weight substances which were easily removed by dialysis.

Molecular weight measurements on crude γ-PGA samples from the two time course studies are given below in Figure 2.

The M_w values of the polymer samples for the control experiment decreased with time. This may be due to the presence of a depolymerase enzyme(s) which is actively degrading chains after day three (or possibly earlier). The molecular weight values obtained for the second feeding time course study were consistently higher than corresponding control samples. This may be due to the synthesis of high molecular weight polymer between days 4 and 5, or to an inhibition in the activity of the depolymerase enzyme(s) due to the addition of nutrients at day 3.

The ^{1}H NMR spectra (200 MHz) and peak assignments for Na-γ-PGA and the corresponding ethyl ester derivative are shown in Figures 3a and 3b, respectively. Assignment of the A,A' and B,B' peaks in Figures 3a and 3b was based on ^{1}H-^{1}H-coupling information obtained from a COSY experiment. Peak integration of the ethyl and propyl ester γ-PGA derivatives show that approximately 90% of the repeat units contain ethyl and propyl ester pendant groups, respectively. The ^{13}C NMR spectrum (67.5 MHz) along with tentative peak assignments for the Na-γ-PGA derivative is displayed in Figure 4. This spectrum is consistent with the proposed polymeric structure though additional NMR experiments are planned to confirm the peak assignments as well as the existence of solely γ-linkages in this polymer.

The DSC thermogram during the first heating scan for the n-propyl ester-γ-PGA derivative is shown in Figure 5. The DSC shows a glass transition temperature, T_g, at approximately 77°C and polymer decomposition (confirmed by visual observation) above 200°C.

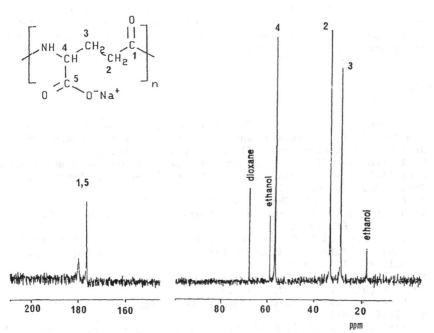

Figure 4. A 67.5 MHz ^{13}C NMR spectrum, recorded in D$_2$O at 25°C, of Na-γ-PGA.

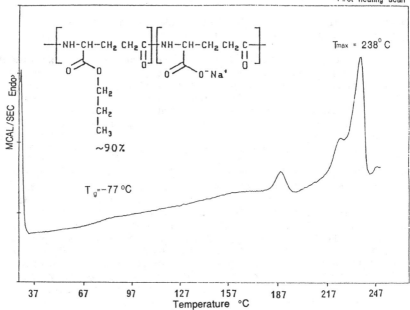

Figure 5. DSC thermogram during the first heating scan of the
propyl-γ-PGA derivative. Heating rate was 20°C/min.

ACKNOWLEDGEMENT

We are grateful for financial support of this work by the U. S. Army
Natick Research Center and The Petroleum Research Fund, Grant # 21657-G7.
We would also like to thank Parasar Dave for carrying out thermal analy-
sis of various γ-PGA derivatives.

REFERENCES

1. R. D. Housewright in: *"The Bacteria: A Treatise on Structure and
   Function,"*, I. C. Gunsalus and R. Y. Stanier, Eds., Academic Press,
   New York and London, Vol. III, 1962, pp. 389-412.
2. F. A. Troy in: *"Peptide Antibiotics: Biosynthesis and Functions,"* H.
   Kleinkauf and H. V. Dohren, Eds., Walter de Gruyter, Berlin and New
   York, 1982, pp. 49-83,.
3. F. A. Troy, J. Biol. Chem., **248** (1), 316 (1973).
4. C. B. Thorne, C. G. Gomez, H. E. Noyes and R. D. Housewright, J.
   Bacteriol., **68**, 307 (1954).
5. C. B. Thorne and C. G. Leonard, J. Biol. Chem., **233**, 1109 (1958).
6. S. Sawa, S. Murao, T. Murakawa and S. Omata, Nippon Nogei Kagaku
   Kaishi, **45**, 124 (1971).
7. S. Sawa, T. Murakawa, T. Watanabe, S. Murao and S. Omata, Nippon
   Nogei Kagaku Kaishi, **47**, 167 (1973).
8. T. Hara, A. Aumayr, Y. Fujio and S. Ueda, Appl. Environ. Microbiol.,
   44, 1456 (1982).
9. J. Kream, B. A. Borek, C. J. Digrado and M. Bovarnick, Archives of
   Biochemistry and Biophysics, 53(2), 333 (1954).

10. M. Torii, J. Biochem., **46**(4), 513 (1959).
11. For the recipe to prepare medium E see: C. G. Leonard, R. D. Housewright and C. B. Thorne, J. Bacteriol., **76**, 499 (1958).
12. V. Bocchi, G. Casnati, A. Dossena and R. Marchelli, Synthesis, 961 (1979).

VERNONIA OIL: A NEW REACTIVE MONOMER

Stoil Dirlikov, Isabelle Frischinger, M. Safiqul Islam, and
T. J. Lepkowski*

Coatings Research Institute
Eastern Michigan University
and
*Paint Research Associates
430 W. Forest Avenue
Ypsilanti, MI 48197

Vernonia galamensis, the plant which produces vernonia
oil, is at a developmental stage in several countries. Vernonia oil is a natural epoxidized vegetable oil containing
three epoxy rings and three carbon-carbon double bonds. It
has several advantages over other epoxidized vegetable oils
produced industrially and appears to be a very attractive raw
material for three large volume industrial applications.
First, vernonia oil is characterized by very low viscosity
(100 cps at 85°F) and low m.p. of about 36°F, with a potential as a reactive diluent for high solids alkyd, epoxy, and
epoxy ester coating formulations by replacing conventional
solvents. Second, vernonia oil can simultaneously improve the
two major disadvantages of commercial epoxy resins: low
toughness and high water absorption. Epoxy resins based on
vernonia oil are elastomers at room temperature with a low
glass transition temperature. Homogeneous mixtures of commercial epoxy resin, diamine, and vernonia oil form thermosets
consisting of a rigid epoxy matrix with randomly distributed
small rubbery "vernonia" spherical particles which dissipate
part of the impact energy. Finally, vernonia oil appears to
be a low-cost substitute of epoxidized soybean oil as a plasticizer and stabilizer of poly(vinyl chloride).

INTRODUCTION

Vernonia oil is a naturally occurring epoxidized vegetable oil which
contains predominantly (~80%) trivernolin, a triglyceride of vernolic
acid (Figure 1). It has three epoxy rings per triglyceride molecule. In
addition, it contains three carbon-carbon double bonds per triglyceride.
There is one epoxy ring and one carbon-carbon double bond per each
vernolic acid residue.

Vernonia galamensis, the source of vernonia oil, is a new potential

Biotechnology and Polymers, Edited by C.G. Gebelein
Plenum Press, New York, 1991

TRIOLEIN:

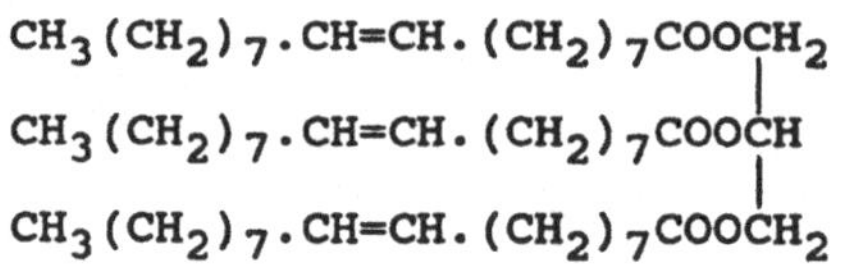

$$CH_3(CH_2)_7.CH=CH.(CH_2)_7COOCH_2$$
$$CH_3(CH_2)_7.CH=CH.(CH_2)_7COOCH$$
$$CH_3(CH_2)_7.CH=CH.(CH_2)_7COOCH_2$$

VERNONIA OIL:

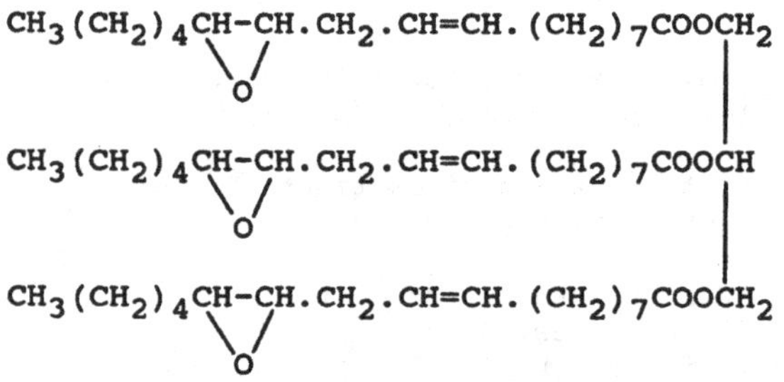

EPOXIDIZED SOYBEAN OIL:

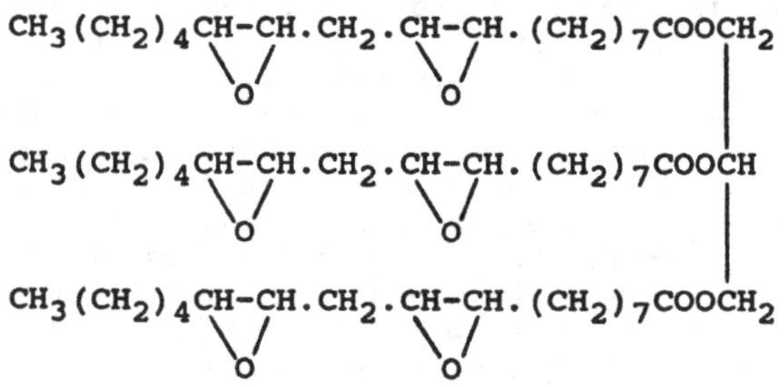

Figure 1. Molecular structure of triolein, vernonia oil, and epoxidized soybean oil.

oil seed crop in Africa, Asia, Central America and in the Southwest U.S.A., in the dry areas of Arizona, New Mexico, etc..[1-6] *Vernonia galamensis* grows in both arid and semi-arid areas of the tropics and subtropics on land that is practically unsuitable for food crops. It is an annual herb, as common as a weed in West Africa, that can be harvested directly. It has good seed retention and the seeds mature uniformly and germinate easily, which is very important for crop utilization. It also has a natural resistance to diseases, nematodes, and insects. Neither wild nor domestic animals consume *Vernonia galamensis*.

"Vernonia" seeds contain about 42% of oil in contrast to soybean seeds, which contain only 17% oil. The maximum seed yield reached at this development stage of *Vernonia galamensis* is 2227 pounds per acre.[2] Unfortunately, it has not been reproduced. Increased yield of vernonia oil, however, is expected by breeding as greater genetic diversity becomes available and by better management of the crop. The best soybean seed yield is 1926 pounds per acre (1979), which was the best soybean oil year in the U.S. after about fifty years of cultivation. Dr. R. Perdue from the U.S. Department of Agriculture expects similar seed yields for soybean and *Vernonia galamensis* after crop maturity.

In summary, it appears that the major agronomic problems for *Vernonia galamensis*' cultivation have been resolved. *Vernonia galamensis* is at a developmental stage at the present moment in several African countries, especially in Zimbabwe and Kenya; and has a potential to soon become an industrial crop. Several major chemical and coatings companies have expressed strong interest in vernonia oil. Hi-Tek Polymers, Inc. has started experimental crop trials for vernonia oil production, and several tons of *Vernonia galamensis* seeds are expected to be harvested next year. Cargill is evaluating the possibilities for cultivation of *Vernonia galamensis* and the potential use of vernonia oil as well.

It appears that the future of *Vernonia galamensis* depends on the development of a large market for application of vernonia oil which will drive its production. Several initial attempts for utilization of vernonia oil in different areas have been reported. Sperling, et al., have reported interpenetrating networks with one component based on vernonia oil.[7-15] Carlson, et al. have been able to obtain good coatings directly from vernonia oil, as well as to further epoxidize vernonia oil.[16,17] Ayorinde et al., have prepared different dibasic acids by oxidation of vernonia oil and discussed their application for preparation of high molecular polymers, i.e. nylons and polyesters.[18-22] Initial evaluation of vernonia oil for modification of poly(vinyl chloride) has been described as well.[23-25] Our research is directed towards development of three large volume industrial applications of vernonia oil: as a reactive diluent for high solids alkyd and epoxy formulations, and for modification of epoxy resins and poly(vinyl chloride).

EXPERIMENTAL

Vernonia oil has been kindly supplied by Dr. K. Carlson from the U.S. Department of Agriculture. Its isolation (extraction) from *Vernonia galamensis* seeds, refining and characterization has been described.[16]

Gel permeation chromatography (GPC) has been measured on a Hewlett-Packard GPC instrument with Waters Associates differential refractometer model R401 and Polymer Laboratory gel columns.

The preparation of the coatings formulations is shown later in Table 3. The characterization of these formulations and their final coatings has been carried out as follows:

1. Viscosity: Stormer viscosimeter (ASTM D562-81) and Brookfield viscosimeter (ASTM 2196-86)
2. Hardness: Pencil hardness (D3363-74) and Rocker Sward hardness (ASTM D2134-66)
3. Adhesion: by tape (ASTM D3359-87)
4. Flexibility: (ASTM D4145-83)
5. Impact strength: by rapid deformation (ASTM D2794-84)
6. Drying: at room temperature (ASTM D1640-83)
7. Package (can) stability: (ASTM D1849-80)
8. UV exposure (Q-UV weathering test): (ASTM D4587-86)

Two-phase epoxy thermosets have been prepared as follows: Vernonia oil was initially B-staged with amines. For this purpose a mixture of 100 g of vernonia oil and 25.63 g of 4,4'-methylenedianiline (DAPM) have been heated at 180°C for 40 hours under nitrogen. Similar results have been observed for vernonia B-staged material obtained from 25.9 g of 1,12-dodecanediamine and 100 g of vernonia oil heated at 180°C for 37 hours. The vernonia B-staged material based on DAPM or 1,12-dodecanediamine is added in 10, 20 or 30 weight percent to a stoichiometric mixture of the diglycidyl ether of bisphenol A (Shell epoxy resin EPON 825) and DAPM or 4-aminophenyl sulfone (DDS). The resulting homogeneous formulation based on DAPM is additionally B-staged at 75°C for 95 minutes under nitrogen, and then poured into the casting mold and vacuumed. Final cure and post-cure are carried out at 75°C for 4 hours and 150°C for two hours respec-tively. The corresponding DDS formulation is B-staged at 150°C for 1 hour under nitrogen and then cured in the mold at 150°C for 2 hours. Both (DAPM or DDS) formulations produce excellent large castings. Electro-micrographs have been taken by an Amray Electron Microscope, Model 1000B, whereas glass transition temperatures have been determined by DuPont Thermal Analyst 2100 instrument, DMA model 983 and DSC model 2910.

RESULTS AND DISCUSSION

1. Properties of Vernonia Oil

   Vernonia oil has several unique features:

   (a). It is a transparent, homogeneous liquid at room temperature with excellent solubility in many organic solvents, diluents and paints.
   (b). Vernonia oil has a low viscosity of 300 cps at 50°F and 100 cps at 85°F.
   (c). It has a low melting point of ~36°F.
   (d). Vernonia oil has a homogeneous molecular structure consisting predominantly of identical triglyceride molecules which have three equal vernolic acid residues. In contrast, all other vege-table oils consist of a heterogeneous mixture of triglycerides with different fatty acid residues.
   (e). It is expected to be available at a low price of approximately one dollar per pound.
   (f). Toxicity of vernonia oil is expected to correspond to that of the epoxidized soybean and linseed oils which are industrially produced.

2. Vernonia vs. Other Epoxidized Vegetable Oils

   We believe that vernonia oil does not compete with other epoxidized vegetable oils, i.e., epoxidized soybean oil and epoxidized linseed oil, which are industrially produced by epoxidation of unsaturated vegetable oils and have structures similar to vernonia oil. Epoxidized soybean and linseed oils are not suitable for coatings applications as reactive di-luents due to their higher viscosity, or at least they are less suitable than vernonia oil.

   Both epoxidized soybean oil and linseed oils have higher viscosity, i.e., the viscosity of the epoxidized soybean and linseed oils is in the range of 1000 - 2000 cps at 50°F, in contrast to 300 cps for vernonia oil. Both are heterogeneous semi-solids even at 75°F and form homo-geneous, clear liquids only above 85°F. Their applications require a warming procedure at 120°F with mild agitation for at least one hour prior to use.

The difference in viscosity of vernonia oil and epoxidized soybean oil is rather surprising even though both oils have similar molecular structures. Initially, we thought that soybean oil partially polymerizes during the epoxidation reaction and the resulting epoxidized soybean oil contains a certain amount of oligomers which increases its viscosity. GPC of vernonia oil and epoxidized soybean oil, however are identical (Figure 2). Evidently, both oils have exactly the same molecular weight distribution and the low viscosity is an inherent characteristic of vernonia oil.

3. Reactive Diluent

The first objective of our project is the application of vernonia oil as a reactive diluent for high solids alkyd, epoxy, and epoxy-ester coating formulations by replacing conventional solvents, which produce volatile organic compound (VOC) emission, with vernonia oil.

There is a strong need for reactive diluents in the reduction of conventional solvents in coating formulations for air pollution control. Recently, further and stricter measurers have been taken for the reduction of air pollution. In 1988, the U.S. Environmental Protection Agency announced a ban on construction of major air-polluting plants in Los Angeles and other U.S. cities in order to reduce air pollution. In the same year the South Coast Air Quality Management District announced a plan for reduction of hydrocarbon emission in Orange County, California by 80-90% over the next five years. This plan requires reduction of VOC emission from paints and varnishes from the current 22.1 tons per day to 2.9 tons per day several years from now. This is an impossible task with today's coatings technology.

Ken Carlson and co-workers, from the U.S. Department of Agriculture, have reported that "vernonia" coatings can be obtained directly from vernonia oil by a standard baking procedure at 150°C for 30 minutes in the presence of different drier systems.[16] Our initial goal is determination of the drying mechanism and compatibility of vernonia oil with alkyd and other resins, characterization of these "vernonia" coatings, evaluation of different drier systems, and eventually, improvement of their properties.

3A. Drying Mechanism

We were interested in the drying mechanism of vernonia oil since it has two functionalities: unsaturated double bonds and epoxy rings. For this purpose, we have compared the drying characteristics of triolein, vernonia oil, and epoxidized soybean oil in the presence of 0.5% cobalt drier under baking conditions at 150°C for one hour. The molecular structures of these three oils are shown in Figure 1.

Epoxidized soybean oil does not form coatings at 185°C for several hours. Obviously, its epoxy groups do not polymerize under these conditions. Both triolein and vernonia oil form good coatings. These results show that vernonia oil drying is based on its unsaturated carbon-carbon double bonds and not on its epoxy functionality. This allows its application as a reactive diluent, not only in epoxy and epoxy-ester formulations, but in alkyd resins as well.

Results also show that vernonia oil dries at a lower temperature and

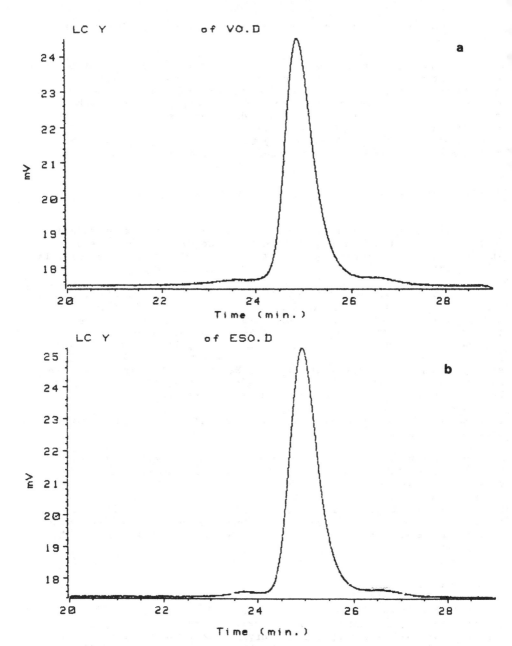

Figure 2. Gel permeation chromatography of vernonia oil (a)
and epoxidized soybean oil (b).

much faster than the triolein. Good coatings have been obtained from
vernonia oil at 150°C for half an hour. Triolein does not form coatings
at 150°C for 1 hour; it forms coatings only at 180°C for 1.5 hours. The
epoxy rings of vernonia oil probably activates the oxidation of the
methylene group between the epoxy ring and the carbon-carbon double bond
and we, therefore, observe its faster drying in comparison to triolein.

3B. Compatibility with Alkyd Resins

Compatibility studies have been carried out with Reichhold Beckosol long oil alkyds 10-560 and 10-060, medium oil alkyd 11-035, and short oil alkyd 12-005. Vernonia oil is compatible with all three types of alkyd resins at any ratio.

3C. Drier Evaluation

We evaluated different driers and drier combinations for vernonia oil and for its 20% mixtures in alkyd resins (Table 1). Cobalt drier at 0.1% concentration forms good coatings at 150°C for 1 hr. Zirconium drier, even at a higher concentration (0.5%), does not form coatings at higher temperature (160°C) and longer time (2 hours). Calcium and manganese driers have intermediate activity, weaker than the cobalt drier, but stronger than the zirconium. Without going into further detail, Manchem manosec CD-44 drier gives the best results. It is more active than all other driers and drier combinations. It contains cobalt (0.6%), calcium (0.2%), and lithium (0.03%) carboxylates in mineral spirits.

Vernonia oil does not air dry with any of the driers at room temperature for several hours. It does dry slowly in about a week. However, alkyd resin mixtures containing 10 to 40% vernonia oil air dry at room temperature within several hours.

3D. Clear Coatings

Table 2 compares the properties of the 2 mil thick coatings obtained under baking conditions in the presence of 0.04% cobalt naphthenate drier at 150°C for one hour on cold rolled steel panels from vernonia oil, pure 10-560 and 10-060 long oil alkyds, and their mixture containing 20% vernonia oil.

All coatings have good adhesion of 4B. The hardness is also good. Pencil hardness is H and Rocker Sward hardness is 42. Both adhesion and

Table 1. Driers for vernonia oil and long oil alkyds.
Cobalt Naphthenate Zirconium Drier Manganese Drier Calcium Drier
Drier Combinations
Mooney Driers: Co, Ca, DRI-Rx Mooney Driers: Co, Mn, DRI-Rx Nuodex Nuxtra Cobalt Nuodex Nuxtra Zirco Nuodex Nuxtra Calcium Nuodex Adr (Blended Metals/Surface Drier) Nuodex Ltd (Complex of Metals/Through Drier) Nuodex Utd (Complex of Metals/Through Drier) Active 8 (Drier Stabilizer and Accelerator) Manchem Manosec CD-44 Cobalt/calcium/lithium carboxylates

Table 2. Characterization of clear coatings prepared under baking conditions: cobalt drier 0.04%, 150°C, 1 hr., on cold rolled steel panel, 2 mil thickness.

	Vernonia Oil	Alkyd (80%) Vernonia Oil (20%)	Alkyd 10-560 10-060
Adhesion	4B	4B	4B
Hardness:			
Pencil	H	2H	H
Rocker	42	42	46
Flexibility	Excellent	Excellent	Excellent
Impact Strength:			
Direct, lb·in.	150	150	160
Reverse, lb·in.	140	140	145

hardness improve at higher baking temperature and longer time. Flexibility, according to a standard test by bending the steel panel 180° around, is excellent. We did not observe any cracks or tape-off under a microscope. The impact strength is also excellent. The direct impact strength is 150 pounds·inch, whereas the reverse is 140 pounds·inch.

All three types of coatings have the same properties which indicates that introduction of vernonia oil does not deteriorate the basic properties of alkyd resin.

3E. Alkyd Paint Formulations

We also prepared a typical paint formulation, as shown in Table 3, and compared it to the paint formulations in which the initial long oil alkyd has been substituted with 20 and 40% vernonia oil.

Table 4 compares the Brookfield viscosity and the RCI drying times, in hours at room temperature, of coatings with 3 mil thickness, and volatile organic compound or VOC in lb. per gallon.

The first paint formulation corresponds to the initial formulation based only on alkyd resin (Table 3). In the second paint formulation, 20% of the alkyd has been substituted with vernonia oil. Lower viscosity is observed for this formulation.

Then, the next two formulations still contain 20% vernonia oil, but their solvent content has been gradually reduced. The goal is to decrease the solvent to the point at which the viscosity of the paint formulation corresponds to the viscosity of the initial paint formulation without vernonia oil. These two formulations have much lower volatiles.

The fifth paint formulation is exactly as the first one with the only difference being that it contains "our" new, more active CD-44 drier. Drying time of the paint formulations or tack-free time changes very little in the presence of vernonia oil, but sharply decreases in the presence of CD-44.

```
Table 3. Composition of paint formulations.
```

	Weight
10-060 Long Oil Alkyd	78.60
Odorless Mineral Spirits	23.59
Byk Bykumen	3.77
Noudex Nuxtra Calcium 10%	1.12
Premix Above	
Dupont R-900 TiO_2	150.97
T & W Atomite	37.74
NL Chemical SD-1 Bentone	2.72
Add Slowly to Cowls	
10-060 Long Oil Alkyd	81.62
Vernonia Oil	
Add to Grind	
Nuodex Nuxtra Cobalt 12%	0.46
Nuodex Nuxtra Zirco 24%	2.34
Odorless Mineral Spirits	93.10
Nuodex Exkin #2	1.62
Premix then Add	

```
Table 4. Characterization of paint formulations (where V.O. ≡
         veronia oil).
```

	Paint Formu-lation	Brookfield Viscosity CPS	RCI Drying Time, hr	VOC lb./gal.
1.	Initial 10-060	2640	13	3.15
2.	V.O. 20%	586	11	3.11
3.	V.O. 20% Less Solvent	2260	15	2.63
4.	V.O. 20% Less Solvent	3476	12	2.52
5.	Initial 10-060 Drier CD-44	3344	3-4	2.50
6.	Drier CD-44 V.O. 20%	780	4	2.60

Finally, the last formulation contains the new drier system and 20% of its alkyd resin has been substituted with vernonia oil. The important result here is that by introducing 20% vernonia oil in this formulation we are able to decrease the volatiles, and thus eventually reduce air pollution, reduce the drying time, and maintain equal or lower viscosity than the initial paint formulation.

A potential problem for direct application of vernonia oil in alkyd formulations is an anticipated reaction between its epoxy groups and the terminal carboxyl groups of the alkyd resins. Our results show that vernonia oil does not change the can stability of the formulations. Obviously, the epoxy groups of vernonia oil have low reactivity and do not react with the free carboxyl groups of the alkyd resins under test conditions for can stability. The low reactivity of the epoxy groups of vernonia oil has been confirmed by additional experiments (see Section 4. Modification of Epoxy Resins).

Q-UV test, which is the decrease in gloss at 20° and 60° after exposure at UV light and humidity for a week, indicates that vernonia oil improves the coatings weatherability.

3F. Epoxy Coatings

Vernonia oil contains reactive epoxy groups and is an attractive reactive diluent for epoxy or epoxy ester formulations as well. We are just starting to explore this area. Another possibility, again for epoxy formulations, is transesterification of vernonia oil with methanol or other higher molecular weight mono-functional alcohols for preparation of very low viscosity diluents. For example, methyl vernolate, shown below, has been prepared in a quantitative yield.

$$CH_3(CH_2)_4CH-CH-CH_2CH=CH-(CH_2)_7COOCH_3$$
$$\backslash \, /$$
$$O$$

Similar esters are industrially produced and the U.S. Navy has used them for coatings applications under water. However, they are very expensive. In summary, our initial research shows that vernonia oil is an attractive reactive diluent for alkyd and, potentially, for epoxy coatings.

4. Modification of Epoxy Resins

The second part of the project is the application of vernonia oil for the modification of commercially available epoxy resins. The project objective is simultaneous improvement of the two main disadvantages of epoxy resins: low toughness and high water absorption.

It is known that a small amount of discrete rubbery particles with an average size of several microns, randomly distributed in a glassy, brittle epoxy resin, dissipates part of the impact energy thus improving crack and impact resistance without deterioration of other properties of the initial epoxy resins.

The epoxy resin toughness is usually achieved by separation of a rubbery phase with a unimodal size distribution from the matrix during the curing process. Different carboxy- or amino- terminated oligomers are used for the formation of the rubbery phase. Some of these oligomers are quite expensive. However, vernonia oil opens new possibilities. The epoxy resins based on vernonia oil and commercial amines are elastomers at room temperature. They have low glass transition temperatures, in the range of -50°C to 0°C, which depends on the nature of the diamine used for B-staging prepolymerization of vernonia oil. They are very suitable for toughening of commercial epoxies.

Our initial attempts for toughening of epoxy resins have been carried out directly with vernonia oil. For this purpose, homogeneous mixtures of commercial epoxy resin (EPON 825) and commercial amines (DDS, DAPM, isophorone diamine, different aliphatic di- and polyamines, etc.), containing 10 to 25% vernonia oil have been cured according to a "standard" curing procedure for epoxy resin. The commercial diamines have higher reactivity with commercial epoxy resins than with the epoxy groups of vernonia oil. As a result, they form a rigid matrix at 70°C in which vernonia oil separates as liquid droplets. The electromicrograph of a formulation based on EPON 825, isophorone diamine and vernonia oil shows a rigid matrix with random distribution of vernonia droplets with diameter in the range of 1 micron (Figure 3). Unfortunately, these droplets do not cure at a high temperature of 150°C with the excess of diamine. Instead, plastification occurs resulting in lower glass transition temperature of the rigid matrix observed in DSC.

We have been able to obtain a two-phase thermoset with rubbery vernonia particles by using vernonia B-staged materials instead of pure vernonia oil as described in the experimental section. Vernonia particles again separate during curing from the initial homogeneous mixture of epoxy resin, diamine and vernonia B-staged material. In this procedure, however, the final thermoset consists of a rigid epoxy matrix with randomly distributed small *rubbery* "vernonia" spherical particles (Figure 4). Diamine molecules on the interface are expected to react with both the epoxy groups of the commercial epoxy resin and the unreacted epoxy

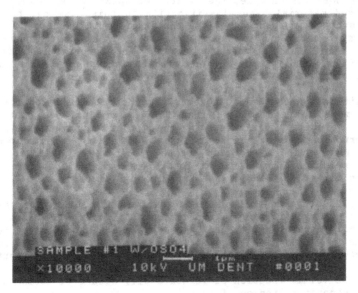

Figure 3. Electromicrograph of two-phase epoxy resin:matrix based on Epon 825 and isophorone diamine and liquid phase of vernonia oil (20%).

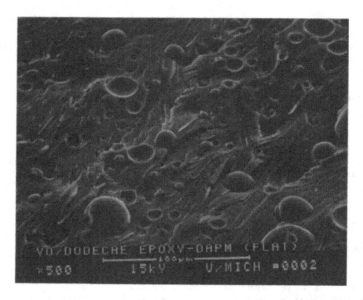

Figure 4. Electromicrograph of two-phase epoxy resin, based on
Epon 825, DAPM, and rubbery particles (20%) based on
vernonia/dodecanediamine B-staged material.

groups of vernonia oil, and will presumably form a chemical bonding
between the rigid matrix and the rubbery particles.

Plastification phenomena at higher temperature (150°C) described
above for pure vernonia oil have been observed here for vernonia B-staged
materials as well, if the solubility parameters of the epoxy matrix and
vernonia rubbery phase are similar. Such formulations are based on less
polar diamines; for example, the mixture of epoxy resin, DAPM, and vernonia B-staged material. The plastification in this case has been avoided
by a two-step curing procedure (see the experimental section). The general rule is to carry out the epoxy resin cure at a lower temperature and
build the rigid matrix, then to increase the temperature and cure the
less reactive vernonia epoxy groups with the excess of diamine into rubbery particles and simultaneously post-cure the matrix.

Plastification phenomena, however, are not observed for formulations
based on more polar diamines (than DAPM), for instance, DDS. In this
case, the solubility parameters of the "polar" epoxy matrix and "nonpolar" vernonia phase are quite different. DDS based formulations, therefore, do not require a two-step curing procedure in order to build the
molecular weight of the matrix at lower temperature. They form two-phase
thermosets directly at 150°C (see Experimental section).

Several factors control phase separation. Miscibility of the initial
mixture of epoxy resin, amine, and vernonia oil or its B-staged material,
is required. Incompatibility and phase separation should start during the
curing process, before gelation, for spherical morphology formation.
Simultaneously, viscosity at this stage should be high to prevent coales-
cence and macroscopic phase separation. The effective rubbery phase depends on the volume fraction of vernonia oil, nature of curing amine, and
time and temperature regime of cure.

The particle diameter is a function of the incompatibility of vernonia B-staged material and the epoxy matrix. We have been able to vary the

particle diameter in the range of the desirable particle size for toughening thermosets, e.g., of 5 to 50 microns. In general, larger particles and even macroscopic total phase separation are observed for less compatible systems with more polar diamine, for instance, DDS for curing of the rigid matrix. The epoxy thermoset based on EPON 825, DAPM, and vernonia/dodecaneamine B-staged material, shown on the electromicrograph, has broad rubbery vernonia oil particles distribution in the range of 5 to 35 microns diameter (Figure 4).

A typical DMA is shown on Figure 5. Glass transition temperatures of the vernonia rubbery phase and the rigid epoxy matrix EPON 825 are -22°C and 160°C, respectively. DAPM is used as diamine curing agent for the matrix, whereas dodecanediamine is used for vernonia oil B-staging. The glass transition of the matrix corresponds to the glass transition temperatures of a pure epoxy resin castings (160°C) prepared under the same conditions. We believe there is no extensive internal plastification.

Our *initial* results indicate improved fracture toughness (K_{IC}) for thus modified epoxy resins. Since vernonia oil is practically resistant to water, the resulting two phase epoxy resins are expected to have lower water absorption. Also, the introduction of vernonia oil into the commercial epoxy resin will eventually result in a price reduction.

5. Poly(vinyl chloride) Modification

The third potential application of vernonia oil is for modification of poly(vinyl chloride) (PVC). Our research in this direction is in the initial stage.

PVC has two major disadvantages. It is inherently brittle and it easily undergoes photodegradation. Upon light or heat treatment, PVC macromolecules release hydrochloride molecules as a primary product, which act as degradation auto-accelerators:

$$\left[CH_2 \text{---} CHCl \right]_n \xrightarrow{\text{light, heat}} \left[CH\text{=}CH \right]_n + HCl$$

Seventy-five to 85% of the epoxidized soybean oil is used for PVC plastification and stabilization. Epoxidized soybean oil decreases PVC's glass transition temperature and makes it tractable and tough. In addition, the epoxy groups of the epoxidized soybean oil react with hydrochloride, preventing the auto-acceleration of PVC degradation, and act as its stabilizer.

$$\underset{O}{CH\text{---}CH} + HCl \longrightarrow \underset{OH \quad Cl}{CH\text{---}CH}$$

Vernonia oil has a molecular structure similar to that of epoxidized soybean oil and initial results have shown that vernonia oil is an excellent PVC plasticizer and stabilizer.[23-25] Since vernonia oil is expected to be less expensive than epoxidized soybean oil, we have little doubt that it will be able to substitute for the epoxidized soybean oil in PVC modification and capture this large market.

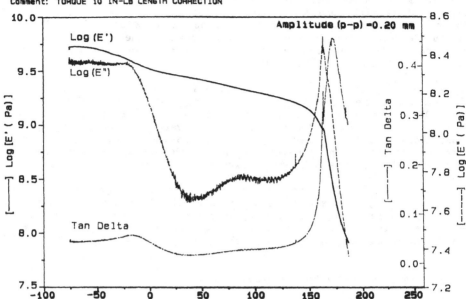

Figure 5. DMA of two-phase epoxy resin:matrix based on Epon
825 and DAPM (Tg = 160°C) and rubbery particles
(20%) of vernonia oil/dodecanediamine B-staged
material (Tg = -22°C).

CONCLUSION

In summary, vernonia oil has a unique structure with reactive epoxy
rings, double bonds, lower viscosity than other epoxidized vegetable
oils, and a low melting point. It appears that the major agronomic problems for *Vernonia galamensis* cultivation have been resolved. Vernonia
oil is an attractive starting raw material for preparation of high solids
alkyds and epoxy coating formulations with low volatile organic compounds, and for epoxy resin and PVC modification. In addition, it is a
potential low-cost substitute of the industrially produced epoxidized
soybean and linseed oils.

ACKNOWLEDGMENT

We would like gratefully to acknowledge the financial support by the
South Coast Air Quality Management District, U.S. Agency for International Development, the Paint Research Associates, and Michigan State. We
would like to also thank Dr. John C. Graham, Coatings Research Institute
and Dr. Robert Perdue from the U.S. Department of Agriculture for advice
and many fruitful discussions. We thank TA Instruments (formerly DuPont
Instruments) for donating DSC and DMA to our institute.

REFERENCES

1. K. Kaplan, Agricultural Research, **37**, 10 (1989).
2. R. Perdue, Jr., Agricultural Engineering, **70**, 11 (1989).
3. R. E. Perdue, Jr., *Vernonia galamensis*, Potential New Industrial Oilseed Crop, USDA, 1987.
4. R. E. Perdue, Jr., K. D. Carlson & M. G. Gilbert, Economic Botany, **40**, 54 (1986).
5. F. R. Earle, J. Amer. Oil Chem. Soc., **47**, 510 (1970).
6. C. F. Krewson, J. Amer. Oil Chem. Soc., **45**, 250 (1968).
7. A. M. Fernandez, J. A. Manson & L. H. Sperling, in: "*Renewable Resource Materials: New Polymer Sources,*" C. E. Carraher, Jr. & L. H. Sperling, Eds., Plenum Press, New York, 1986, p. 177.
8. G. M. Jordhamo, J. A. Manson & L. H. Sperling, Polym. Eng. Sci., **26**, 517 (1986).
9. S. Qureshi, J. A. Manson, J. C. Michel, R. W. Hertzberg & L. H. Sperling, in: "*Characterization of Highly Cross-Linked Polymers,*" ACS Symposium Series Book No. 243, S. S. Labana & R. A. Dickie, Eds., ACS, 1984, p.109.
10. L. H. Sperling, J. A. Manson & M. A. Linne, J. Polym. Mater., **1**, 54 (1984).
11. M. A. Linne, L. H. Sperling, A. M. Fernandez, S. Qureshi, & J. A. Manson, in: "*Rubber-Modified Thermoset Resins,*" ACS Advances in Chemistry Series, No. 208, C. Keith Riew & J. K. Gillham, Eds., ACS, 1984, Chapter 4, p.37.
12. A. M. Fernandez, C. J. Murphy, M. T. DeCosta, J. A. Manson & L. H. Sperling, in: "*Polymer Science & Technology,*" Vol. 17, C. E. Carraher, Jr. & L. H. Sperling, Eds., Plenum Press, New York & Toronto, 1983, p.273.
13. S. Qureshi, J. A. Manson, L. H. Sperling & C. J. Murphy, in: "*Polymer Applications of Renewable Resource Materials,*" C. E. Carraher, Jr. & L. H. Sperling, Eds., Plenum Press, New York, 1983, p.249.
14. L. H. Sperling & J. A. Manson, J. Amer. Oil Chem. Soc., **60**, 1887 (1983).
15. L. H. Sperling, J. A. Manson, S. Qureshi & A. M. Fernandez, Ind. Eng. Chem. Prod. Res. Dev., **20**, 163 (1981).
16. K. D. Carlson, W. J. Schneider, S. P. Chang & L. H. Princen, in: "*New Sources of Fats & Oils,*" AOCS Monograph No. 9, E. H. Pryde, L. H. Princen & K. D. Mukherjee, Eds., Amer. Oil Chemists' Soc., Champaign, IL 1981, Chapter 21, p. 297.
17. K. D. Carlson & S. P. Chang, J. Amer. Oil Chem. Soc., **62**, 934 (1985).
18. M. O. Ologunde, F. O. Ayorinde & R. L. Shephard, J. Amer. Oil Chem. Soc., **67**, 92 (1990).
19. O. A. Afolabi, M. E. Aluko, G. C. Wang, W. A. Anderson & F. O. Ayorinde, J. Amer. Oil Chem. Soc., **66**, 983 (1989).
20. F. O. Ayorinde, F. T. Powers, L. D. Streete, R. L. Shepard & D. N. Tabi, J. Amer. Oil Chem. Soc., **66**, 690 (1989).
21. F. O. Ayorinde, G. Osman, R. L. Shepard & F. T. Powers, J. Amer. Oil Chem. Soc., **65**, 1774 (1988).
22. F. O. Ayorinde, J. Cliffon, Jr., O. A. Ofolabi & R. L. Shepard, J. Amer. Oil Chem. Soc., **65**, 942 (1988).
23. G. R. Riser, R. W. Riemenschneider & L. P. Witnauer, J. Amer. Oil Chem. Soc., **43**, 456 (1966).
24. C. F. Krewson, G. R. Riser & W. E. Scott, J. Amer. Oil Chem. Soc., **43**, 377 (1966).
25. G. R. Riser, J. J. Hunter, J. S. Ard & L. P. Witnauer, J. Amer. Oil Chem. Soc., **39**, 266 (1962).

INTERPENETRATING POLYMER NETWORKS BASED ON FUNCTIONAL TRIGLYCERIDE OILS

AND OTHER NOT YET COMMERCIAL RENEWABLE RESOURCES

L. H. Sperling, C. E. Carraher,[a] S. P. Qureshi,[b]
J. A. Manson[c], and L. W. Barrett

Materials Research Center
Center for Polymer Science and Engineering
Department of Chemical Engineering
Department of Materials Science and Engineering
Whitaker Laboratory #5, Lehigh University
Bethlehem, PA 18015

(a) Department of Chemistry
    Florida Atlantic University
    Boca Raton, FL 33431

(b) Current address: Amoco Performance Products, P. O. Box
    409, Bound Brook, NJ 08805

(c) Posthumously

A plethora of chemicals are available from natural feed-stocks for the production of polymers and related materials. These natural resources range from seed oils to a spider's silk. While the wide use of these materials may not be economically viable at present, as oil prices continue to rise, these materials will become more attractive commercially. In this paper, the use of triglyceride oils to form novel interpenetrating networks is reviewed, and other natural sources of reactive chemicals and their uses in polymerizations are discussed.

INTRODUCTION

Many triglyceride oils such as castor oil, lesquerella oil, and vernonia oil have special functionalities on each of their three fatty acid chains in addition to double bonds. For castor and lesquerella, the acid residues have hydroxyl groups, while vernonia oil has oxirane groups. Besides these naturally occurring functionalized oils, linseed, crambe, lunaria, and lesquerella oil have been epoxidized to yield the oxirane functionality wherever a double bond existed. These special functionalities permit the formation of crosslinked polyesters or polyurethanes, which are usually soft elastomers with glass transition temperatures of about -50°C. The castor oil polyester with sebacic acid is especially interesting, since sebacic acid itself is derived commercially from castor oil. Similarly, vernonia oil polyesters may be made with suberic acid, which can be obtained from vernonia oil.[1,2] Thus, these elastomers are based on renewable resources, rather than petrochemical sources.

These soft elastomers can be swelled by other monomers such as styrene, which together with crosslinkers such as divinylbenzene, can be polymerized *in situ* to form a sequential interpenetrating polymer network, IPN. Alternately, and in some cases preferably, the two polymerizations can be carried out simultaneously, since the esterification of the oil and the vinyl polymerization do not significantly interfere with one another. These latter products are called simultaneous interpenetrating networks, SINs.

Most of the IPNs and SINs so formed are phase separated, as are most polymer blends. For IPNs and SINs, however, if gelation of both networks precedes phase separation, the crosslinks will tend to hold the phases together and limit their sizes, forming a microheterogeneous morphology. However, if phase separation precedes gelation, then the crosslinks will tend to keep the phases apart, and gross phase separation will occur with concomitant loss of physical and mechanical properties.

This paper will review the synthesis of oil-based IPNs and SINs, and describe the morphology and viscoelastic behavior of some representative materials.[3-8] Other renewable resource materials will also be briefly described, especially lignin, starch, and other hydroxyl bearing natural chemicals.

IPNs BASED ON FUNCTIONAL TRIGLYCERIDE OILS

1. Synthesis of Triglyceride Oil Networks

For reaction purposes, the composition of castor oil is considered as the triglyceride of ricinoleic acid. The average hydroxyl functionality of castor oil is around 2.7, since it is about 90% pure triglyceride of ricinoleic acid, with the remaining 10% being diluent oils. The structure of the castor oil triglyceride is shown in Equation 1 below:

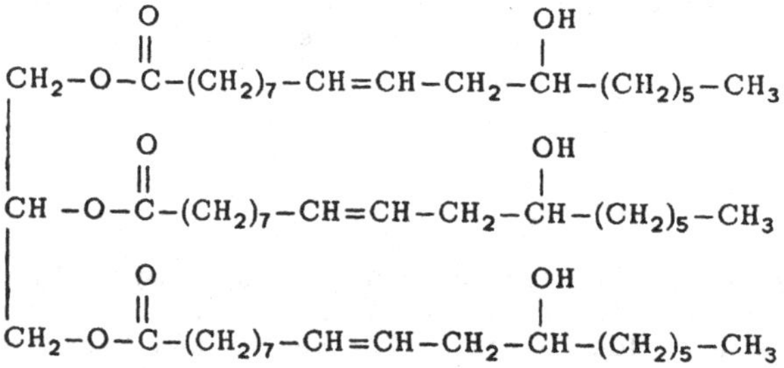

$$
\begin{array}{l}
\overset{\displaystyle O}{}\overset{\displaystyle \|}{}\overset{\displaystyle OH}{}\overset{\displaystyle |}{}\\
CH_2-O-C-(CH_2)_7-CH=CH-CH_2-CH-(CH_2)_5-CH_3\\
|\quad\quad O\quad\quad\quad\quad\quad\quad\quad\quad OH\\
|\quad\quad \|\quad\quad\quad\quad\quad\quad\quad\quad\ |\\
CH-O-C-(CH_2)_7-CH=CH-CH_2-CH-(CH_2)_5-CH_3\\
|\quad\quad O\quad\quad\quad\quad\quad\quad\quad\quad OH\\
|\quad\quad \|\quad\quad\quad\quad\quad\quad\quad\quad\ |\\
CH_2-O-C-(CH_2)_7-CH=CH-CH_2-CH-(CH_2)_5-CH_3
\end{array}
$$

Lesquerella oil differs from castor in having nine methylene units before the double bond, rather than seven. Similarly, the structure of the predominant vernonia oil triglyceride, trivernolin, is shown below in Equation 2, where the functionality is around 2.4, depending on the exact species of vernonia, corresponding to a purity of about 80% trivernolin.

$$
\begin{array}{l}
\text{CH}_2-\text{O}-\overset{\displaystyle\text{O}}{\overset{\|}{\text{C}}}-(\text{CH}_2)_7-\text{CH}=\text{CH}-\text{CH}_2-\overset{\displaystyle\text{O}}{\overset{/\ \backslash}{\text{CH}-\text{CH}}}-(\text{CH}_2)_4-\text{CH}_3 \\[2mm]
\text{CH}\ -\text{O}-\overset{\displaystyle\text{O}}{\overset{\|}{\text{C}}}-(\text{CH}_2)_7-\text{CH}=\text{CH}-\text{CH}_2-\overset{\displaystyle\text{O}}{\overset{/\ \backslash}{\text{CH}-\text{CH}}}-(\text{CH}_2)_4-\text{CH}_3 \\[2mm]
\text{CH}_2-\text{O}-\overset{\displaystyle\text{O}}{\overset{\|}{\text{C}}}-(\text{CH}_2)_7-\text{CH}=\text{CH}-\text{CH}_2-\overset{\displaystyle\text{O}}{\overset{/\ \backslash}{\text{CH}-\text{CH}}}-(\text{CH}_2)_5-\text{CH}_3
\end{array}
$$

In addition to these naturally occurring functionalized oils, any unsaturated triglyceride oil may be epoxidized, by converting double bonds into oxirane groups via reaction with a peroxy acid. Linseed oil is a particularly interesting oil to epoxidize, as it contains six double bond sites which may all be epoxidized to yield an oil of exceptionally high functionality. By reacting acetic acid with hydrogen peroxide in the presence of the proper catalyst, Equation 3, peroxyacetic acid is prepared, which will react with carbon to carbon double bonds to form the epoxy group and return acetic acid.

$$
\text{H}_3\text{C}-\overset{\displaystyle\text{O}}{\overset{\|}{\text{C}}}-\text{OH} \quad + \quad \text{H}_2\text{O}_2 \quad \rightarrow \quad \text{H}_3\text{C}-\overset{\displaystyle\text{O}}{\overset{\|}{\text{C}}}-\text{O}-\text{OH} \ + \quad \text{H}_2\text{O}
$$

$$
\text{H}_3\text{C}-\overset{\displaystyle\text{O}}{\overset{\|}{\text{C}}}-\text{O}-\text{OH} \ + \quad \sim\text{CH}=\text{CH}\sim \quad \rightarrow \quad \sim\overset{\displaystyle\text{O}}{\overset{/\ \backslash}{\text{CH}-\text{CH}}}\sim + \quad \text{H}_3\text{C}-\overset{\displaystyle\text{O}}{\overset{\|}{\text{C}}}-\text{OH}
$$

$$(3)$$

Epoxidized lunaria oil, which will be the subject of later discussion, has the structure shown in Equation 4.

$$
\begin{array}{l}
\text{CH}_2-\text{O}-\overset{\displaystyle\text{O}}{\overset{\|}{\text{C}}}-(\text{CH}_2)_{14}-\overset{\displaystyle\text{O}}{\overset{/\ \backslash}{\text{CH}-\text{CH}}}-(\text{CH}_2)_7-\text{CH}_3 \\[2mm]
\text{CH}\ -\text{O}-\overset{\displaystyle\text{O}}{\overset{\|}{\text{C}}}-(\text{CH}_2)_{14}-\overset{\displaystyle\text{O}}{\overset{/\ \backslash}{\text{CH}-\text{CH}}}-(\text{CH}_2)_7-\text{CH}_3 \\[2mm]
\text{CH}_2-\text{O}-\overset{\displaystyle\text{O}}{\overset{\|}{\text{C}}}-(\text{CH}_2)_{14}-\overset{\displaystyle\text{O}}{\overset{/\ \backslash}{\text{CH}-\text{CH}}}-(\text{CH}_2)_7-\text{CH}_3
\end{array}
$$

In all these oils, because the average functionality is greater than two, a crosslinked product can be made using difunctional reactants. When castor (or lesquerella) oil reacts with difunctional sebacic acid, a three dimensional esterification reaction occurs between the hydroxyl and acid functionalities, forming a network polymer. Oxirane (epoxy) groups can also react to form crosslinked polyesters with dibasic acids, albeit by different routes. The reaction between a dibasic acid, such as sebacic acid, and an epoxy bearing oil is shown in Equation 5.

$$
\begin{array}{ccc}
\underset{\displaystyle \sim CH=CH\sim}{\overset{\displaystyle \overset{O}{/\ \backslash}}{}} & & \underset{\displaystyle \sim CH-CH\sim}{\overset{\displaystyle \overset{OH}{|}}{}} \\
& & | \\
& \underset{\displaystyle (CH_2)_8}{\overset{\displaystyle C}{\overset{\displaystyle \backslash\!/}{}}} & O \\
+ & | & C=O \\
& C & (CH_2)_8 \\
\underset{\displaystyle \sim CH=CH\sim}{\overset{\displaystyle \overset{O}{/\ \backslash}}{}} & \overset{\displaystyle /\!/\backslash}{O\ \ OH} & C=O \\
& & O \\
& & \sim CH-CH\sim \\
& & | \\
& & OH
\end{array}
$$

Simultaneous with the formation of ester groups, free hydroxyl groups are formed in the reaction above, without the release of a small molecule such as water, which is common to most step polymerizations. Given favorable reaction conditions, these hydroxyl groups constitute a new locus for continued reactions. In fact, some care must be exercised for highly unsaturated triglycerides, because too large a functionality may be introduced.

The analogous reaction between castor (or lesquerella) oil and sebacic acid also forms an aliphatic polyester. Unlike the epoxy-acid reaction, water is formed as a by-product, which must be removed from the reaction medium in order to obtain complete reaction and a void-free material. The reaction may be carried out with deliberate stoichiometric imbalance to form a high molecular weight hydroxy endcapped prepolymer, which may be thoroughly dried, and then crosslinked with a diisocyanate, which form urethane crosslinks without the release of water. Of course, hydroxy functionalized oils can also be reacted from the start with a diisocyanate, forming a complete polyurethane network.

IPNs can be formed by two routes. In one, sequential IPNs may be formed by swelling the fully reacted polyester (or polyurethane) network with a mixture of styrene (or other appropriate monomer) and the cross-linking agent divinylbenzene, and polymerizing at 70-80°C. In the other, both monomers are polymerized simultaneously, forming a simultaneous interpenetrating network, SIN. The SIN route is illustrated in Figure 1.[8] The triglyceride oil is made into a prepolymer, which is then stirred into styrene and divinylbenzene along with the difunctional reactant. The difunctional reactant might be a diacyl chloride or a difunctional isocyanate, which crosslinks the trifunctional oil prepolymer. This is followed by more or less simultaneous polymerization of both components via non-interfering routes. The degree of polymerization of the prepolymer can have marked effects on the properties of the resulting material, as will be discussed below.

2. Viscoelastic and Morphological Behavior

The eight methylene units in sebacic acid, plus the numerous methylene units in the triglyceride all contribute to the low glass transition temperature, T_g, of the final product. The ten-second modulus vs. temper-

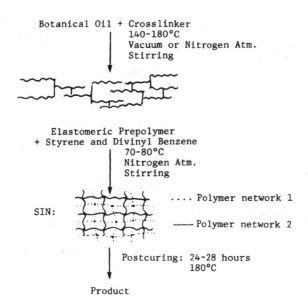

Figure 1. Simultaneous interpenetrating network (SIN) synthesis based on botanical oils.[8]

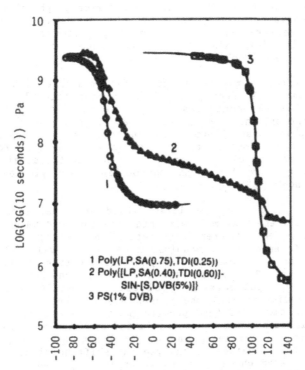

Figure 2. Modulus-temperature behavior of lesquerella oil-polystyrene SINs.[8]

ature for lesquerella oil polymer, crosslinked polystyrene, and their SIN is shown in Figure 2.[8] Here, the T_g of the lesquerella polymer is seen to be about -50°C, suggesting applications in the field of rubber-toughened plastics. The plateau modulus between the two glass transitions depends on the overall composition, so that either impact resistant plastics or tough elastomers can be made, the latter when the lesquerella oil polymer is the major component.

Morphology of the interpenetrating network depends on the method of synthesis. In Figure 3, the remaining double bonds in a partially epoxidized lesquerella oil-polystyrene IPN were stained with osmium tetroxide, and appear as the dark areas in the transmission electron microscopy (TEM) micrographs.[8] Simultaneous polymerization led to phase separation before the networks could form, and thus the domain size of the phases is rather large. The sequential route to the IPN formed very small domains, since the rubber is already crosslinked before the styrene is polymerized. The size of the phase domains controls the clarity of the product, as well as the mechanical behavior. Larger domains result in more opaque materials, with less strength and resistance to impact. The optimum size of the rubber domains for impact resistance is 0.1-0.3 micrometer.

The extent of oil prepolymerization also effects the morphology, so that the SIN depicted in Figure 3 can be made to have nearly the same characteristics as the IPN by adjusting the prepolymer molecular weight. The extent of oil prepolymerization may be measured by the acid value (ASTM D1639), which measures the carboxylic acid end groups on the polymer chain (from sebacic acid). A high acid value corresponds to a low degree of polymerization. In the epoxidized lunaria oil-dimer acid system, the degree of polymerization is unity (no reaction) at an acid value of about 100, and the gel point (incipient infinite network) corresponds to an acid value of 40. In Figure 4, the TEM micrographs show that as the epoxidized lunaria oil prepolymer acid value decreases (degree of polymerization increases), the final morphology changes from large oil-polyester domains dispersed in the polystyrene matrix, to a cellular structure with the oil-polyester rubber as the continuous phase.

Initially, styrene monomer is completely miscible with the oil prepolymer, but as the styrene polymerizes to high molecular weight polystyrene, the two components phase separate from each other. At this point, it is thought that the oil-rich phase is continuous, and the polystyrene-rich phase discontinuous. If the oil is not prepolymerized, a phase inversion will occur, as in Figures 4-A and 4-B. For extensive prepolymerization, the oil remains continuous, but the domains are smaller. As described below, the morphology of Figure 4-C yielded the highest impact resistance.

The dynamic moduli for the 15/85 epoxidized lunaria oil dimer acid (ELuODAN)/polystyrene SINs made with high and low acid value are shown in Figure 5. The effect of prepolymerization on the continuous phase explains the difference in the modulus. At high acid value (low degree of prepolymerization), the loss modulus peaks (indicative of the glass transition temperature, T_g) are shifted towards one another as compared to the respective homopolymers (oil-polyester T_g -30°C; polystyrene T_g 108°C). This indicates that the phases are not pure homopolymers, but contain a small amount of the opposite polymer. The amount of the opposite polymer in each phase may be estimated by using Equation 6, the Fox equation, where W_i represents the weight fraction of component in the phase. Use of this equation to analyze the shifting of T_gs leads to the conclusion that each phase contains about three percent of the opposite polymer. The amount of opposite polymer contained in each phase is a result of the phase diagram position at 25°C for the pair of polymers.

Poly [(ELP,SA(0.70))-SIN-(S,DVB(5%))]
50/50

Poly [(ELP,SA(0.70))-IPN-)S,DVB(5%)]
50/50

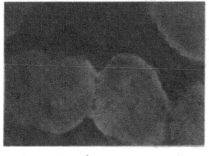 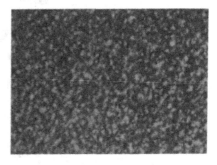

1μm

Figure 3. Morphology of epoxidized lesquerella oil-polystyrene SINs.[8]

$$1/T_g = W_1/T_{g1} + W_2/T_g$$

The pure polystyrene phase in the highly prepolymerized system should
have no shift in T_g, since no miscibility is expected between the high
molecular weight oil and polystyrene, which is observed (Figure 5, A.V. =
47). The oil-polyester phase in this case contains more polystyrene than
in the non-prepolymerized case, causing a more pronounced shift in the
oil-polyester T_g. This is due to the trapping effect of the gel, since
some of the polystyrene to polymerize inside of the oil-polyester phase
may not be able to nucleate a phase domain, or diffuse to an already
formed domain.

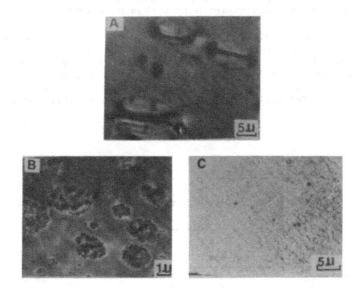

Figure 4. Effect of oil prepolymerization on morphology of
15/85 epoxidized lunaria oil-dimer acid/polystyrene
network SINs. Acid values: A, 98; B, 75; C, 55.

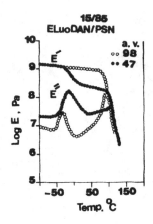

Figure 5. Effect of oil prepolymerization (morphology) on dynamic moduli of epoxidized lunaria oil SINs.

As alluded to earlier, the mechanical properties are a direct result of morphology. For the same series of 15/85 epoxidized lunaria oil dimer acid/polystyrene SINs discussed previously, stress-strain curves are depicted in Figure 6. As the acid value decreases, toughness as defined as the area under the stress-strain curve increases. This behavior is a direct result of the formation of the cellular structure of Figure 4-C, where the oil-polyester rubber becomes the continuous phase, with higher elongation than pure polystyrene, while being reinforced by the 85% dispersed polystyrene, resulting in higher tensile strength than the pure rubber. Notched Izod impact strength followed similar behavior (Table 1), with the highly prepolymerized oil-polyester material achieving nearly five times the impact strength of pure polystyrene. This result is comparable to commercial high-impact polystyrene materials.

At the present time, only castor oil based polyurethane elastomers are commercial, the others are not. None of the polyesters are commercial, although they are particularly promising for developing tropical countries due to the ability to grow the oil seeds locally, and produce the dibasic acid from the oil itself. Some of the current literature on the subject includes work from India, the Peoples Republic of China, and Taiwan.[9-13]

Table 1. Effect of oil prepolymerization (morphology) on notched Izod impact strength of epoxidized lunaria oil SINs.

Material	Acid Value		Impact Strength
Polystyrene			16 J/m
15/85 ELuODAN/PSN	98	Increasing	20
	75	Prepolymerization	64
	55	↓	77

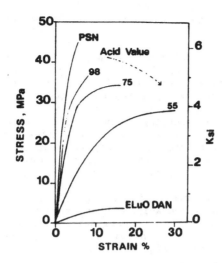

Figure 6. Effect of oil prepolymerization (morphology) on stress-strain behavior of epoxidized lunaria oil SINs.

OTHER RENEWABLE RESOURCE MATERIALS

There are many renewable resource materials worthy of consideration which are not yet commercial. Although current economics are the main driving force keeping them out of commercial production, problems of petroleum shortages, waste disposal, biodegradability, and pollution may bring them to the fore in the near future. Naturally occurring materials, because they are natural, have "natural" mechanisms for their degradation. These mechanisms utilize moisture, heat and specialized enzymes. Under the proper circumstances, namely exposure to the needed mode of degradation, these polymers are biodegradable. In our quest for biodegradable materials, selected natural biomasses should be considered as likely candidates. Again, it must be noted that such materials will only suffer biodegradability under the appropriate conditions. Compacted newspaper has been retrieved from within landfills and dumps that are over forty years old and remains essentially unchanged, illustrating the importance of the correct biodegradation conditions.

Examples of such naturally occurring polymeric materials include spider web silk, which is extremely strong and elastic, and mucus glycoproteins, which serve as nature's ball bearings. Sources of the latter include earthworm surfaces, snails and slugs, and fish and eels. The glycoproteins are spheroidal in shape, which is thought to yield their ball bearing characteristics. Silk from the golden orb spider has been mass produced through biotechnology to be used in bullet-resistant vests for the Army.[14]

1. Natural Monomers

While the major components of nature's biomass are polymeric in nature, there exists many small-molecule materials that are present in sufficient amount with appropriate or modifiable functional group sites which may allow their consideration as feedstocks for the polymer industry. Examples include the triglyceride oils delineated in the first half of this paper.

Those materials containing more than one appropriate functional group or an active, sterically suitable vinyl group can be employed directly. Thus terephthalic acid and phthalic anhydride, both available from natural sources, are employed in polyester and alkyd resin synthesis. Furfural, and one of its most important derivatives, furfuryl alcohol, reacts with phenol forming thermosetting resins utilized in molding applications and in the manufacture of coated abrasives. Furfuryl alcohol reacts with formaldehyde and urea forming a variety of industrially important products. Furfural also reacts with selected ketones forming polymeric materials used to form resin aggregate mixtures.

Materials containing sites of unsaturation can be polymerized using these sites. Many unsaturated acids, on exposure to air, undergo a series of reactions giving crosslinked thermosetting materials. An important example of such a material is linseed oil, a triglyceride of unsaturated acids which is used to make paints.

Polymerizations involving terpenes can be initiated through a wide variety of routes including high energy radiation, free radical, cationic, and Ziegler-Natta polymerizations. Terpine resins are used as pressure sensitive adhesives and coatings. Naturally occurring small molecules can also serve as starting materials in the synthesis of monomers. Thus, furfural is a precrsor to 1,6-hexanediamine and adipic acid, employed in the synthesis of nylon-66.[15] The distillation of castor oil yields sebacic acid, also employed in the polyamide synthesis of nylon-6,10.

The following are additional examples where naturally occurring small molecules are employed in the synthesis of polymers. In truth, the number of actual and potential natural materials able to be used for polymer synthesis is quite large and the following is only to be considered as illustrative. Only examples a decade or more old are cited here to encourage readers to investigate the older literature for potential leads.

Abietic acid from the oleoresin of *P. paulstris* has been reacted with poly(glycerol-diglycerol) eventually to form varnishes.[16]

Eleostearic acid from the bagilum-bang nut is polymerized forming an elastic gel.[17,18]

Glutaric acid from green sugar and beet juice reacts with xylitol forming polyesters.[19]

Tartaric acid esters from chimnsis, shizardra, etc., polymerize forming thermoplastic resins.[20]

Isamic acid from angokea oil has been used as a drying oil.[17]

Glutamine from sugar beets forms polyamino acids.[21,22]

Furoic acid from eronymus atropurpureus has been polymerized forming a cation exchange resin.[23,24]

Cardanol from anacardium occidentale gave vinyl resins used as additives, sealants, in brake linings and in coatings applications.[25,26]

2. Hydroxyl-containing Natural Materials

Hydroxyl-containing materials comprise the vast majority of natural materials. These include the carbohydrates (or saccharides), lignin, many steroids and fused ring systems (such as cholesterol, luciferin, anhalamine, folic acid, Vitamin E, riboflavin, and tetracycline). With the possible exception of the saccharides, these materials are either vastly underused (such as lignin) or unused as commercial polymers or as feedstock within the chemical industry.

Inclusion of hydroxyl-containing "smaller molecules" will vary according to the number of functional sites present. Focusing only on the hydroxyl groups, those materials containing only one hydroxyl, such as the alkaloid berberine, can be reacted with polymers, such as poly(acryl chloride), attaching the berberine through an ester linkage. This product could be used as an antimalarial agent, a febrifuge, or in the treatment of ulcers. Dihydroxyl-containing materials, such as bishydroxycoumarin can be reacted with acid chlorides forming polyesters that may be used as an anticoagulant assisting to increase coronary flow. These uses are limited because of the specificness and availability of the hydroxyl-containing agent.

The following will focus on polyhydroxyl compounds. Some of these, such as sucrose, are monomeric while others, such as the polysaccharides and lignin, are polymeric. Examples are only given to illustrate possible uses.

Recently, Zaffaroni[27] and Usamani and Salyer[28] have exploited the basic concept of polysugar. Zaffaroni's approach involves reacting a monosaccharide sugar with a reactive compound before coupling the reactive sugar intermediate with a high polymer carrier molecule. The Usamani-Salyer approach is simpler and employs etherification of sucrose with poly(vinyl alcohol). The resulting product is largely stable to boiling water (permitting its use in cooking) and boiling dilute HCl (corresponding to the pH present in the human stomach; allowing its use as a sweetener without addition of digestable calories). Some of the products were sweet without a bitter aftertaste. These polysugars are candidates for satisfying our "sweet tooth" while not adding inches to our middles. Most sugars can be polymerized through reaction with di- and polyfunctional Lewis acids giving a variety of products. Most of these products are biodegradable under the proper conditions. Carraher and coworkers have synthesized a variety of products from organostannane dihalides.[29,30] These products are generally bioactive inhibiting selected mildew and rot causing organisms. Similar reactions were effected involving numerous organometallic mono and di-halo tin, antimony, titanium, zirconium, and hafnium containing reactants, and include the polysaccharides dextran, cellulose, xylan, and amylose.[31-34] The modification of polysaccharides producing industrially valuable and potentially useful materials is well known with research continuing.[35,36]

Some starch-based materials have found commercial application, however. As shown in Table 2, starch compositions permit controlled release of pesticides in agricultural settings, while kenaf cellulose is a potential source of paper fiber.[37]

Some other materials have been tried commercially, but are lacking in some way. An excellent example is lignin. Although lignin has seen some commercial utilization, most of it is burned as an energy source. Since the amount of lignin produced in pulping a tree is almost as large as the resulting pulp, very significant amounts of material are involved. Lignin

Table 2. Newly industrialized renewable resource polymers.	
Polymer	Application
ALREADY ARRIVED	
Starch-based	Encapsulated pesticides for controlled release
Enzyme production of sugar from starch	Ethanol fermentation for gasoline
Crosslinked starch-based "Superslurper"	Food thickener, water absorber from gasoline
Resins from dimer acids	Polyurethane adhesives and gels
STILL IN PROGRESS	
Polymer	Application
Kenaf cellulose	Newsprint
Guayule Poly(cis-isoprene)	Aircraft tires
Rape seed triglyceride oil	Nylon, paints

constitutes probably the most underutilized material available in the biomass in large volume. Wood is composed of cellulose and lignin in roughly equal amounts. Lignin approximates a three-dimensional network and is often called nature's glue. It is recovered in a variety of forms that are dependant on the precise procedures employed in its isolation. Kraft lignin is the lowest molecular weight commercially available fraction with number-average molecular weight of about 3500 Daltons. The major functional groups are aromatic and aliphatic hydroxyls. (Sulfonated lignins are widely used in drilling muds and as industrial surface active agents.)

Carraher and coworkers have extended their work on polysaccharides to lignin, forming a wide variety of biologically active materials derived from dihalo and monohaloorganostannanes. These materials are brittle.[38] Flexible products that can be formed into sheets are formed from the copolymerization of lignin and hydroxyl-terminated poly(ethylene oxide) with the organostannanes. Glasser and coworkers made a series of tough resins employing lignin as a prepolymer.[39-42] Kraft lignin was reacted with propylene oxide and ethylene oxide forming hydroxyalkyl lignin. These lignins were then reacted with diisocyanates to yield polyurethane products. Young's modulus ranged from 1 to 2 GPa, in the range of soft plastics, and the glass transition temperature ranged from 70 to 190°C.

Lignin has been reacted with chlorophosphazenes producing materials with decent flame resistance and thermal stability.[43-48] Ultrazine NAS lignosulfonates have the greater reactivity towards the chlorophosphazenes. The products showed good flame resistance while Kraft lignin sustained a flame. The modified materials also exhibited good hydrolytic stability to aqueous acids and bases, and good resistance to organic solvents. The increased phosphorous content is at least partially responsible for the increased flame resistance.

Organic acid chlorides have also been employed in the modification of lignin. Diacid chlorides form crosslinked products with the formation of ester groups.[45,49-52] For products derived from terephthaloyl chloride, the products show good hydrolytic and chemical resistance, increased thermal stability, a white to yellow color, and can be mixed in the melt with other polymers. Lignin has also been employed to produce controlled release materials. Thus lignin sulfonates have been reacted with 2,4-dichlorophenoxyacetic acid (2,4-D), with some of the products being good controlled release agents for this herbicide.[53]

In summary, much work has been done utilizing the hydroxyl groups present on lignin with the production of numerous materials with potential industrial uses. The above results and the ready availability of vast quantities of lignin encourage further development of lignin as a renewable resource feedstock. As our sources for synthetic and oil-based feedstocks decreases, the renewable biomass must be considered as an alternative.

ACKNOWLEDGEMENT

The authors wish to thank the National Science Foundation Program on Alternate Biological Sources of Materials for financial support through Grant No. PFR 7827336.

REFERENCES

1. O. A. Afolabi, M. E. Aluko, G. C. Wang, W. A. Anderson, and F. O. Ayorinde J. Amer. Oil Chem. Soc., 66, 7 (1989).
2. F. O. Ayorinde, G. Osman, R. L. Shepard, and F. T. Powers J. Amer. Oil Chem. Soc., 65, 11 (1988).
3. A. M. Fernandez, J. A. Manson, and L. H. Sperling, in: "Renewable Resource Materials: New Polymer Sources," C. E. Carraher, Jr. and L. H. Sperling, Eds., Plenum Press, New York, 1986.
4. L. H. Sperling and J. A. Manson, J. Amer. Oil Chem. Soc., 60, 11 (1983).
5. A. M. Fernandez, C. J. Murphy, M. T. DeCrosta, J. A. Manson, and L. H. Sperling, in: "Polymer Applications of Renewable Resource Materials," C. E.Carraher, Jr., and L. H. Sperling, Eds., Plenum, New York, 1983.
6. L. H. Sperling, J. A. Manson, and M. A. Linne, J. Polym. Mater., 1(1), 54 (1984).
7. C. E. Carraher, Jr. and L. H. Sperling, in: "The Encyclopedia of Polymer Sci. Eng.", 2nd Ed., Vol. 12, 1988, J. Kroschwitz, Ed.
8. M. A. Linne, L. H. Sperling, A. M. Fernandez, S. P. Qureshi, and J. A. Manson in: "Rubber Modified Thermoset Resins," C. Keith Riew and John K. Gillham, Eds., ACS Advances in Chemistry Series #208, American Chemical Society, Washington, DC, 1984, reprinted with permission.
9. P. Patel and B. Suthar, Polym. Eng. Sci., 28, 901 (1988).
10. P. Rajalingam and G. Radhakrishnan, Abstract for poster session, "International Symposium on Multiphase macromolecular Systems," ACS Div. Polym. Chem., 14th Biennial Meeting, San Diego, California, November 19-23, 1988.
11. Ma Song and Zhang Donghua, Plastics Industry (P. R. C.), 2, 42 (1987).
12. P. Tan and H. Xie, Hecheng Xiangjiao Gongye, 1, 180 (1984).
13. J. L. Liang, H. T. Liu, W. H. Ku, and G. M. Wang, Abstracts, IUPAC

International Symposium on Polymers for Advanced Technologies, Jerusalem, Israel, August, 1987.
14. Associated Press news release in The Morning Call, Allentown, PA. January 27, 1990.
15. R. Seymour and C. Carraher "*Polymer Chemistry*," Dekker: New York, 1989.
16. E. Symmers U. S. Pat. 1,696,337; Dec. 25, 1929 (CA 23: 848.8)
17. A. W. Ralston, "*Fatty Acids and their Derivatives*," Wiley, New York, 1948.
18. J. Boedtker J. Pharm. Chem. **29**, 313 (1924).
19. B. Zhubanov and K. Mirfaizou Vesth. Akad. Nauk. Kaz. SSR 7, 3 (1974).
20. B. Garvey, C. Alexander, F. Kung, and D. Henderson Ind. Eng. Chem. **33**, 1060 (1941).
21. C. Morris, J. Thompson, and S. Asen J. Biol. Chem. 239, 1833 (1964).
22. Farbwerke Hoechst, A.G. Neth. Pat. 6508183; Dec. 27, 1965 (CA **64**: 17747f)
23. F. M. Dean, "*Naturally Occurring Oxygen Ring Compounds*," Butterworth: London, 1963.
24. R. Oda and H. Shimizu Chim. High Polymers **6**, 14 (1947).
25. J. Tyman J. Amer. Chem. Soc. **55**, 663 (1978).
26. S. Verneker Ind. J. Tech. **18**, 170 (1980).
27. A. Zaffaroni Fr. Pat. 2204369 (assigned to Dynapol), May 24, 1974.
28. A. M. Usmani and I. O. Salyer, in: "*Modification of Polymers*," C. E. Carraher and J. Moore, Eds.; Plenum: New York, 1983.
29. Y. Naoshima, C. Carraher, S. Hirono, T. Bekele, and P. Mykytiuk, in: "*Renewable Resource Materials*," C. Carraher and L. Sperling, Eds., Plenum, New York, 1986.
30. Y. Naoshima, S. Hirono, and C. Carraher J. Polymer Materials 2, **43** (1985).
31. Y, Naoshima, C. Carraher, G. Hess, and M. Kurokawa, in: "*Metal-Containing Polymeric Systems*," J. Sheats, C. Carraher, and C. Pittman, Eds., Plenum, New York, 1985.
32. Y. Naoshima, C. Carraher, S. Iwamoto, and H. Shudo Appl. Organometal. Chem. 1, **245** (1987).
33. Y. Naoshima, C. Carraher, T. Gehrke, M. Kurokawa, and D. Blair J. Macromol. Sci.- Chem. **A23(7)**, 861 (1986).
34. C. Carraher, T. Gehrke, D. Giron, D. Cerutis, and H. M. Molloy J. Macromol. Sci.-Chem., **A19**, 1121 (1983).
35. C. Carraher and L. Sperling, Eds., "*Polymer Applications of Renewable Resource Materials*," Plenum, New York, 1983.
36. C. Carraher and L. Sperling, Eds., "*Renewable-Resource Materials*," Plenum, New York, 1986.
37. USDA presentation at "First International Conference on New Industrial Crops and Products" Peoria, Ill. 1989.
38. C. Carraher and D. Sterling, unpublished work and chapter in this book.
39. L. Wu and W. G. Glasser J. Appl. Polym. Sci., 29, 1111 (1984).
40. W. G. Glasser, C. Barnett, T. Rials, and V. Saraf J. Appl. Polym. Sci., 29, 1815 (1984).
41. V. Saraf and W. G. Glasser J. Appl. Polym. Sci., 29, 1831 (1984).
42. W. G. Glasser and S. Sarkaner "*Lignin*," ACS, Washington, D.C., 1989.
43. H. Struszczyk and J. Laine Polish Pat. 125877, 1981.
44. H. Struszczyk, Polish Pat. Appl. P-265167, 1987.
45. H. Struszczyk and K. Krajewski Polish Pat. 134256, 1982.
46. H. Struszczyk and J. Laine J. Macromol. Sci.- Chem. A17(8), 1193 (1982).
47. H. Struszczyk, Fire and Materials 6(1), 7 (1982).
48. H. Struszczyk, J. Macromol. Sci.-Chem., **A23(8)**, 973, (1986).
49. K. Sarkanen and C. Ludwig, Eds., "*Lignins*," Wiley-Interscience, New York, 1971.
50. H. Struszczyk and K. Wrzesniewska-Tosik Proc. Inter. Sym. on Fiber Sci. and Technol., Hakone, Japan, 1985.

51. G. Allen, W. Balaba, J. Dutkiewicz, and H. Struszczyk, *"Chemicals from Western Hardwoods and Agricultural Residues; Semiannual Report,"* April 1979, NSE-7708979, Univ. of Washington, Seattle.
52. G. Van der Klashort, C. Forbes, and K. Psotta Holzforschung, **37(6)**, 279 (1983).
53. H. Struszczyk and K. Wrzesniewska-Tosik Polish Pats. 141253, 141254, and 141255, 1985.

STRUCTURAL CHARACTERIZATION OF ORGANOSTANNANE - KRAFT LIGNIN

Charles E. Carraher, Jr.,[a] Dorothy C. Sterling,[a] Thomas H. Ridgway[b] and J. William Louda[a]

(a) Department of Chemistry
    Florida Atlantic University
    Boca Raton, FL 33431
    and
(b) Department of Chemistry
    University of Cincinnati
    Cincinnati, Ohio 45221

Lignin is the second most abundant natural, renewable material. It is produced at an annual rate of about 4×10^{13} lbs. Organostannane moieties were successfully incorporated into Kraft lignin (Indulin AT and C) through the classical interfacial condensation of organotin chlorides with lignin. IR, MS and other physical data are consistent with the presence of organotin-modified lignin formed through condensation with the lignin-hydroxyl groups.

INTRODUCTION

Lignin is second only to cellulose as a natural, renewable material. It is produced at an annual rate of about 2×10^{10} tons and is present in the biosphere at a level of 3×10^{11} tons.[1] Based on pulp and paper production, lignin is produced worldwide from woody plants, in mills, at an annual rate of 5×10^7 tons.[1-3] USA production of lignin from mills is estimated to be about one half of this or about 200 lbs. per each person in the USA. Only about 2% of the lignin present in spent pulping solutions of pulp mills worldwide is isolated for further use. The remainder is discarded or burned. Burning is typically carried out to supply energy needed for the recovery of inorganic pulping chemicals. Lignin sulfonates rank first among industrial sulfactants.[1] Even so, lignin is a greatly under-used resource.

Lignin appears to serve two major functions. It is an adhesive, connecting the various portions of the plants. It is also active in the conduction of fluids and energy. Whatever its' true role(s) in plants is and whether these roles vary between plants, lignin is present in land-based vascular plants. The exact structure, proportion and distribution of lignin varies between and within plants and probably even during the life of the plant. The precise nature of the lignin also varies with the processing. Thus, specific companies will specialize in lignin obtained from a common source and processed in a specific manner. Lignin is associated with the vascular system of plants and solar enery collection as well as serving as an adhesive to bind together the various components of wood.

Lignin is macromolecular with a defined general structure that varies throughout a matrix. It is three dimensional, but is often considered two dimensional since its molecular thickness is small compared to its width and breadth. It forms an integral part of a polysaccharide-lignin composite. Thus, the particular composition of a lignin sample is dependent on the source as well as the particular conditions employed in its isolation. Lignin is then a mixture of polydisperse "sheet-like" molecules that vary in size and exact chemical structure. Even so, lignin is commonly treated as consisting of a C_9 repeat formula unit wherein the superstructure contains aromatic and aliphatic alcohols and ethers, aliphatic aldehydes and vinyl units.

All lignins industrially employed for nonfuel applications are chemically modified copolymers.[1] The major copolymer is formed through the sulfonation of lignin and is the major industrial surfactant. Sulfonated lignin is generally water soluble. Numerous structural networks have been formed using lignin as a basic component. Lignin sulfonates have been crosslinked using hydrogen peroxide.[1] Lignin has been incorporated into a phenol formaldehyde resin,[4-6] reacted with isocyanates,[7-12] alkylene oxides,[13-16] and epichlorohydrin forming epoxy resins.[17,18]

The trend today is away from the production of sulfonated lignin due to environmental considerations. Thus nonsulfonated lignin was employed for the current study.

EXPERIMENTAL

1. Materials Used

Chemicals were used as received without further purification. Indulin AT and C were obtained as gifts from Westva Company, Charleston, S.C. Indulin C is the sodium salt of Kraft Pine lignin. Indulin AT is a purified form of lignin from the Krafting process.

2. Procedures

The following reagents, instruments and procedures were employed to determine the hydroxyl content of the lignin.

2A. Reagents - Hydroxyl Content Determination

Acetic anhydride, 99% minimum
Ethyl Acetate, 99% undenatured
Acetylating reagent (prepared as per instructions,[25] and stored in a freezer to prolong shelf-life)
Standard 0.5 N potassium hydroxide in methanol

2B. Procedure - Hydroxyl Content Determination

Instructions for preparing samples and reagent blanks were followed according to the modified procedure, with the following changes.[23] A sample size containing about 10 mequiv. of hydroxyl compound was required. However, this was cut by half for Indulin AT to ensure that the acetylating agent penetrated the sample. A 10 mequiv. sample formed a solid cake in the reaction flask with no supernatant fluid. A reaction time of 30 minutes in 20 mL of acetylating reagent was used since the slowest acting alcohol (1,2,6-hexanetriol) required this period with the modified derivatization procedure. 0.5N KOH was used to titrate to a potentiometric endpoint, using a Fisher 301 pH/Potentiometer with a glass electrode, instead of the visual phenolphthalein endpoint, as indicated in the modified procedure.

3. Reaction Procedure

A one quart Kimex emulsifying jar, placed on a Waring Blendor (Model 1120) with a no-load stirring rate of about 1,000 rpm, was employed as the reaction system vessel. The aqueous solution containing the lignin and added base is placed in the reaction jar. The blender is turned on and the organic phase containing the organostannane halide is introduced through a funnel penetrating the lid into the reaction jar. After a designated time, stirring is stopped and the blender jar contents poured into a separatory funnel. The product is collected as a precipitate, washed repeatedly with water and the organic liquid, transferred to a preweighed glass petri dish and allowed to air dry.

4. Instrumentation

Infrared spectra were obtained using potassium bromide pellets employing an Alpha Centeri FTIR. Mass spectroscopy (DIP) was carried out employing a Kratos MS-50 Mass Spectrometer operating in the EI mode, 8 KV acceleration and a ten second per decade scan rate with variable probe temperature (Midwest Center for Mass Spectrometry, Lincoln, Nebraska) and a DuPont 21-491 Mass Spectrometer at 1.8kv. Elemental analysis was carried out by Galbraith Labs. (Knoxville, TN). Solubilities were determined by placing about 3 mg of sample in approximately 5 mL of liquid.

RESULTS AND DISCUSSION

For structural considerations, empirical C_9 units are typically employed. For the present study two samples of lignin were employed. As noted, these two samples were analyzed for hydroxyl content. The Kraft Pine C material, which is about 60% lignin as the sodium salt, has a total hydroxyl content of 1.74 meg/gram of which 0.44 meg/gram is phenolic hydroxyls, 8% catechol and other aromatic hydroxyls and the remainder aliphatic hydroxyls. This corresponds to 31 hydroxyls per 100 C_9 units. The Kraft Pine AT is about 99 to 100% lignin in the "acid" form with 1.14 meg/gram being phenolic, 8% catechol and other aromatic hydroxyls, the remainder aliphatic hydroxyls, with a total hydroxyl content being 3.30 meg/gram. This corresponds to a hydroxyl concentration of 5.9 hydroxyls for 100 C_9 units. Included in the analysis scheme were two model compounds, representative of a theoretical monomeric unit of lignin (an aromatic alcohol with at least one methoxy group ortho to the phenolic

113

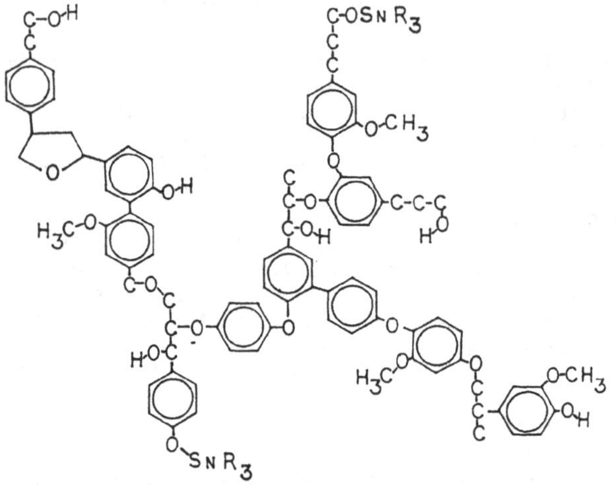

Figure 1. Hypothetical lignin structure corresponding to a molecular weight of 3500 Daltons and containing appropriately placed organostannane moieties.

hydroxyl). Results appear in Table 1. The general C_9 unit for Kraft lignin is taken to have 178 Daltons with a general formula: $C_9H_{7.9}O_{2.1}S_{0.1}(OCH_3)_{0.82}$.

Kraft lignin has the lowest molecular weight of any large scale lignin with a number average molecular weight of 1600 and a weight average molecular weight of 3500. Figure 1 shows an approximate lignin structure corresponding to a molecular weight of 3500 Daltons (except containing appropriately placed moieties).

Reaction was rapid, with a decent yield of modified material produced within seconds (Table 2). Percentage yield was based on a hypothetical repeat unit containing a single hydroxyl group. For instance, the average molecular weight for a single C_9 unit is 178 Daltons. The Kraft Pine C material contains 0.31 hydroxyls per C_9 unit and thus corresponds to a hydroxyl-based repeat unit molecular weight of 574 Daltons. For the AT material, the hydroxyl-based repeat unit molecular weight is 302 Daltons. For the dihaloorganostannanes the repeat unit [2] is assumed to have one organostannane moiety for every two hydroxyl-based repeat units, whereas for the monohaloorganostannane products, percentage yield is based on one organostannane for each hydroxyl-based repeat unit [1].

Table 1. Results for hydroxyl content.

Sample	Milli-eqiv/gram	Theoretical
Indulin C	1.741	–
Indulin AT	3.3065	–
2, 6-Dimethoxyphenol	9.3259	10.0000
Vanillyl Alcohol	11.5403	10.03048
(4-Hydroxy-3-methoxybenzyl alcohol)		

Table 2. Yield as a function of organostannane (1 mmole lignin (OH)).

Lignin	Organostannane (mmole)	Yield (g)	Yield (%)
C	Bu_2SnCl_2 (0.5)	0.5311	38
C	Bu_3SnCl (1.0)	0.4005	46
C	Lu_2SnCl_2 (0.5)	1.8211	100
C	Ph_3SnCl (1.0)	0.2737	29
AT	Bu_3SnCl (1.0)	0.2134	36

Reaction Conditions: The organic phase consisting of 100 mL of $CHCl_3$ and the organostannane is added to rapidly (18,000 rpm, no load) stirred aqueous solutions (100 mL) containing lignin and added sodium hydroxide (1 mmole) for 30 seconds stirring time at room temperature, ca. 25°C.

[1]   [2]

Purification of the product is required to remove unreacted organostannane. For instance, simple washing of the precipitated product gives material that exhibits ion fragments at m/e 36 and 38 (HCl) in about a 3 to 1 ratio, consistent with the presence of chlorine. The following discussion focuses on the product from tributyltin chloride. Ion fragments are also found about m/e 154 (SnCl), 177 (SnBu), 212 (SnBuCl) and 269 (SnBu$_2$Cl) present in the correct ratios to reflect the natural abundances of tin and chlorine for both the tributyltin chloride itself and the nonpurified product (Table 3). These ion fragments are present in the ratios predicted from the natural isotope abundance ratios for tin (Table 4). Purification is accomplished by washing the precipitated material with acetonitrile, whereas the modified product is only slowly, partially soluble or insoluble in acetonitrile. The purified products give no significant ion fragments that are associated with tin-containing moieties (for instance Table 4).

Ion fragments derived from the hydrocarbon portion of the organostannane and the lignin are found in the mass spectra of purified products (Table 5). Ion fragments characteristic of the lignin[19] are present including ion fragments at m/e 121-124 (guaiacol), 136-138 (methylguaiacol), 147-152 (isoeugenol (149), acetoguaiacone (149), 4-vinylguaiacol (150), ethylquaiacol (152), and vanillin (152)) and 162-166 (trans-coniferyl alcohol (163), and homovanillic acid (165)). Thus, degradation of the lignin is characterized by a breakup of the side chains, beta-guaiacyl ether bond cleavage, some demethylation and formation of stilbene structures (for instance,[19-22]). Numerous sulfur-containing fragments are formed including those found at m/e 34 (H_2S), 48 (CH_3SH), 62 (C_2H_5SH and/or CH_3SCH_3), 64 (SO_2), 96 (CH_3SSCH_3) and 128 (CH_3SSSCH_3).

The absence of evolved tin-containing ion fragments is consistent

Table 3. Major ion fragments for uncleaned Bu₃Sn-Lignin material (m/e >50; >10%) where Bu = a butyl derived fragment and L = a lignin derived fragment.

m/e	%-Relative Abundance		m/e	%-Relative Abundance	
55	62	Bu	173	14	
56	43	Bu	175	33	SnBu
57	82	Bu	177	41	
60	44	L	209	23	
69	42	L	211	36	
71	36	L	212	45	BuSnCl
73	22	L	213	22	
81	20	L	215	20	
83	20	L	266	49	
100	36	L	267	30	
118	20	Sn	268	82	
120	22		269	56	
151	48	L	270	100	Bu₂SnCl
152	36		271	21	
153	24	SnCl	272	53	
155	43		274	23	
157	14				

with other organostannane-containing polymers connected through the ether linkage, i.e., Sn-O-R (for instance,[23,24]). Presumably a thermally stable oxygen-associated intermediate is formed in such materials (SnO decomposes at 1080°C and SnO₂ has a melting point of 1127°C; the pyroprobe is taken to only about 600° for production of evolved moieties for the mass spectrophotometry).

Unmodified lignin exhibits a broad, intense infrared band from 3400 to 2800 (all infrared bands are given in cm⁻¹), assigned to OH stretching in alcohols. Purified samples show only a small band centering about 3300. The presence of C-H stretching is hidden in the lignin but clearly present for the modified products and confirms the presence of the organostannane moiety from its shape.

The fraction of substitution is determined from the percentage yields of tin (Table 6). For instance, for a product derived from dilauryltin

Table 4. Relative ion intensities for the crude Bu₃Sn-lignin sample for ion fragments associated with various tin-isotopes for Bu-Sn.

Sn	%-Natural (Predicted)	m/e	%-Relative (Found)
116	14	173	12
117	8	174	10
118	24	175	27
119	9	176	7
120	33	177	32
122	5	179	8
124	6	181	4

Table 5. Major ion fragments from lignin and Bu₃Sn-lignin products (m/e >50; >10% relative abundance).

m/e	Lignin Rel. Abundance	Bu₃Sn-Lignin Rel. Abundance
51	14	
53	18	70
55	15	72
56		60
57	20	100
60	100	
62		40
64	54	
77		15
81	38	15
91		15
94	11	
100		13
102		16
122		12
123		20
124	74	18
137		82
138		12
151	24	22

dichloride, the percentage tin present was 6.7%. Total substitution, assuming the products to be form [2], would give 15.0% tin. To reduce this to near the observed value requires only one dilauryltin moiety for each five hydroxyls.

Table 6. Percentage substitution of organotin lignins.

Organostannane Moiety	%-Sn Found	%Sn Calc.-100%	Corresponding Substitution
Bu₃Sn	8.25	20.1	1:4 (8.0%)
Bu₂Sn	17.28	22.4	1:1.5 (17.5%)
Lu₂Sn	6.72	15.8	1:5 (6.1%)

REFERENCES

1. W. Glasser and S. Kelley, in: "*Encyclopedia of Polymer Science and Engineering,*" Vol. 8, 2nd. Ed., John Wiley, NY, 1988, P. 795.
2. L. Sperling and C. Carraher, in: "*Encyclopedia of Polymer Science and Engineering,*" Vol. 12, 2nd. Ed., John Wiley, NY, 1988 p. 658.
3. L. Sperling and C. Carraher, "*Renewable-Resource Materials,*" Plenum, NY, 1986.
4. P. Muller and W. Glasser, J. Adhes., **17**, 157 (1984).
5. P. Muller and S. Kelley and W. Glasser, J. Adhes., **17**, 185 (1984).
6. A. Dolenko and M. Clarke, Forest Prod. J. **28C8**, 41 (1978).

7. K. Kratzl, K. Buchtela, J. Gratzl, J. Zauner and O. Ettingshausen, Tappi, **45(2)**, 117 (1962).
8. O. Hsu and W. Glasser, Wood Sci., **9(2)**, 97 (1976).
9. W. Glasser, O. Hsu, D. Reed, R. Forte and L. Wu, ACS Symp. Ser., **172**, 311 (1981).
10. V. Saraf and W. Glasser, J. Appl. Polym. Sci., **29**, 1831 (1984).
11. V. Saraf, W. Glasser, G. Wilkes, J. McGrath, J. Appl. Polym. Sci., **30**, 2207 (1985).
12. V. Saraf, W. Glasser and G. Wilkes, J. Appl Polym. Sci., **30**, 3809 (1985).
13. D. Christian, M. Look, A. Nobell and T. Armstrong, U. S. Pat. 3,546,199 (Dec. 8, 1970).
14. G. Allan, U. S. Pat. 3,476,795 (Nov. 4, 1969).
15. W. Glasser, C. Barnett, T. Rials and V. Saraf, J. Appl. Polym. Sci., **29**, 1815 (1984).
16. L. Wu and W. Glasser, J. Appl. Polym. Sci., **29**, 1111 (1984).
17. S. Tai, J. Nakano, and N. Migita, Mokuzai Gakkaishi, 13, 257 (1967).
18. S. Tai, M. Nagata, J. Nakano and N. Migita, Mokuzai Gakkaishi, 13, 102, (1967).
19. C. Saiz-Jimenez and J. W. de Leeuw, Org. Geochem., **6**, 417 (1984).
20. J. Marton, Tappi, **47**, 713 (1964).
21. E. Adler, I. Falkehag, J. Marton and H. Halvarson, Acta Chem. Scand., **18**, 1313 (1964).
22. K. Lundquist, T. Kirk and W. Connors, Arch. Microbiol., **112**, 291 (1977).
23. C. Carraher, T. Gehrke, D. Giron, D. Cerutis and H. M. Molloy, J. Macromol. Sci.-Chem., **A19**, 1121 (1983).
24. C. Carraher, W. Burt, D. Giron, J. Schroeder, M. L. Taylor, H. M. Molloy and T. O. Tiernan, J. Appl. Polymer Sci., **28**, 1919 (1983).
25. F. Critchfield, *"Organic Functional Group Analysis,"* MacMillan, New York, (1963), p. 82.

EXAMPLES OF ANALYTICAL APPROACHES TO INDUSTRIALLY IMPORTANT

POLY(SACCHARIDES)

J. F. Kennedy,[1] V. M. Cabalda,[1] K. Jumel,[2] and E. H. Melo[1]

(1) Research Laboratory for the Chemistry of Bioactive
    Carbohydrates and Proteins
    School of Chemistry
    The University of Birmingham
    P. O. Box 363
    Birmingham B15 2TT, England

(2) Chembiotech, Ltd.
    Institute of Research and Development
    Vincent Drive
    Edgebaskon, Birmingham B15 2SQ, England

Poly(saccharides) are extensively used in a large variety
of industries as emulsifiers, thickeners and stabilizers.
Therefore, a need exists for the determination of purity
and/or molecular weight of these materials. This paper uses
some examples to describe different approaches to the problem. Carboxymethyl cellulose (CMC) is a thickening agent
which is used in a large number of industries, including the
food industry. A universal chemical method for the determination of CMC in all types of formulations has previously not
been available; the development of such a method is described
here. Also described is the development of a method for the
determination of starch purity in starches. This measurement
is a particular problem for the member countries of the European Community where high grade starches are subsidized. The
method of choice is based on total enzyme hydrolysis and is
applicable to all starches. There are a number of ways for
the determination of molecular weights and molecular weight
distribution of poly(saccharides). Most of these are based on
some form of calibration and can therefore only give values
relative to a particular standard. Low angle laser light
scattering provides a means of obtaining absolute weight
average molecular weights and molecular weight distributions
and examples for its use are shown.

INTRODUCTION

Poly(saccharides) are used in all types of industries from the construction industry to the food industry. In order to fulfill the differ-

ent needs of the end user, manufacturers produce the materials in different grades and formulations. However, this increases the difficulties in determining the purity of the actual poly(saccharide) since a large number of impurities, often undefined, will interfere with analyses, and different methods have to be found for different types of materials. This paper describes some approaches to dealing with these problems.

The derivatization of cellulose renders it more soluble and gives it different properties depending on the type of substituent and the degree of substitution (DS). Derivatization of cellulose is therefore popular on account of the cheap cost of the original polymer. Cellulose derivatives are used in paints, thickeners, adhesives, gels, food materials, adsorbents, detergents, etc. Although normally the cellulose derivatives would be added to a formulation to a specific, predetermined level, there are many instances where cellulose derivatives need to be measured in a formulation of which they compromise a minority amount. Methodology for the estimation of carboxymethyl cellulose in industrial formulations is reported here as an example.

Starch is present in almost all plants, with its major source being the seeds of maize, wheat, rice, etc. and the tubers and roots of potato, arrowroot and cassava. It consists of a mixture of linear and branched homopolymers of D-glucose. Amylose is made up of linear chains of α-D-glucopyranse units linked by (1$\rightarrow$4)-bonds, and has a degree of polymerization of between 1×10^2 and 4×10^5. Amylopectin contains short linear chains (between 19 and 23 units long) of (1$\rightarrow$4)-linked α-D-glucopyranose residues which are joined by (1$\rightarrow$6) linkages to form a highly branched structure which has a degree of polymerization between 1×10^4 and 4×10^7. The degree of branching and the ratio of amylose:amylopectin vary with the source, age, etc. of the starch.

Such was the demand for starch by the food industry, that food-grade starches became the sole product of most, if not all, starch manufacturers. Understandably, this subsequently led to a surplus in food-grade starches, most especially in the EC. To help reverse the situation and to encourage the use of starch as a feedstock in the chemical industry, substantial rebates are given to the end-user by the EC for the utilization of food-grade starches, which have a minimum purity of 97%, in non-food applications.

Therefore, this requires a rapid and reliable method for the measurement of starch in high purity starches and several enzymic and non-enzymic methods have been proposed, some of which have been adopted for official testing, although none have been found to be truly satisfactory. In this paper an enzymic method using an enzyme regime for the hydrolysis of starch to glucose is described.

The determination of molecular weights of polymers has become more and more important to industry since molecular weight determines the properties of a polymer to a large extent. However, most methods of molecular weight determination, such as viscosity measurements or gel permeation chromatography (GPC), are based on calibrations with standards similar to the samples being tested. However, these standards can only represent the behavior of the sample to various degrees, and molecular weight values obtained can only be quoted relative to the standard. Low angle laser light scattering (LALLS) is an absolute method which can be used in a stand-alone (static) mode and in conjunction with GPC for the determination of absolute molecular weight distribution. The use of this method is demonstrated for several poly(saccharides).

DETERMINATION OF INDUSTRIAL GRADE CARBOXYMETHYL CELLULOSE BY A CHEMICAL METHOD

1. Introduction

Carboxymethyl cellulose is used in a large range of DS and for many different applications. To measure it is not an easy task since the glucose repeating unit should be measured, but this is complicated by its random derivatization which affects reagent reaction for qualitative recognition and uniformity of calculations for quantitation. Enzymatic hydrolysis is only effective up to a DS of 0.7 and acid hydrolysis produces undesirable by-products. Therefore, a method had to be found which could cope with CMCs of DS higher than 0.7 and also separate any interfering by-products. A manual and an automated method are described here, by which the crude samples are subjected to acid hydrolysis and then separated by dialysis (manual method) or column chromatography (automated method) prior to colorimetric measurement.

EXPERIMENTAL

1. Materials

Crude CMC (DS = 0.55) - samples A, B, and C - as a solid powder and in a concentration of 1% in solid (powder) and liquid detergent formulations were used in this study. All the samples were dried for 24 hours in a methanol drier before use.

2. Analysis of Crude CMC by Manual and Automated L-Cysteine Assays

Carboxymethyl cellulose (1 mg) - samples A, B, and C - was dissolved in distilled water (1 mL). Aliquots of 20 µl were added to 180 µL of distilled water to give aqueous solutions with final concentrations of 20µg/200µL. The L-cysteine-sulfuric acid (86%) reagent (1 mL) was added to these solutions, the mixture stirred and heated at 95°C for 3 min, cooled and the absorbencies measured spectrophotometrically at 415 nm. An aliquot of 100 µL was loaded on to the chromatographic column packed with Biogel^R P6 which was connected to an automated L-cysteine sulfuric acid assay. The same procedure was performed for these samples after 24 hours dialysis in order to remove the impurities present in the preparations.

3. Analysis of Crude CMC in the Form of Detergent Formulations by Manual and Automated L-Cysteine Assays

Powdered detergent formulations (87 mg to give 1 mg CMC) samples D, E and F were dissolved in 1 mL of distilled water. Aliquots (20 µL) were taken and added to 180 µL of distilled water to give aqueous solutions with final concentrations of 20µg/200µL. The L-cysteine-sulfuric acid (86%) reagent (1 mL) was added to these solutions, the mixture stirred and heated at 95°C for 3 min., cooled and the absorbencies measured spectrophotometrically at 415 nm. An aliquot of 100 µL was taken and loaded on to the chromatographic column packed with Biogel^R P6 which was connected to an automated L-cysteine-sulfuric acid assay.

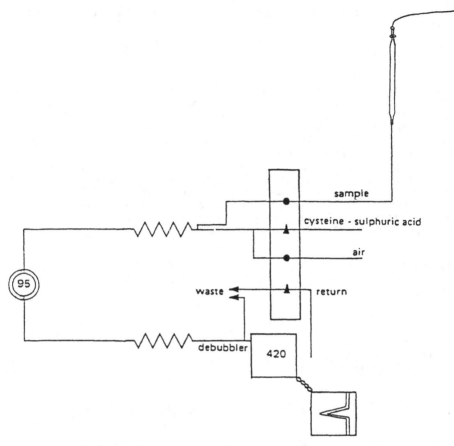

Figure 1. Automated L-cysteine-sulfuric acid assay.

Liquid detergent formulations (0.5 mL to give 1 mg CMC) samples G, H and I were added to distilled water (0.5 mL) to give aqueous solutions (1mg/mL). Aliquots of 20 µL were taken and the same procedure as described for the crude CMC was followed for both dialyzed and non-dialyzed materials (Figure 1).

RESULTS AND DISCUSSION

The chemical analysis used for measuring the total sugar content of crude CMC samples by manual L-cysteine sulfuric acid assay was affected by the impurities present in the preparations since higher percentages of glucose were obtained when compared to the stated values of approximately 30%. However, after dialysis of the preparations, these percentages came down to the expected values (Table 1).

The percentages of glucose produced from crude CMC samples were the same before and after dialysis when assayed by an automated L-cysteine-sulfuric acid system which was connected to a chromagraphic column packed with Biogel[R] P6. This is due to the capacity of the chromatographic system to remove the impurities present in the CMC samples (Table 2).[1] The

Table 1. Analysis of crude CMC by direct manual L-cysteine assay.

Sample No.	DS	Glucose Produced (%)
A*	0.55	52.4 ± 0.1
B*	0.55	53.7 ± 0.1
C*	0.55	60.2 ± 0.1
A+	0.55	32.5 ± 0.0
B+	0.55	32.1 ± 0.1
C+	0.55	31.4 ± 0.1

* Before dialysis + After dialysis

same behavior was observed when a manual and automated L-cysteine-sulfuric acid assay was used for estimating the total sugar content of the crude CMC samples present as 1% CMC in powdered and liquid detergent formulations (Tables 3 and 4). This proves that crude CMC and CMC in the form of detergent formulations can be quanitatively determined by measuring their total sugar content. Either manual or automated L-cysteine-sulfuric acid assays can be used for the quantitative determination of CMC, but in the former case dialysis is required prior to the assay to avoid interference of the contaminants present in the samples. However, the dialysis step was not necessary when a chromatographic column packed with Biogel^R P6 was connected to the automated L-cysteine-sulfuric acid system. This is due to the capability of the chromatographic column to purify the CMC thus avoiding interference in the automated L-cysteine assay.

DETERMINATION OF STARCH PURITY IN HIGH PURITY STARCHES

1. Introduction

The enzyme methods for the measurement of starch purity are based on the total hydrolysis of starch to glucose, by starch degrading enzymes, and the subsequent analysis of the glucose produced to give the starch

Table 2. Analysis of crude CMC by column chromatography and automated L-cysteine assay.

Sample No.	DS	Glucose Produced (%)
A*	0.55	31.2 ± 0.1
B*	0.55	30.8 ± 0.1
C*	0.55	30.2 ± 0.1
A+	0.55	30.4 ± 0.1
B+	0.55	31.6 ± 0.1
C+	0.55	31.5 ± 0.0

* Before dialysis + After dialysis

Table 3. Analysis of CMC in liquid detergent formulations using direct manual and chromatography automated L-cysteine assay.

| Sample No. | System | Glucose Produced (%) | |
		Before dialysis	After dialysis
G	manual	0.85 ± 0.01	0.35 ± 0.01
H	manual	0.72 ± 0.01	0.39 ± 0.01
I	manual	0.66 ± 0.01	0.40 ± 0.01
G	automated	0.29 ± 0.01	0.26 ± 0.01
H	automated	0.29 ± 0.01	0.26 ± 0.01
I	automated	0.28 ± 0.01	0.27 ± 0.01

purity value. One or more of the following enzymes are commonly used. Amyloglucosidase (EC 3.2.1.3) is an exo-acting enzyme capable of hydrolyzing the successive (1→4)-linkages between α-D-glucopyranosyl residues in starch with the production of β-D-glucose. The enzyme is not specific for (1→4)-linkages since it hydrolyzes (1→6)- and (1→3)-linkages, although these linkages are hydrolyzed 15 to 30 times more slowly. The nature of the linkage adjacent to that being hydrolyzed and the molecular weight of the substrate both affect the rate of hydrolysis. Amyloglucosidase is the most important among the enzymes used in enzymic methods for the determination of starch purity since it hydrolyzes more than 90% of the starch to glucose.

α-Amylase (EC 3.2.1.1) is an endo-acting enzyme and randomly hydrolyzes any of the (1→4)-linkages within the poly(saccharide) chain resulting in the rapid reduction in the molecular size of starch. α-Amylase is used in some enzymic methods for the solubilization of starch before the addition of amyloglucosidase.

Pullulanase (EC 3.2.1.41) is an endo-acting enzyme which hydrolyzes the poly(saccharide) pullulan almost entirely to maltotriose and is capable of acting on the (1→6)-linkages in starch, amylopectin, and limit dextrins to give products containing no branch points. All pullulanases require that each of the two chains linked by a (1→6)-linkage contains at least two (1→4)-linked α-D-glucopyranosyl residues.

Table 4. Analysis of CMC in solid detergent formulations using direct manual and chromatography automated L-cysteine assay.

| Sample No. | System | Glucose Produced (%) | |
		Before dialysis	After dialysis
D	manual	0.62 ± 0.01	0.38 ± 0.01
E	manual	0.64 ± 0.01	0.27 ± 0.01
F	manual	0.56 ± 0.01	0.36 ± 0.01
D	automated	0.32 ± 0.0	0.32 ± 0.0
E	automated	0.33 ± 0.0	0.30 ± 0.0
F	automated	0.32 ± 0.01	0.35 ± 0.01

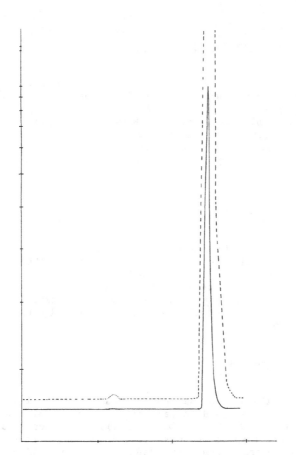

Figure 2. Gel permeation chromatography profile of starch
hydrolysate using multi-enzyme regime of Birmingham
2 method.

Isoamylase (EC 3.2.1.68) is a starch degrading enzyme which hydrolyzes all the (1→6)-linkages in amylopectin and glycogen, but which has
very little or no action on pullulan.

An enzyme Method using only amyloglucosidase for the conversion of
starch to glucose has been developed,[2] but gel permeation studies of the
hydrolysate have shown that approximately 5% of most starches is not
hydrolyzed to glucose (Figure 2), probably due to the branched points of
starch which amyloglucosidase would have difficulty in hydrolyzing. This
led to the development of an enzyme method which utilizes a multienzyme
regime consisting of α-amylase, amyloglucosidase and pullulanase which
ensures total hydrolysis.

EXPERIMENTAL

1. Determination of enzyme activity

In this paper, enzyme activity is defined as the amount of enzyme
which produces 1mg/mL of dextrose equivalent in one minute from 5 mL of
1% soluble starch solution at 60°C and pH 5.0

2. Gel Permeation Chromatography[2,3,4,5]

Aliquots (75 µl of 1% solution) of the supernatants of the enzymic hydrolysates of starches were injected into a water-jacketed column (50 x 1.0 cm i.d.) packed with Biogel P2 (400 mesh particle size) and maintained at 60°C. The mobile phase was 0.1 M sodium chloride with a flow rate of 0.16 mL/min. The column eluent was continuously monitored using an automated L-cysteine sulfuric acid assay. The system was standardized by injecting a mixture of glucose and malto-oligosaccharides of known composition.

3. Determination of Glucose

GOD/POD reagent (2 mL) and distilled water (2 mL) were added to solutions of glucose (1 mL) containing 10 to 50 µg of glucose. The reaction mixtures were left in the dark at ambient temperature for exactly 30 minutes and their optical density was measured at 560 nm using water (3 mL) and the GOD/POD solution (2 mL) as blank. Standard glucose solutions containing 10 to 50 µg of glucose were similarly treated to get a calibration curve.

4. Hydrolysis of the Starch Samples

α-Amylase (0.125 units), amyloglucosidase (22.5 units) and pullulanase (0.625 units) were added to each of the starch solutions. The hydrolyses were carried out at 60° for 45 minutes with continuous stirring. The solutions were then cooled down to 25° and transferred quanitatively into 250 mL volumetric flasks and made up to the mark with water. The hydrolysates were then filtered and the first 50 mL filtrate was discarded. Portions (10 mL) of the solutions were further diluted to 250 mL with water and aliquots (1 mL) of these final solutions were assayed for their glucose (and thereafter starch) content.

RESULTS AND DISCUSSION

A critical point for the development of a reliable enzyme method is the total enzymic hydrolysis of starch to glucose. In order to obtain total hydrolysis of any starch including amylopectinaceous material to glucose, a debranching enzyme (pullulanase) together with α-amylase and amyloglucosidase was deemed necessary. The efficiency of this multi-enzyme regime to hydrolyze starch to glucose was tested on a wide range of starches and was monitored by gel permeation chromatography. Table 5 and Figure 2 show the almost total conversion (99.5-100.00%) of the starches to glucose except for potato starches which give less than 99.25% conversion and a rye starch which was only 98% hydrolyzed to glucose. Potato starches are known to contain phosphorylated glucose units which can not be recognized by the amyloglucosidase, and/or heavily phosphorylated portions might offer steric hindrance to enzyme attack. The presence of a brown residue after the enzyme hydrolysis of a rye starch might suggest that this particular starch is not a very high purity starch and that the more than 1.5% high molecular weight portion in its starch hydrolysate could very well be soluble hemicellulose.

Table 5. Data on the percentage composition of some starch hydrolysates using Birmingham 2 Enzyme Method.

| Starch Type | % Composition of starch hydrolysates | | |
	G1	G2-IMW	HMW
Wheat	99.67	0.33	0.00
Wheat	100.00	0.00	0.00
Maize, regular	99.75	0.25	0.00
Maize, regular	99.70	0.30	0.00
Maize, waxy	99.82	0.18	0.00
Maize, waxy	99.82	0.18	0.00
Maize, 50% amylose	99.84	0.16	0.00
Maize, 50% amylose	99.67	0.33	0.00
Maize, 70% amylose	99.86	0.14	0.00
Maize, 70% amylose	99.74	0.26	0.00
Potato	99.15	0.85	0.00
Potato	98.54	1.46	0.00
Rice	99.60	0.40	0.00
Rice	99.87	0.13	0.00
Rye	97.98	0.20	1.82
Sago	99.58	0.42	0.00
Tapioca	99.77	0.23	0.00

Approximately 0.5% of most of the starches is still not hydrolyzed to glucose. This could be attributed to the densely branched portion of the starch which is not susceptible to enzyme attack due to steric hindrance. However, earlier studies showed that further increases in the amount of pullulanase resulted in an even lower conversion, suggesting enzyme inhibition which would be due to the molecular size of the pullulanase. Alternatively, the 0.5% unhydrolyzed portion could be products from the reverse reaction catalyzed by amyloglucosidase.

In order to optimize the method described above, studies concerning the solubilization of the starch and optimization of the glucose measurement have also been carried out.

ABSOLUTE MOLECULAR WEIGHT AND MOLECULAR WEIGHT DISTRIBUTIONS BY LALLS AND GPC/LALLS

1. Introduction

The technique of low angle laser light scattering (LALLS) when applied to molecular weight determination is absolute in that no standardization or external calibration is required since the method of calibration is based on an internal calibration system using the actual sample.[9] It is equally applicable to linear and branched polymers and offers high sensitivity to trace fractions of high molecular weight species. When coupled to a GPC column system and concentration selective detector (such as refractive index), LALLS removes the need for column calibrations. The absolute molecular weight distribution is calculated directly from LALLS measurement of excess Rayleigh factor and the refractive index response using a data acquisition system equipped with GPC/LALLS software. The use of LALLS and GPC/LALLS for poly(saccharides) ia demonstrated by three examples.

EXPERIMENTAL

1. Static LALLS

At least four concentrations of the polymer to be analyzed are prepared. The intensities of the scattered and incident light (P_θ and P_0) are measured for the solvent and each polymer solution. A plot can then be constructed of Kc/R_θ vs. c, where K is the polymer constant, c the solute concentration and R_θ the excess Rayleigh factor. The weight average molecular weight M_w is given by the reciprocal of the intercept and the slope is equal to $2A_2$ which is a measure of the solvent solute interactions.

2. GPC/LALLS

To obtain absolute molecular weight distributions in GPC effluents, the LALLS detector is connected in series with the column and the concentration-selective detector (Figure 3). The sample is prepared and injected as for GPC. Two elution curves are generated, one by the LALLS detector showing the intensity of the scattered light at each point on the curve, and one by the concentration detector, showing the corresponding concentration. The molecular weight at each point is then calculated similarly to the calculations described above for static LALLS. M_w and M_n are calculated in the conventional manner.

3. Determination of molecular weight of hyaluronic acid

Hyaluronic acid is a poly(saccharide) electrolyte having $\rightarrow$4)-(β-D-glucopyranosyluronic acid)-(1$\rightarrow$3)-(2-acetamido-2-deoxy-β-D-glucopyranosyl)-(1 as the repeating disaccharide unit (Figure 4).

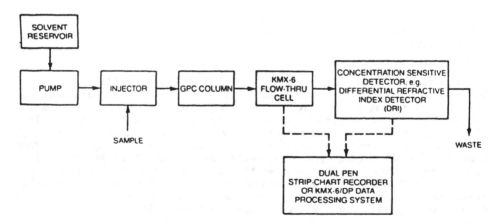

Figure 3. Block diagram of the GPC/LALLS system.

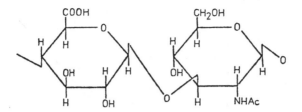

Figure 4. Repeating unit of hyaluronic acid.

It occurs in cartilage, in eye vitreous humor and in synovial fluid. It forms highly viscous solutions and its viscosity decreases with increasing pressure, a property important for its ability to act as a "shock-absorber" in joints. In contrast to other glycosaminoglycans, hyaluronic acid exhibits molecular weights of up to 8×10^6 daltons, a highly regular structure and no sulfur containing groups.

The material can now be produced by a microbiological process which will increase the number of applications possible. These applications depend to a large extent on the molecular weight of hyaluronic acid. It is, therefore, important to have the means of measuring this accurately. Figures 5A and 5B show a plot of Kc/R_θ vs c and the elution profiles for hyaluronic acid respectively. As can be seen, results obtained by CPC/LALLS are in fairly good agreement with those obtained by the static method.

4. Determination of Molecular Weight of Alginate by GPC/LALLS

Alginates are naturally occurring poly(saccharides) which are extracted from brown seaweed and are used mainly in the food industry, in the pharmaceutical industry, for textile processing, paper making and many other applications where they act as gelling and thickening agents.

The poly(saccharide) is a linear co-polymer of β-D-mannuronic acid and a α-L-guluronic acid linked by (1→4) bonds to give blocks of poly-(mannuronic acid) (M-blocks), blocks of poly(guluronic acid) (G-blocks), and blocks in which the two acidic monosaccharides occur in mixed sequences (MG-blocks).

Since alginates are most commonly used as thickeners and gelling agents, measurements of molecular weight and M/G ratios which determine properties such as viscosity are extremely important.

Figure 6 shows the chromatograms obtained for a low viscosity sodium alginate using GPC/LALLS. Both trances show that the material is very pure and the low molecular weight would be expected from a low viscosity alginate.

5. Determination of Molecular Weight of a Low DE Maltodextrin

Maltodextrins are polymers of D-glucose in which the individual α-D-glucopyranosyl residues are joined by (1→4) linkages to give linear chains with a degree of chain branching being brought about by (1→6) linkages.

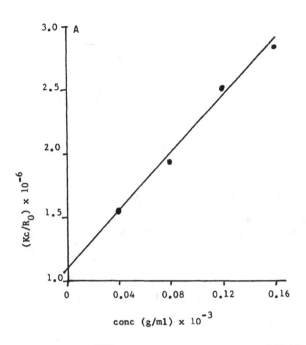

Conditions:	GPC	LALLS	
eluent	0.25MNaCl/0.05M sodium phosphate buffer (pH7.0)	Annulus	6 to 7°
flowrate	1.066ml/min	Fieldstop	0.15mm
temperature	20°C	dn/dc	0.166ml/g
detector	refractive index, lalls		
columns	30cm, TSK G6000PW + G5000PW + G4000PW + guard cartridge		

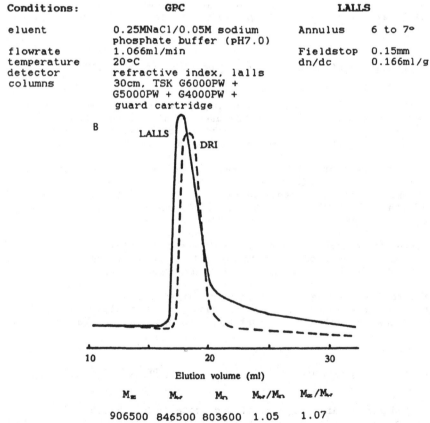

M_{Ξ}	M_{wr}	M_n	M_{wr}/M_n	M_{Ξ}/M_{wr}
906500	846500	803600	1.05	1.07

Figure 5. A. Static LALLS analysis of hyaluronic acid.
B. GPC/LALLS analysis of hyaluronic acid.

Conditions:

	GPC		LALLS	
eluent	0.1M KMO₃		Annulus	6 to 7°
flowrate	0.531ml/min		Fieldstop	0.15mm
temperature	20°C		dn/dc	0.150ml/g
detector	refractive index, lalls			
columns	30cm, TSK G6000PW + G5000PW + G4000PW + guard cartridge			

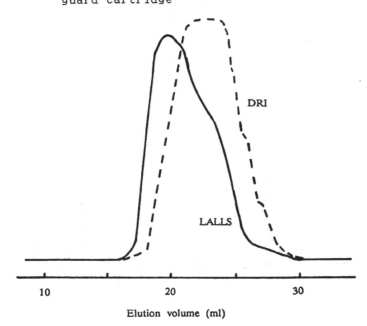

M$_z$	M$_w$	M$_n$	M$_w$/M$_n$	M$_z$/M$_w$
320300	244800	175300	1.40	1.31

Figure 6. GPC/LALLS analysis of sodium alginate.

Commercially, maltodextrins are produced by acid hydrolysis,[6] enzyme hydrolysis,[7] or combined acid and enzyme hydrolysis[8] of starches.

Maltodextrins have a range of industrial applications, particularly in the food and pharmaceutical industries, as fillers, stabilizers, thickeners, pastes, glues etc. and solutions of maltodextrins have been used in the after care treatment of chronic renal failure, liver cirrhosis, disorders of amino acid metabolism and conditions which require high energy but low fluid volume and electrolyte content diet.

Figure 7 shows the RI and LALLS traces of a maltodextrin sample. The LALLS trace shows a small amount of high molecular weight materials which is not detected by the concentration detector and is probably due to some form of retrogradation. This demonstrates the sensitivity of LALLS to high molecular weight materials and its use in detecting these.

Conditions:	GPC	LALLS	
eluent	0.25MNaCl/0.05M sodium phosphate buffer (pH7.0)	Annulus	6 to 7º
flowrate	0.528ml/min	Fieldstop	0.15mm
temperature	20ºC	dn/dc	0.147ml/g
detector	refractive index, lalls		
columns	30cm, TSK G4000PW plus guard cartridge		

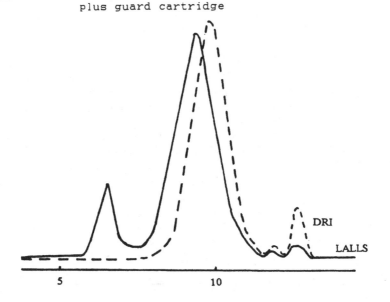

Elution volume (ml)

M_z	M_w	M_n	M_w/M_n	M_z/M_w
42700	41200	39400	1.05	1.04

Figure 7. GPC/LALLS analysis of low DE maltodextrin.

OVERALL CONCLUSION

The above examples show that different approaches of analysis are required for different types of poly(saccharides). Whereas enzymic hydrolysis of CMC was not appropriate for many CMC formulations and a chemical method had to be developed, the enzymic method for the determination of starch purity in high purity starches could be significantly improved.

Molecular weight determinations using LALLS and GPC/LALLS have been shown to give absolute and accurate measurements of molecular weights and molecular weight distributions and prove sensitive to low concentrations of high molecular weight materials.

REFERENCES

1. J. F. Kennedy and E. H. M. Melo, in: "*Cellulose: Structural and Functional Aspects,*" J. F. Kennedy, C. O. Philips and P. A. Williams Eds., Ellis Horwood, Chichester, 1990, p. 2.
2. J. F. Kennedy, D. L. Stevenson and K. Jumel, Starke, **42**, 8-12 (1990).
3. OJEC L158 3-5 (1986).
4. Analytical Working Party of the Starch Experts Group (STEX) of the European Starch Association (ESA), Starke, **39**, 414-416 (1987).
5. J. F. Kennedy, V. M. Calalda and K. Jumel, Starke, in press, (RP 372).
6. T. J. Palmer, In: "*Glucose Syrups and Related Carbohydrates,*" G. G. Birch, L. T. Green and C. B. Coulson, Eds., Applied Science, London, 1970, p. 23.
7. W. M. Fogarty, P. J. Griffin and A. M. Joyce, Process Biochem. **19** (6), 11 (1974).
8. D. Howling, In: "*Sugar Science and Technology,*", G. G. Birch and K. J. Palmer, Eds., Applied Science, London, 1979, p. 263.
9. W. Kaye and J. B. McDaniel, Appl. Opt., **13**, 1934 (1974).

BACTERIAL POLYSACCHARIDES FOR USE IN FOOD AND AGRICULTURE

V. J. Morris

AFRC Institute of Food Research
Norwich Laboratory
Colney Lane, Norwich NR4 7UA, U.K.

Polysaccharides secreted by bacteria provide a new range
of polymers with interesting functionality for use by the
food and agricultural industries. One application involves
mass production, purification and sale of such polymers as
additives. Recent research on potential new additives will be
illustrated by discussions of the gelation mechanism of gellan gum, a broad spectrum gelling agent, and early studies on
the production, purification and application of cyclosophorans, a new family of cyclic polysaccharides with potential
use as encapsulating agents. A second area of research involves developing and optimizing natural functional properties of bacterial polysaccharides. In the food industry this
involves developing self-textured fermentation based foods.
Early research in this area will be illustrated by studies on
polysaccharides produced by lactic acid bacteria and *Acetobacter*. In the agricultural area the biological and possible
ecological importance of the extracellular polysaccharides
produced by the soil bacterial species of *Rhizobium* and *Agrobacterium* will be discussed.

INTRODUCTION

Extracellular polysaccharides secreted by bacteria provide a source
of new polymers which may have potential use in the food and agricultural
industries. Likely food applications for such polysaccharides include
encapsulation or use as gelling, thickening or suspending agents. Agricultural applications also include encapsulation, viscosity enhancement,
suspension of solids, stabilization of emulsions and control of droplet
size during spraying. In most instances it is envisaged that the polysaccharides will be fermented in bulk, purified and sold as functional additives. An example of such a commercially successful bacterial polysaccharide is xanthan gum.[1] An alternative to such an approach is to develop
natural functionalities of bacterial polysaccharides. In the food industry this can involve optimizing the role played by extracellular polysaccharides in fermented food products produced by bacterial starter cultures. In the area of agriculture it may be possible to optimize and
exploit the role played by polysaccharides produced by soil bacteria.

Biotechnology and Polymers, Edited by C.G. Gebelein
Plenum Press, New York, 1991

This article describes recent studies at IFR on a range of new bacterial polysaccharides for use in food and agriculture, either as additives, or through exploitation of their natural functions.

GELLAN GUM

Gellan gum is produced by the aerobic fermentation of *Pseudomonas elodea* in batch culture.[2] Kelco have shown that gellan gum can be used as a broad spectrum gelling agent in a variety of food applications.[3] The additional advantage is that gellan can be used in such applications at concentrations substantially lower than those required for the use of traditional algal or plant polysaccharides. Successful toxicity trials have been carried out and gellan has received food approval in Japan in 1988. Food approval is currently being sought in the UK, Europe and the USA. Potential agricultural application lies in the demonstrated use of gellan as plant tissue medium,[1] and the use of gellan-gelatin mixtures in encapsulation.[4]

At IFR studies have been concentrated on the molecular basis for gelation. Gellan gum is a linear anionic heteropolysaccharide possessing a tetrasaccharide repeat unit (Figure 1).[5,6] The native polysaccharide is esterified and forms soft elastic gels.[7,8] The commercial product (Gelrite) is deesterified by subjecting the broth to alkaline conditions. This material forms hard brittle gels.[8] X-ray diffraction studies,[9-13] of fibers prepared from oriented stretched gels, are consistent with a three-fold helix showing an axial rise per chemical repeat unit equal to one-half the extended length of the repeat unit; suggestive of a double helical structure. Oriented deesterified gellan gum gels give crystalline fiber diffraction patterns consistent with regular side-to-side packing of segments of the gellan helices. Esterification inhibits crystallization thus increasing the elasticity and reducing the brittleness of the gels.[9-13] Latest modeling studies suggest that gellan forms a left-handed three-fold double helix.[14,15] Acetylation of the gellan helix does not prevent these substituted helices packing into the unit cell characteristic of deesterified gellan, but L-glycerate substitution does inhibit packing into this unit cell.[15] Previous X-ray diffraction studies have shown that the unit cell dimensions are the same for the native and deesterified gellan,[9-13] suggesting that the bulky L-glycerate esters appear to restrict crystallization to unsubstituted regions of the helix.

Rheological studies have shown that gelation is dependent upon the type and concentration of associated cations.[13,16] Bulky tetramethyl ammonium (TMA) ions prevent gelation resulting in viscoelastic fluids. Studies using a variety of physical chemical techniques,[17-22] suggest a conformational change for TMA gellan upon heating and cooling. Optical

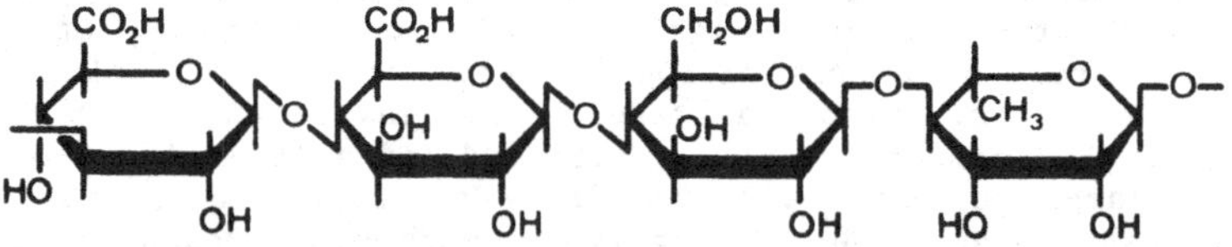

Figure 1. The chemical repeat unit of gellan gum. (The native polysaccharide is partially substituted at C2 with L-glycerate and contains ~50% C6 substitution with acetate, both on the (1→3) linked glycosyl residue).

rotation data have been taken to suggest a helix-to-coil transition upon heating and cooling.[17-19] Light scattering studies on TMA gellan in TMACl, under conditions favoring the suggested helical structure, have been taken to favor a double helical structure.[21,22]

Preliminary light scattering studies carried out in our laboratory suggest that the structures observed in such fluids may be sensitive to the type of clarification procedure. The TMA form of gellan was prepared and absence of gel promoting cations demonstrated by X-ray microprobe analysis of the solid and atomic absorption spectroscopy of dilute solutions (<100ppm Na^+, K^+, Ca^{2+}). Light scattering studies were carried out using a FICA 50 operating at 436 nm. The Rayleigh ratio for benzene was taken to be 45.6×10^{-6} cm^{-1},[23] dn/dc was taken to be 0.156 cm^3/g at 436 nm,[21] and measured as 0.136 cm^3/g at 633 nm, and the depolarization ratio (ρ_u) was measured as $\rho_u = 0.016$ giving a Cabannes-Rocard factor $(6+6\rho_u)/(6-7\rho_u) = 1.04$.[23] Stock solutions of 1 mg/mL TMA gellan in 0.075M TMACl, conditions said to favor the ordered helical form, were prepared. Sample concentrations were measured using a differential refractometer at 633 nm.

Table 1 shows the results of a Zimm plot analysis of TMA gellan in 0.075M TMACl solutions after filtration through either 0.45 μm filters or 3μm filters.[23] These large differences in scattering suggest that the clarification procedure is either removing or disrupting aggregates. Carbohydrate analysis of TMA gellan gives a composition of 96% carbohydrate.[24] Table 2 shows the measured solution concentrations before and after filtration. The absence of any marked loss of TMA gellan upon filtration suggests that filtration disrupts rather than removes TMA gellan aggregates. To examine the nature of these aggregates Holtzer plots (Figure 2) have been obtained for 1 mg/mL stock TMA gellan in 0.075M TMACl after filtration through different pore size filters.[25] The 0.45 μm filtered sample shows a Holtzer plot characteristic of short rod-like molecules whereas, with increasing filter pore size the scattering behavior becomes increasingly characteristic of an extended more flexible worm-like coil molecule. All the curves seem to tend to a similar asymptote at high wavevector (q) suggesting a similar mass per unit length, and fibrils formed by end-to-end association.

These data have been combined to suggest a model for gellan gelatin (Figure 3). In the TMA form inter-chain crystallization of gellan fibrils is taken to be suppressed. Helix formation on cooling is considered to promote end-to-end association, via double helix formation, into fibrils. Thickening or bifurcation of these fibrils could arise by a chain end

Table 1. Zimm plot analysis results.

Method of Clarification	Molecular weight (daltons)	Radius of gyration (nm)	Virial Coefficient (cm^3 mol./g^2)
3 μm	4.5×10^6	159	0.6×10^{-4}
0.45 μm	1.06×10^5	42	8.6×10^{-4}
0.45 μm*	4.3×10^5	159	22×10^{-4}

Data measured using a FICA 50 at a wavelength of 436 nm. Rayleigh ratio for benzene (R_θ) = 45.6×10^{-6} cm^{-1}, (dn/dc) = 0.156 cm^3/g, depolarization ratio (ρ_u) = 0.016. * Data reported in reference 17.

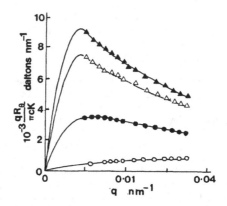

Figure 2. Holtzer plots for 1mg/mL TMA gellan in 0.075M
TMACl. Different curves represent clarification
through different pore size filters: -o- 0.45 μm,
-o- 0.6 μm, -△- 3 μm and -▲- 5 μm.

linking to a central region of another chain. Gel promoting cations are
considered to induce inter fibril crystallization and gelation.

CYCLOSOPHORANS

The cyclosophorans are a family of cyclic β(1→2) D-glucans produced[26]
by the soil bacterial species *Agrobacterium* and *Rhizobium*. Preliminary
studies have suggested potential applications as a new range of encapsulating agents.[27] Complex formation has been claimed[27] with indomethacin,
fluorescein, steroids and a number of vitamins. Molecular modeling
studies suggest a minimum ring size corresponding to a degree of polymerization (dp) of 15.[28] Different bacterial species produce different spectra of ring sizes.[26]

Preparation of cyclosophorans is restricted because the bacteria
secrete a highly viscous extracellular polysaccharide (EPS).[29] This
extracellular slime restricts aeration of the broth, hinders removal of
bacterial cells, and hence the separation of the glucan. Transposon muta-
genesis has been used to produce EPS-*Rhizobium leguminosarum* which produce cyclosophorans but no longer produce EPS. The bacterial broths can

Table 2. Polymer solution concentrations.	
Filter size	Polymer concentration after filtration mg/cc
unfiltered	0.82
3.0 μm	0.81
1.2 μm	0.81
0.45 μm	0.78

Concentrations of TMA gellan samples after filtration through
different pore size filters. Concentrations measured using a
differential refractometer at 25°C.

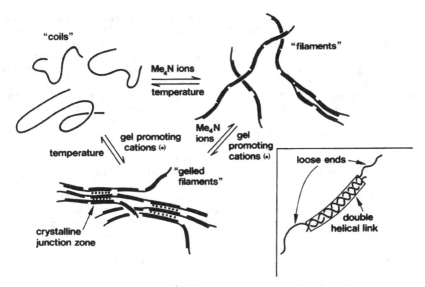

Figure 3. Schematic model for the gelation of gellan gum.

be clarified by filtration to remove bacterial cells. Cyclosophorans can be removed from the clarified broths by passage through an ion exchange column, to remove pigment contaminants, and then selective absorption onto a charcoal column. The cyclosophorans can then be selectively de-sorbed and eluted from the column. Purification into individual ring sizes can be carried out by HPLC. Figure 4 shows the major ring sizes produced by an EPS *R. leguminosarum* strain. The dp values assigned were determined by FAB mass spectrometry. The spectrum of ring sizes shown in figure 4 is typical for that obtained from a native EPS+ *R. leguminosarum* strain.[26]

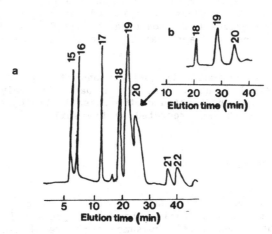

Figure 4. Distribution of molecular sizes (degree of polymerization - dp) for cyclosophorans produced by an
EPS-*Rhizobium leguminosarum* species. Separation on
Zorbax ODS, 3000 nm, 9.4 x 250 mm column (a) Eluant
3% CH_3OH, 1.5 mL/min increased to 2.5 mL/min, 10 mg
loading, (b) Eluant 2% CH_3OH, 2.5 - 2.7 mL/min, 0.8
mg loading.

Present studies at IFR are concerned with the development of methods for examining potential complex formation with purified cyclosophorans. Preliminary studies have been carried out on the macrolide antibiotic Amphotericin B (AmB). This material is a potent systemic fungicide but its considerable toxicity and poor solubility have restricted its therapeutic applications.[30,31] Attempts to improve water solubility, and thus lessen the aggregation of AmB, by derivatization have resulted in decreasing activity.[33-36] Recently it has been reported that AmB can be solubilized by complexation with γ-cyclodextran and that the complex retains the full anti-fungal activity.[37-40]

AmB was dissolved in dimethylsulphoxide (DMSO). Micellar suspensions were prepared by adding water to the DMSO solution. The Micellar solution shows a strong absorption band between 340-350 nm and the circular dichroism spectrum (Figure 5) shows an intense exciton couplet at the same wavelength. Both features arise because the AmB first dimerises and then associates into large helical aggregates. Loss of these bands can therefore be used to monitor dissolution of the aggregate.[37] Figure 5 shows the effect of adding a mixture of cyclosophorans with the size distribution shown in Figure 4. The reduction in the circular dichroism signal is taken as evidence for dissolution of the helical aggregates due to solubilization of AmB by the cyclosophorans.

ADDITIVES OR NATURAL FUNCTIONALITY

Market opportunities still exist for the introduction of new bacterial polysaccharides as additives. In the food industry the cost for clearance for food use is extremely expensive and there is increasingly the requirement to justify need in addition to safety or superior properties.

An alternative may be in the use of starter cultures to produce certain fermentation based foods. Is it possible to select or engineer these starter cultures to optimize the texture of the final product or develop new products? Are such products natural and can they be used as natural ingredients to texturize other food products. Studies at IFR have concentrated upon two types of bacteria: the lactic acid bacteria and *Acetobacter*.

In the case of agricultural applications it may be possible to improve on the natural functions of certain bacterial polysaccharides. At Norwich studies have concentrated on the possible biological and ecological roles played by the EPS secreted by the soil bacteria *Agrobacterium* and *Rhizobium*.

LACTIC ACID BACTERIA

Lactic acid bacteria polysaccharides are obvious candidates for study because many of the polysaccharides have been present in fermented food products which have been safely consumed for several centuries. Certain yogurt preparations contain polysaccharides which thicken the yogurt. Very few of these polysaccharides have been isolated, purified and evaluated as thickening agents. Products such as sugary kefir contain a gelling polysaccharide.[41] The starter culture contains both bacteria and yeasts and it is the bacterium *Lactobacillus hilgardii* which secretes the polysaccharide. Chemical analysis suggests the polymer is a dextran containing additional α(1→3)D-Glucose residues.[42] Gelation of dextrans con-

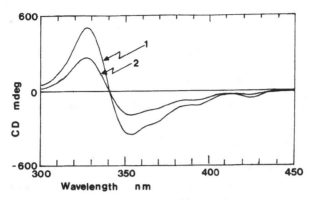

Figure 5. Circular dichroism spectrum for a micellar suspension of AmB in water. Data collected on a JASCO J-
600. Curve 1 shows the original spectrum and curve
2 the effect of adding cyclosophorans to the AmB
dispersion.

taining such residues normally occurs because these residues are present
as insoluble ribbon-like blocks, insoluble at normal pH, which serve to
cross-link the dextran.[43,44] Clearly there is scope for modifying gelation by increasing the α(1→3)D-Glucose content, or enhancing polymer
yield, through selective or genetic modification.

ACETOBACTER BACTERIA

Acetobacter strains are used in the production of vinegar. A major
use of UK produced vinegar is as a base for acid-based foods (sauces,
salad dressings, spreads, etc.). Xanthan gum is often added to such foods
as a thickening and suspending agent and it imparts a unique thixotropic
property to such foods.

Xanthan consists of a cellulosic backbone substituted on alternate
glucose residues with a trisaccharide sidechain (Figure 6).[45,46] The
sidechain alters the solid-state conformation of xanthan from a two-fold
ribbon-like structure to a helix with five-fold symmetry.[47-49] Thixotropy
of xanthan samples has been shown to result from association of xanthan
molecules into a "weak gel" network.[50] This structure is stabilized by
the formation of the xanthan helix. To develop self-thickening vinegars
it would be necessary to identify Acetobacter which are capable of producing both acetic acid and an extracellular polysaccharide whose rheological properties match or exceed those of xanthan. X-ray fiber diffraction studies of xanthan and xanthan derivatives deficient in terminal
mannose, and both terminal mannose and glucuronic acid, have demonstrated
the crucial features of the xanthan structure.[51,52] The cellulosic backbone and the (→2)Man are crucial for establishing the five-fold helical
structure. The glucuronic acid serves to stabilize the polymer and controls the helix-coil transition temperature. Accepting these as the crucial elements it is then possible to screen the chemical structures of
bacterial polysaccharides produced by Acetobacter for those structures
which satisfy the criteria. At present the most acceptable polysaccharide
appears to be acetan (Figure 6).

Acetan is a bacterial polysaccharide secreted by Acetobacter xylinium

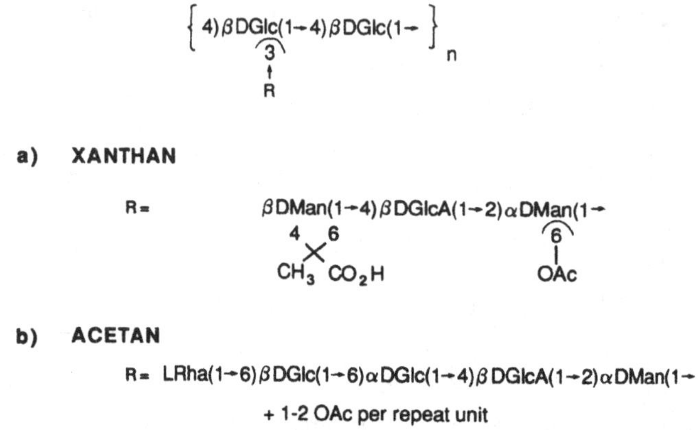

$$\left\{ 4)\beta DGlc(1\rightarrow4)\beta DGlc(1\rightarrow \atop \underset{\underset{R}{\uparrow}}{3} \right\}_n$$

a) XANTHAN

R= $\beta DMan(1\rightarrow4)\beta DGlcA(1\rightarrow2)\alpha DMan(1\rightarrow$
        $\underset{CH_3}{4}\underset{CO_2H}{\overset{6}{\times}}$ $\underset{OAc}{\overset{6}{|}}$

b) ACETAN

R= $LRha(1\rightarrow6)\beta DGlc(1\rightarrow6)\alpha DGlc(1\rightarrow4)\beta DGlcA(1\rightarrow2)\alpha DMan(1\rightarrow$

+ 1-2 OAc per repeat unit

Figure 6. The chemical repeat unit for the bacterial polysaccharides (a) xanthan and (b) acetan.

NRRL B42. The partial chemical structure for acetan (Figure 6) consists
of a cellulosic backbone substituted on alternate glucose residues with a
pentasaccharide sidechain.[53] The backbone, backbone-sidechain linkage,
and the first two sugars in the backbone are all identical to those of
xanthan. The ester distribution is still undetermined. X-ray fiber diffraction studies have shown that acetan forms a helix with five-fold
symmetry and similar pitch to xanthan.[53,54] Optical rotation and circular
dichroism studies are consistent with a reversible helix-to-coil transition on heating.[54] In the helical conformation acetan forms thixotropic
fluids (Figure 7) characteristic of xanthan samples.

Thixotropic vinegars can be prepared by dispersing purified acetan in
vinegar or by blending cell-free, clarified *A. xylinum* broths with vinegar. This type of product can be developed by biotechnological methods in
a variety of ways depending upon industrial interest and preference. The
likely similar biosynthetic mechanisms for production of acetan and xanthan suggest that it could be possible to engineer *Acetobacter* which
produce xanthan rather than acetan.

RHIZOBIUM AND *AGROBACTERIUM* POLYSACCHARIDES

The acidic heteropolysaccharides secreted by various species of *Rhizobium leguminosarum* possess the same backbone but have different sidechains (Figure 8). Aqueous solutions of these polysaccharides form viscoelastic fluids. Recently it has been discovered that upon addition of
sufficient amount of electrolyte these polymer solutions form thermore-
versible gels.[55] Divalent cations have been found to be more effective at
inducing gelation than monovalent cations, although the moduli of the
gels are similar at comparable ionic strength.[55]

R. leguminosarum species invade, induce module formation, and fix
nitrogen in leguminous plant roots. The mechanisms of the symbiosis are
largely unknown. Experimental evidence suggests the involvement of the
EPS in the early stages of the interaction.[56-58] Recent studies suggest
that EPS is unlikely to be involved in specific host recognition.[59,60]
However, gelation of the EPS may provide a non-specific binding mechanism. A two-stage binding mechanism has been proposed in which the initial
attachment to plant roots involves a calcium-dependent adhesion,[61] al-

3%
2%
1%
0·5%
0·25%
log10{η (mPas-1)}
log10{γ̇ (s-1)}
0
1
2
3
4
5
-1
-0·5
0
0·5
1·0
1·5
2·0
2·5

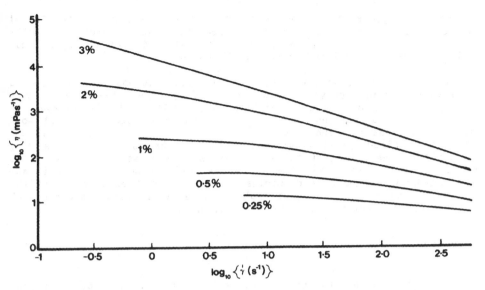

Figure 7. Viscosity (η) versus shear rate (γ) plots for aqueous acetan samples. The curves are reproducible upon raising or lowering the shear rate. Note the data is plotted on a log-log scale.

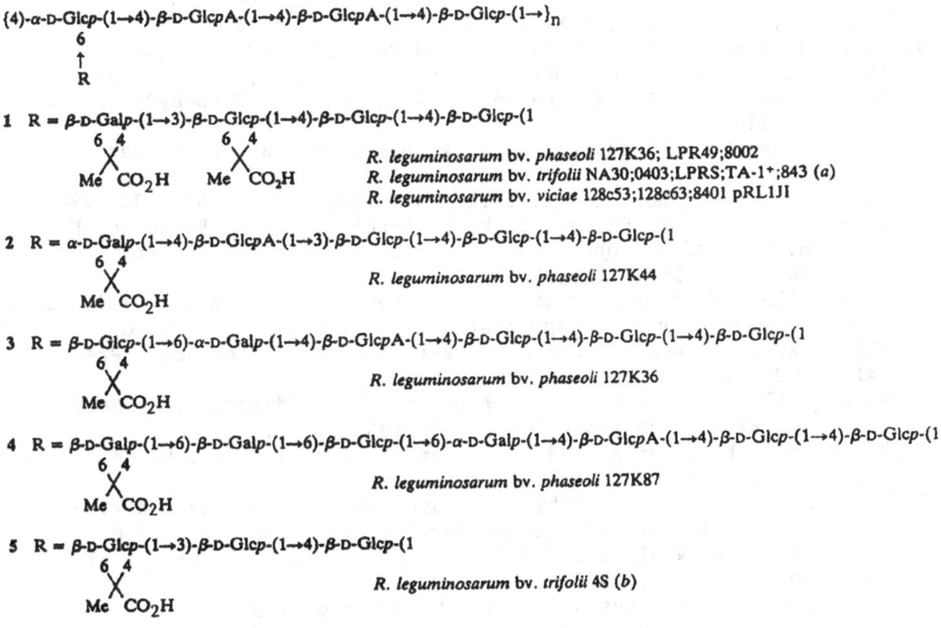

{4)-α-D-Glcp-(1→4)-β-D-GlcpA-(1→4)-β-D-GlcpA-(1→4)-β-D-Glcp-(1→}n
6
↑
R
1 R = β-D-Galp-(1→3)-β-D-Glcp-(1→4)-β-D-Glcp-(1→4)-β-D-Glcp-(1
6 4 Me CO2H 6 4 Me CO2H
R. leguminosarum bv. phaseoli 127K36; LPR49;8002
R. leguminosarum bv. trifolii NA30;0403;LPRS;TA-1+;843 (a)
R. leguminosarum bv. viciae 128c53;128c63;8401 pRL1JI
2 R = α-D-Galp-(1→4)-β-D-GlcpA-(1→3)-β-D-Glcp-(1→4)-β-D-Glcp-(1→4)-β-D-Glcp-(1
6 4 Me CO2H
R. leguminosarum bv. phaseoli 127K44
3 R = β-D-Glcp-(1→6)-α-D-Galp-(1→4)-β-D-GlcpA-(1→4)-β-D-Glcp-(1→4)-β-D-Glcp-(1→4)-β-D-Glcp-(1
6 4 Me CO2H
R. leguminosarum bv. phaseoli 127K36
4 R = β-D-Galp-(1→6)-β-D-Galp-(1→6)-β-D-Glcp-(1→6)-α-D-Galp-(1→4)-β-D-GlcpA-(1→4)-β-D-Glcp-(1→4)-β-D-Glcp-(1
6 4 Me CO2H
R. leguminosarum bv. phaseoli 127K87
5 R = β-D-Glcp-(1→3)-β-D-Glcp-(1→4)-β-D-Glcp-(1
6 4 Me CO2H
R. leguminosarum bv. trifolii 4S (b)

Figure 8. Family of bacterial polysaccharide structures produced by certain *Rhizobium leguminosarum* species. The chemical structures were reported by McNeill, et al (ref. 60) except for (a) due to Hollingworth, et al. (ref. 67) and Kuo and Mort (ref. 68) and (b) due to Amemura, et al. (ref. 69).

though other group IIA cations may also suffice. Gelation of the EPS of bacteria growing at the surface of the root tip, due to calcium (or other divalent cations) present in the soil, or released from the plant cell wall, may aid the initial binding of the bacteria.

Recently similar gelation behavior has also been observed in a second family of bacterial polysaccharides secreted by other *Rhizobium* and certain *Agrobacterium* species.[62] Gelations of such EPS within soils would be important in controlling water retention in soils and could aid soil particle aggregation. Indeed there is strong evidence that the beneficial effects of ploughing-in leguminous crops,[63] known since BC times, may result from the action of the microorganisms associated with this plant material. Indeed there is direct evidence implicating soil bacteria and their EPS in binding silt particles together.[64-66] Thus gelation of the EPS may provide a mechanism to explain these observations. Whether such effects can be exploited remains to be determined.

ACKNOWLEDGMENTS

The author wishes to thank G. R. Chilvers and A. P. Gunning for providing experimental data prior to publication.

REFERENCES

1. G. T. Colegrave, Ind. Eng. Chem. Prod. Res. Dev., **22**, 456 (1983).
2. K. S. Kang, G. T. Veeder, P. J. Mirrasoul, T. Kanecko and I. W. Cottrell, Appl. Environ. Microbiol., **43**, 1086 (1982).
3. G. R. Sanderson and R. C. Clark, Food Technol., **37**, 63 (1983).
4. G. R. Chivers and V. J. Morris, Carbohydr. Polym., **7**, 111 (1987).
5. M. A. O'Neill, R. R. Selvendran and V. J. Morris, Carbohydr. Res., **124**, 123 (1983).
6. P. E. Jansson, B. Lindberg and P. A. Sandford, Carbohydr. Res., **124**, 135 (1983).
7. M. S. Kuo, A. J. Mort and A. Dell, Carbohydr. Res., **156**, 173 (1986).
8. R. Moorhouse, G. T. Colegrave, P. A. Sandford, J. K. Baird and K. S. Kang in: "*Solution Properties of Polysaccharides*," D. A. Brant, Ed., ACS Symp. Ser., **150**, 111 (1981).
9. M. J. Miles, V. J. Morris and M. A. O'Neill in: "*Gums and Stabilizers for the Food Industry: 2 Application of Hydrocolloids*," G. O. Phillips, D. J. Wedlock and P. A. Williams, Eds., Pergammon Press, Oxford, U. K., 1984, p. 485.
10. V. Carroll, G. R. Chilvers, D. Franklin, M. J. Miles, V. J. Morris and S. G. Ring, Carbohydr. Res., **114**, 181 (1983).
11. V. Carroll, M. J. Miles and V. J. Morris, Int. J. Biol. Macromol., **4**, 432 (1982).
12. P. T. Attwool, E. D. T. Atkins, C. Upstill, M. J. Miles and V. J. Morris in: "*Gums and Stabilizers for the Food Industry: 3*," G. O. Phillips, D. J. Wedlock and P. A. Williams, Eds., Elsevier Appl. Sci., London, U. K., 1986 p. 135.
13. P. T. Attwool, PhD Thesis, Bristol University (1987).
14. R. Chandrasekaran, L. C. Puigjaner, K. L. Joyce and S. Arnott, Carbohydr. Res., **181**, 23 (1988).
15. R. Chandrasekaran and V. G. Thailambal, Carbohydr. Polym., **12**, 431 (1990).
16. V. J. Morris in: "*Food Biotechnology*," R. D. King and P. S. J. Cheetham, Eds., Elsevier Appl. Sci., London, U. K., 1987, p. 193.
17. H. Grasdalen and O. Smidsrod, Carbohydr. Polym., **7**, 371 (1987).

18. V. Crescenzi, M. Dentini, T. Coviello and R. Rizzo, Carbohydr. Res., **149**, 425 (1986).
19. V. Crescenzi, M. Dentini and I. C. M. Dea, Carbohydr. Res., **160**, 283 (1987).
20. H. D. Chapman, G. R. Chilvers, M. J. Miles and V. J. Morris in: "*Gums and Stabilizers for the Food Industry: 4,*" G. O. Phillips, D. J. Wedlock and P. A. Williams, Eds., IRL Press, Oxford, U. K., 1988, p. 147.
21. M. Dentini, T. Covello, W. Burchard and V. Crescenzi, Macromol., **21**, 3312 (1988).
22. P. Denkinger, W. Burchard and M. Kung, J. Phys. Chem., **93**, 1428 (1989).
23. M. B. Huglin "*Light Scattering from Polymer Solutions,*" Academic Press, London, U. K., 1972.
24. M. Dubois, K. A. Gilles, J. K. Hamilton, P. A. Rebers and F. Smith, Anal. Chem., **28**, 350 (1956).
25. M. Schmidt, G. Paradossi and W. Burchard, Makromol. Chem. Rapid Commun., **6**, 767 (1985).
26. M. Hisamatsu, A. Amemura, K. Kaizumi, T. Utamura and Y. Okada, Carbohydr. Res., **121**, 31 (1988).
27. K. Koizumi, Y. Okada, S. Horiyama, T. Utamura, T. Higashiura and M. Ikeda, J. Inclusion Phenomena, **2**, 891 (1984).
28. A. Palleschi and V. Crescenzi, Gaz. Chem. Ital., **115**, 243 (1985).
29. L. P. T. M. Zevenhuizen, FEMS Microbiol. Letts., **35**, 43 (1986).
30. G. Medoff, J. Brajtburg, G. S. Kobayashi and J. Bolard, Ann. Rev. Pharmacol Toxicol., **23**, 303 (1983).
31. G. Medoff and G. S. Kobayashi, in: "*Antifungal Chemotherapy,*" D. C. E. Speller, Ed., Wiley, New York, USA., 1980, p. 3.
32. W. Martindale in: "*The Extra Pharmacopoeia,*" 27th Edition, Pharm. Press, London, U. K., 1973, p. 637.
33. C. P. Schaffner and E. Borowski, Antibiot. Chemother., **11**, 724 (1961).
34. W. Mechlinski and C. P. Schaffner, J. Antibiot., **25**, 256 (1972).
35. G. S. Kobayashi, J. R. Little and G. Medoff, Antimicrob. Agents Chemother., **27**, 302 (1985).
36. G. R. Keim Jr., P. L. Sibley, Y. H. Yoon, J. S. Kulesza, I. H. Zaidi, M. M. Miller and J. W. Poutsiaka, Antimicrob. Agents Chemother., **27**, 302 (1985).
37. M. Kajtar, M. Vilmom, E. Morlin and J. Szejtli, Biopolymers, **28**, 1585 (1989).
38. N. Rajagopalan, S. C. Chen and W.-S. Chow, Int. J. Pharmaceut., **29**, 161 (1986).
39. J. Szejtli, M. Vikmom, A. Stradler-Szoke, S. Piukovits, I. Intzefy, G. Kulcsar, M. Jaray and G. Zlatos, European Patent Application, 841163.71 (1984).
40. M. Vikmom, A. Stadler-Szoke and J. Szejtli, J. Antibiot., **38**, 1822 (1985).
41. M. Pidoux, J. M. Brillouet and B. Quenemer, Biotech. Letts., **10**, 415 (1988).
42. M. Pidoux, G. A. DeRuiter, B. E. Brooker, I. J. Colquhoun and V. J. Morris, Carbohydr. Polym., **13**, 351 (1990).
43. R. L. Sidebotham, Adv. Carbohydr. Chem. Biochem., **30**, 371 (1974).
44. R. H. Marchessault and Y. Deslandes, Carbohydr. Polym., **1**, 31 (1981).
45. P. E. Jansson, L. Kenne and B. Lindberg, Carbohydr. Res., **45**, 275 (1975).
46. L. D. Melton, L. Mindt, D. A. Rees and G. R. Sanderson, Carbohydr. Res., **46**, 245 (1976).
47. R. Moorhouse, M. D. Wilkinshaw and S. Arnott in: "*Extracellular Microbial Polysaccharides,*" P. A. Sandford and A. Laskin, Eds., ACS Symp. Ser., **45**, 90 (1977).
48. R. Moorhouse, M. D. Walkinshaw, W. T. Winter and S. Arnott in:

"*Cellulose Chemistry and Technology*," J. D. Arthur, Ed., p. 133 (1977).

49. K. Okwyama, S. Arnott, R. Moorhouse, M. D. Walkinshaw, E. D. T. Atkins and C. H. Wolf-Ullish in: "*Fiber Diffraction Methods*," A. D. French and K. D. Gardener, Eds., ACS Symp. Ser., 141, 411 (1980).

50. S. B. Ross-Murphy, V. J. Morris and E. R. Morris, Faraday Symp. Chem. Soc., 18, 7 (1983).

51. R. P. Millane and T. V. Narasaiah, Carbohydr. Polym., 12, 315 (1990).

52. R. P. Millane, T. V. Narasaiah and S. Arnott in: "*Recent Developments in Industrial Polysaccharides: Biomedical and Biotechnological Advances*," V. Crescenzi, I. C. M. Dea and S. S. Stirala, Eds., Gordon and Breach, New York, USA (in press).

53. R. O. Couso, L. Ielfi and M. Dankert, J. Gen. Microbiol., 133, 2123 (1987).

54. V. J. Morris, G. J. Brownsey, P. Cairns, G. R. Chilvers and M. J. Miles, Int. J. Biol. Macromol., 11, 326 (1989).

55. V. J. Morris, G. J. Brownsey, J. E. Harris, A. P. Gunning, B. J. H. Stevens and A. W. B. Johnston, Carbohydr. Res., 191, 315 (1989).

56. W. D. Bauer, T. V. Bhuvaneswari, A. J. Mort and G. Turgeon, Plant Physiol., 63, 5 (1979).

57. C. Napoli and P. Albersheim, J. Bacteriol., 141, 1454 (1980).

58. R. E. Sanders, R. W. Carlson and P. Albersheim, Nature, 271, 240 (1978).

59. B. K. Robertson, P. Aman, A. G. Darvill, M. McNeill and P. Albersheim, Plant Physiol., 67, 389 (1981).

60. M. McNeill, J. Darvil, A. G. Darvil, P. Albersheim, R. van Veen, P. Hooykaas, R. Schilperoort and A. Dell, Carbohydr. Res., 146, 307 (1986).

61. G. Smit. J. W. Kline and B. J. J. Lugtenberg, J. Bacteriol., 169, 4294 (1987).

62. V. J. Morris, G. J. Brownsey, A. P. Gunning and J. R. Harris, Carbohydr. Polym., (in press).

63. J. M. Lynch and E. Bragg, Adv. Soil Sci., 2, 133 (1985).

64. T. Santoro and G. Stotzky, Cand. J. Microbiol., 14, 299 (1968).

65. G. Stotzky and V. Bystricky, Bacteriol. Proc., A93 (1969).

66. R. C. Fehrman and R. W. Weaver, Soil Sci. Soc. Am. J., 42, 279 (1978).

67. R. I. Hollingsworth, F. B. Dazzo, K. Hallenga and B. Musselman, Carbohydr. Res., 172, 97 (1988).

68. M. S. Kuo and A. J. Mort, Carbohydr. Res., 145, 247 (1986).

69. A. Amemura, T. Harada, M. Abe and S. Higashi, Carbohydr. Res., 115, 165 (1983).

DEGREE OF SUBSTITUTION OF DEXTRAN MODIFIED THROUGH REACTION WITH

ORGANOSTANNANE CHLORIDES AND GROUP IV-B METALLOCENE DICHLORIDES

Charles E. Carraher, Jr.[a] and Yoshinobu Naoshima[b]

(a) Department of Chemistry
    Florida Atlantic University
    Boca Raton, FL 33431

(b) Department of Biological Chemistry
    Okayama University of Science
    Ridai-cho, Okayama 700, Japan

Dextrans represent a potential market of 2 x 10⁵ tons per year. Reaction with Groups IV-A and B organometallics show distinct trends with respect to loading or presence of organometallic moiety per repeat sugar unit. For Group IV-B metallocene dichlorides, loading can be easily varied from less than one metallocene per unit to full loading, presumably due to the lack of a tendency to form cyclic products. By comparison, loading of the organostannane halides is typically of the order of 0.25 to 0.5 per repeat unit. For the tin dihalides internal cyclic formation probably occurs to a major extent limiting loading. Use of selected PTAs allows loading of organostannanes to be increased to two tins per repeat unit.

INTRODUCTION

Saccharides are weight-wise the most abundant class of compounds upon the face of the earth. They form the basis for foods (sucrose, starch) and building (wood). Polysaccharides are one of three fundamental macromolecular building blocks of life (along with nucleic acids and proteins). They serve as structural materials, in energy storage and can offer specific biological properties.

Dextrans are α-(1→6) glucans produced by bacteria. They are largely linear with varying degrees of branching. Linear, water-soluble dextrans have many uses. One dextran is employed in viscous water-flooding for secondary recovery of petroleum, with a potential market of about 2 x 10⁵ tons per year. This dextran is superior to carboxymethyl cellulose when employed in high calcium drilling muds.[1,2] These dextran-muds show superior stability and performance at high pH and in saturated brines.[3-5] Other dextrans show good resistance to deterioration in soil and the ability to stabilize aggregates in soils.[6] They can also be used in binding collagen fibers into surgical sutures.[7] Water insoluble dextran flavored with fruit liquids constitute the Philippine dessert hata.[8] Dextran is employed in food processing based on their ability to:

> stabilize syrup confections against crystallization;
> stabilize texture in ice creams and sherbets;
> act as adhesives and humectants; and
> form viscous liquid sugars.[9]

Chemically, dextrans are similar to one another. The activation energy for acid hydrolysis is about 30-35 kcal/mole.[10] The C-2 hydroxyls appear to be the most reactive in most Lewis base and acid-type reactions. A wide variety of esters and ethers have been described as well as carbonates and xanthates.[11,12] In alkaline solution, dextran forms various complexes with a number of metal ions.[13]

We have modified polysaccharides employing a variety of Group IVA and B and Group VA organometallic dihalides.[14-25] The modified polysaccharides include those with two (xylan) and three (cellulose, dextran and amylose) hydroxyls per repeat ring unit. From such studies some generalities are emerging with respect to the limiting factors affecting amount of substitution.

Here the nature of the organometallic reactant will be emphasized. The lower limit of loading or substitution is achieved through use of limited amounts of the organometallic reactant. The following variables will be briefly considered: (a) nature of the organo moiety, (b) size of the metal, (c) nature of the metal and (d) number of halides.

The products consist of repeat units containing no substitution, one substitute at any of the hydroxyls, two substitutes, etc. These organometallic moieties may be internally cyclicized, form crosslinks (leading to formation of insoluble materials), or possess unreacted metal-halide groupings or end groups resulting from reaction with water or organic base.[26-32] Thus, only generalized structures are typically reported.

EXPERIMENTAL

Biscyclopentadienyltitanium dichloride (Aldrich, Milwaukee, WI), biscyclopentadienylzirconium dichloride (Aldrich, Milwaukee, WI), biscyclopentadienylhafnium dichloride (Alfa, Danvers); dextran (molecular weight 2 to 3 X 10^5, United States Biochemical Corporation, Cleveland, OH); dibutyltin dichloride (Fisher, Fairlawn, NJ); diphenyltin dichloride (Metallomer Labs., Maynard, MA); diethyltin dichloride (Ventron-Alfa Inorganics, Beverly, MA); tri-n-butyltin chloride (Alfa); tri-n-propyltin chloride (Alfa); tri-n-butyltin chloride (Aldrich, Milwaukee, WI); dioctyltin dichloride (Alfa); dilauryltin dichloride (Metallomer); dimethyltin dichloride (Alfa); tricyclohexyltin bromide (Alfa); triethylamine (Eastman); piperdine (Matheson, Coleman and Bell, Cincinnati, OH); N,N-dimethylaniline (Fisher); 3,5-latidine (Aldrich); and pyridine (J. T. Baker Chemical Co., Phillipsburg, NJ) were used as received.

Reactions were conducted using a one quart Kimex emulsifying jar placed on a Waring Blender (Model 7011 G) with a "no load" stirring rate of 18,000 rpm. The product is obtained by suction filtration. Repeated washings with the organic liquid and water-assisted in the purification of the product.

Elemental analyses for tin were conducted employing the usual wet analysis procedures except that one drop of water and concentrated nitric

Table 1. Fraction of substitution as a function of stirring time employing dibutyltin dichloride.				
Reaction Time (sec)	Yield (%)	Yield (g)	Sn (%)	#Sn/ Units
15	76	1.6	26	0.6
30	81	1.7	26	0.6
60	81	1.7		
120	76	1.6	26	0.6

acid (10 drops) are added subsequent to the sodium fusion and this mixture is heated until brown fumes are evolved. Also, the stannic sulfide is heated in air converting it to stannic oxide which is weighed and this weight utilized in determining the percentage tin. This process is not suitable for volatile organostannanes such as dibutyltin dichloride, but is suitable for tin polyesters, polyethers, and polyamines. Tin polyesters were utilized as control compounds. Thus the polyester derived from disodium adipate and dibutyltin dichloride has been found to contain 31% tin with the calculated value being 31%. Analysis for titanium was accomplished by treating the samples with $HClO_4$ and then heating the samples until a white product (titanium dioxide) was formed. Other analysis procedures and results are given in our previous papers.[14-25]

RESULTS AND DISCUSSION

For the organostannane dibutyltin dichloride reaction is rapid and essentially complete after 15 seconds with the degree of substitution being about 0.6 dibutyltin moieties per repeat sugar unit (Table 1). Dextran (4.00 mmole) and sodium hydroxide (12.0 mmole), in 40 mL of water, was added to a stirred (18,000 rpm, no load) $CHCl_3$ solution containing dibutylin dichloride (6.00 mmole) at 25°C.

The corresponding reaction with the titanocene dichloride gives almost complete substitution with the number of Cp_2Ti units per sugar unit being three (Table 2). Reaction conditions - Dextran (2.00 mmole) and TEA (6.00 mmole), dissolved in 50 mL water, were added to a stirred (18,500 rpm, no load) solution of Cp_2TiCl_2 (3.00 mmole) in 50 mL $CHCl_3$ at 25°C.

The reaction with dibutyltin dichloride was accomplished at varying stirring rates. Percentage yield and amount of substitution are low for low stirring rates but achieve a plateau around a stirring rate of 10,000

Table 2. Fraction of substitution as a function of stirring time.				
Stirring Time (sec)	Yield (%)	Yield (g)	Ti (%)	#Ti/ Units
10	0	0	–	
15	7	0.05	20	3
30	12	0.09	19	3
45	4	0.03	19	3

Table 3. Fraction of substitution as a function of stirring
rate employing dibutyltin dichloride.

Stirring Rate (rpm-no load)	Yield (%)	Yield (g)	Sn (%)	#Sn/ Unit
4,900	42	0.9	9.9	0.2
7,350	42	0.0	23	0.5
11,600	85	1.8	24	0.5
15,500	85	1.8	24	0.5
18,000	91	1.7	23	0.5
19,500	85	1.8	23	0.5

rpm where the number of dibutyltin units per unit is about one half
(Table 3). Dextran (4.00 mmole) and sodium hydroxide (12.0 mmole) in 40
mL of water are added to stirred CCl_4 (40 mL) solutions containing dibutyltin dichloride (6.00 mmole) at 25°C. for a stirring time of 30
seconds.

The amount of organotin included in the modified dextran appears to
be somewhat independent of the size of the organo-substituents (Table 4)
with the order of substitution being R = C_8H_{17} > CH_3 > C_4H_9 > $C_{12}H_{23}$ with
all of the values for organotin substitution being within the range of
0.9 to 0.5. The dextran and sodium hydroxide ratio was held constant at
3:1 NaOH:dextran unit in 40 mL of water added to stirred (18,000 rpm)
chloroform (40 mL) solutions containing the organotin halide at 25°C for
a stirring time of 30 secs.

For the Group IV-B metallocene dichlorides degree or fraction of
substitution increases as the ratio of metallocene to hydroxyl increases
with substitution increasing from about one metallocene per ring to three
metallocene's per unit (Table 5). This approach to approximately total
substitution appears independent of the Group IV-B metal and is mainly
dependent on the ratio of metallocene to available hydroxyls.

A similar study was made except employing dibutylin dichloride where
the ratio of organostannane and dextran were varied (Table 6). While
yield varied greatly, the degree of substitution remained approximately
constant corresponding to 0.3 to 0.5 dibutyltin moieties per sugar unit.
As noted before, for the Group IV B metallocene dichlorides, the trends
with respect to degree of organometallic inclusion is not greatly depen-

Table 4. Results as a function of organostannane.

Organotin Halide	Organotin (mmole)	Dextran (mmole)	Yield (%)	Sn (%)	#Sn/ Unit
Ph_3SnCl	12	4	95	14	0.3
$(C_6H_5CH_2)_3SnCl$	12	4	13	14	0.3
$(C_6H_{11})_3Sn\ Br$	6	2	11	16	0.4
Ph_2SnCl_2	6	4	77	12	0.25
Me_2SnCl_2	6	4	65	33	0.7
Bu_2SnCl_2	6	4	81	15	0.5
$(C_{12}H_{25})_2SnCl_2$	6	4	76	15	0.5
$(C_8H_{17})_2SnCl_2$	6	4	92	23	0.9

Table 5. Percentage metal content as a function of monomer reactant for the reaction of Cp_2MCl_2 with dextran.

Cp_2MCl_2 (mmole)	Dextran (mmole)	Ti (%)	#Ti/ Unit	Zr (%)	#Zr/ Unit	Hf (%)	#Hf/ Unit
0.5	3.0	7	0.3	-		-	
1.0	3.0	9	0.5	20	0.7	32	0.7
2.0	3.0	12	0.9	24	1.	31	0.7
3.0	3.0	21	3.	30	2.	37	1.
4.0	3.0	18	2.	33	3.	48	3.
5.0	3.0	-				49	3.
6.0	3.0	-				49	3.
3.0	0.5	21	3.				
3.0	1.0	15	1.5			49	3.
3.0	2.0	18	2.	30	2.	39	1.
3.0	3.0	21	3.	30	2.	37	1.
3.0	4.0	17	1.8	21	0.8	37	1.

dent on the nature of the metal though the crystal ionic radii do vary by about 15% [Ti(+4) = 0.68A and Zr(+4) and Hf(+4) = 0.78 A]. Dextran and sodium hydroxide (in a ratio of 1:3 moles NaOH/unit dextran) in 40 mL of water was added to stirred solutions (18,000 rpm) containing dibutyltin dichloride at 25°C for a stirring time of 30 secs.

Dextran modification was also studied as a function of the type of solution condensation process employed.[16] The previously described results are all for modifications carried out employing the classical aqueous interfacial process. For a two solvent solution system (H_2O/DMSO) the percentage of titanium was 13% corresponding to about one titanocene moiety per repeat sugar unit. For the analogous system except employing dibutyltin dichloride, the percentage tin was 17% or about 0.4 organotin moieties per repeat unit. For aqueous solution systems employing titanocene dichloride, the percentage of titanium varied from 15 to 21%, similar to the corresponding classical interfacial systems. The inverse interfacial systems with titanocene dichloride had 17% titanium content. The nonaqueous interfacial product had a tin content of 16 to 21%. Thus, for each of these varying systems, the metallic-moiety content remained

Table 6. Results as a function of dibutyltin dichloride to dextran molar ratios.

Bu_2SnCl_2 (mmole)	Dextran (mmole)	Yield (%)	Sn (%)	#Sn/ Unit
12	4	1	-	
8	4	5	-	
6	4	81	23	0.5
2	4	55	20	0.5
1	4	45	14	0.3
6	12	45	22	0.5
6	8	56	17	0.4
6	4	81	23	0.5
6	2	0	0	
6	1	0	0	

approximately constant. (The ratio of dextran to organometallic dihalide was held constant in these studies - titanocene dichloride to 1 mmole to 1 mmole of dextran and for dibutyltin dichloride 4 mmole to 6 mmole of dextran.)

An extensive study was carried out employing PTAs. PTAs may well affect the abundance and rate of arrival within the "reaction zone" of the various organometallics. This is indicated in the case of reactions involving dibutyltin dichloride where the percentages of tin reached 37% for selected PTAs consistent with the presence of two dibutyltin moieties per repeat unit.[23]

The reaction between dibutyltin dichloride and dextran was carried out under a wide variety of reaction conditions (for instance,[15,19,24]) but in almost all cases the percentage tin was within the range of 23 to 26% corresponding to about one half dibutyltin moiety per each repeat sugar unit.[19] These reactions were carried out employing 6 mmoles of dibutyltin dichloride and 4 mmoles of dextran (with 12 mmoles of sodium hydroxide). Only at very slow stirring rates (ca. 5,000 rpm) was the tin content significantly different (ca. 10% Sn). It is believed that for reactions run below about 10^4 rpm precipitation probably occurs before optimum substitution can occur. Higher substitution was only achieved employing selected PTAs, but even here full substitution was not achieved.It is probable that formation of the cyclic five-membered ring occurs leading to the functionality of the repeat sugar unit being two instead of three. The formation of such a five-membered unit is well established and is employed in sugar chemistry for the protection of the hydroxyls (for instance,[33-39]).

Alternately, the fraction of metallocene substitution is easily and widely variable. Further, full substitution is achieved. The difference in bond angle and length for the metallocene probably minimizes cyclic formation allowing full substitution to occur.

REFERENCES

1. P. Monaghan and J. Gidley, Oil Gas J., **57**, 100 (1959).
2. G. Dumbauld and P. Monagham, U.S. Pat. 3,065,170 (1962).
3. E. Mueller, Z. Angew. Geol., **9**, 213 (1963).
4. W. Owen, Sugar, **47** (7), 50 (1952).
5. W. Owen, U.S. Pat. 2,602,082 (1952).
6. L. Novak, E. Witt and M. Hiler, Agr. Food Chem., **3**, 1028 (1955).
7. L. Novak, U.S. Pat. 2,748,774 (1956).
8. F. Adriano, J. Oliverso and E. Villanueva, Philippine J. Ed., **16**, 373 (1933).
9. D. Wadsworth and M. Hughes, U.S. Pat. 2,409,816 (1946).
10. E. Antonini, L. Bellelli, M. Bruzzesi, A. Caputo, E. Chiancone and A. Rossi-Fanelli, Biopolymers, **2**, 35 (1964).
11. P. J. Baker, "*Dextrans,*" Academic Press, N.Y. 1959.
12. P. Flodin, "*Dextran Gels and Their Applications in Gel Filtration,*" Halmstead, Uppsala, 1962.
13. C. E. Rowe, Ph.D. Thesis, Univ. Birmingham, 1956.
14. C. Carraher, D. Giron, J. Schroeder and C. McNeely, U. S. Patent 4,312,981 Jan. 26, 1982.
15. C. Carraher, T. Gehrke, D. Giron, D. Cerutis and H. Molloy, J. Macromol. Sci-Chem., **A19**, 1121 (1983).
16. Y. Naoshima, C. Carraher, T. Gehrke, M. Kurokawa and D. Blair, J. Macromol. Sci.-Chem., **A23**;861 (1986).
17. C. Carraher, W. Burt, D. Giron, J. A. Schroeder, M. L. Taylor, H. M.

Molloy, and T. Tiernan, J. Applied Polymer Sc., **28**, 1919 (1983).

18. Y. Naoshima, S. Hirono and C. Carraher, J. Polymer Materials, 2, 43 (1985).

19. C. Carraher and T. Gehrke, in: "*Modification of Polymers,*" C. Carraher and J. Moore, Eds., Plenum, NY, 1983, Chapter 18.

20. Y. Naoshima, C. Carraher and K. Matsumoto, in: "*Renewable Resource Materials,*" C. Carraher and L. Sperling, Eds., Plenum, NY, 1986.

21. C. Carraher, C. Butler, Y. Naoshima, V. Foster, D. Giron and P. Mykytiuk, in: "*Applied Bioactive Polymeric Materials,*" C. G. Gebelein, C. E. Carraher and V. Foster, Eds., Plenum, NY, 1988, p. 198.

22. Y. Naoshima, H. Shudo, M. Uenishi and C. Carraher, J. Macromol. Sci.-Chem., **A25**, 895 (1988).

23. Y. Naoshima, C. Carraher, S. Iwamoto and H. Shudo, Applied Organo-metalic Chem., **1**, 245 (1987).

24. C. Carraher and T. Gehrke, in: "*Polymer Applications of Renewable-Resource Materials,*" C. Carraher and L. Sperling, Eds., Plenum, NY, 1983, Chapter 4.

25. Y. Naoshima, C. Carraher, G. Hess and M. Kurokawa, in: "*Metal-Containing Polymeric Systems,*" J. Sheats, C. Carraher and C. Pittman, Eds., Plenum, NY, 1985.

26. C. Anchisi, A. Maccioni, A. M. Maccioni and G. Podda, Gazzetta Chimica Italiana, **113**, 73 (1983).

27. R. M. Munavu and H. H. Szmant, J. Org. Chem., **41**, 1832 (1976).

28. Y. Tsuda, M. E. Haque and K. Yoshimoto, Chem. Pharm. Bull., **31**, 1612 (1983).

29. T. Ogaea and M. Matsui, Carbohydr. Res., **56** C1 (1977).

30. D. Wagner, J. P. H. Verheyden and J. G. Moffatt, J. Org. Chem., **39**, 24 (1974).

31. M. A. Nashed and L. Anderson, Tetrahedron Lett., 3503 (1976).

32. M. A. Nashed, Carbohydr. Res., **60**, 200 (1978).

33. R. M. Munavu and H. H. Szmant, J. Org. Chem., **41**, 1832 (1976)

34. D. Wagner, J. Verheyden, and J. Moffatt, J. Org. Chem., *Ibid*, 39, 39(1), 24 (1974).

35. M. Nashed and L. Anderson, Tetrahedron Lett., **39**, 3503 (1976).

36. M. Nashed, Carbohydrate Res., **60**, 200 (1978).

37. T. Ogawa and M. Matsui, *Ibid*, 56, C1 (1977).

38. Y. Tsuda, M. Hague and K. Yoshimoto, Chem. Pharm. Bull., **31**(5), 1612 (1983).

39. C. Anchisi, A. Maccioni and G. Podda, Gazz. Chim. Ital., **113**, 73 (1983).

ACIDIC POLYSACCHARIDES: THEIR MODIFICATION AND POTENTIAL USES

Robert J. Linhardt, Ali Al-Hakim and Jian Liu

Division of Medicinal and Natural Products Chemistry
College of Pharmacy
University of Iowa
Iowa City, IA   52242

Acidic polysaccharides are known for their wide range of biological activities. These natural products are obtained from animal sources (i.e., glycosaminoglycans) or are prepared by the chemical sulfation of naturally occurring microbial and plant polysaccharides. For medical applications, heparin is the most commonly used acidic polysaccharide.

Heparin is a polydisperse, highly sulfated, linear polysaccharide of repeating uronic acid and glucosamine residues. Although heparin has been used clinically as an anticoagulant for the past 50 years, its precise structure remains unknown. Heparin's primary application is as an anticoagulant, however, it is more appropriate to consider heparin as a polyelectrolytic drug having a multiplicity of biological activities. Virtually any cationic protein and many anionic proteins are capable of binding to heparin under physiological conditions. Within the past decade, a growing number of biological activities have been demonstrated to be regulated by heparin, ranging from its effect on angiogenesis to the regulation of the immune response. Low molecular weight heparins have been undergoing clinical trials as antithrombotic agents for use in a wide variety of diseases ranging from deep vein thrombosis to non-hemorrhagic stroke. This chapter discusses new agents including natural products chemically or enzymatically derived from heparin, synthetic acidic oligosaccharides and polysaccharides, and heparin covalently immobilized to polymers.

INTRODUCTION

Glycosaminoglycans are a group of highly sulfated acidic polysaccharides.[1] One or more polysaccharide chains are typically found attached to a protein core rich in serine-glycine repeating sequences. Unlike the highly branched oligosaccharides found in glycoproteins, these glycosaminoglycans are linear (Figure 1). Although these linear polysaccharide are typically O-linked to a serine residue in their core protein, they

Biotechnology and Polymers, Edited by C.G. Gebelein
Plenum Press, New York, 1991

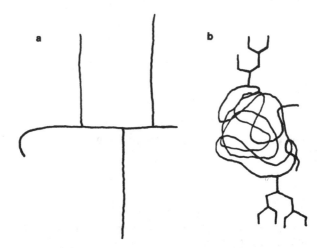

Figure 1. Proteoglycan and glycoprotein structures. Represen-
tations are shown of linear glycosaminoglycan
chains extending from the protein core of a proteoglycan (a) and the branched oligosaccharides of a
glycoprotein (b).

are also found N-linked similar to glycoprotein oligosaccharides.[2]

There are several major classes of glycosaminoglycans.[1] These include
hyaluronic acid, chondroitin sulfates A and C, dermatan sulfate, heparan
sulfate, heparin and keratan sulfate. The structure of the major repeating disaccharide unit in each glycosaminoglycan is shown in Figure 2.
Hyaluronic acid contains no sulfate and is not found linked to a core
protein.[2] Its structure is (→3)-β-D-N-acetylglucosamine-(1→4)β-D-glucuro-
nic acid (1→). Chondroitin sulfates are another class of PGs which can be
broken down into two major classes: chondroitin sulfate A, (→3)-β-D-N-
acetylgalactosamine-4 sulfate-(1→4)-β-D-glucuronic acid (1→); chondroitin
sulfate C, (→3)-β-D-N-acetylgalactosamine-6 sulfate (1→4)-β-D-glucuronic
acid (1→); Dermatan sulfate B is also a galactosaminoglycan and is primarily a polymer of (→3)-β-D-N-acetylgalactosamine-4 sulfate (1→4)-α-L-

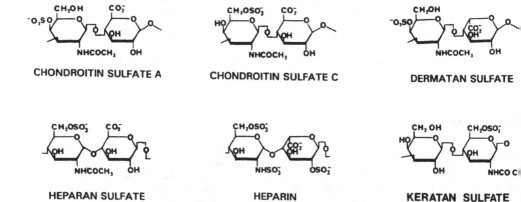

CHONDROITIN SULFATE A

CHONDROITIN SULFATE C

DERMATAN SULFATE

HEPARAN SULFATE

HEPARIN

KERATAN SULFATE

Figure 2. Major repeating disaccharide units found in various
glycosaminoglycans.

iduronic acid (1→).[2] Heparin and heparan sulfate are structurally similar glucosaminoglycans[3] comprised of(→4)-β-D-N-acetyl or N-sulfated glucosamine (6-O-sulfation is possible), (1→4)-α-L-iduronic acid or β-D-glucuronic acid (with possible 2-sulfation) (1→). Keratan sulfate, the final class of glycosaminoglycans, has a major disaccharide repeating unit: (→3) β-D-galactose (1→4)-β-D-N-acetylglucosamine-6 sulfate (1→).[2]

Although heparin's primary application is as an anticoagulant, it is more appropriately considered a polyelectrolytic drug having a multiplicity of biological activities.[4,5] Most cationic proteins (pI > 7), and many anionic proteins, are capable of binding to heparin under physiological conditions. In the past decade, a number of new biological activities have been demonstrated to be regulated by heparin.[1,6] These range from its effect on angiogenesis to the regulation of the immune response.[7,8] Low molecular weight heparins, heparin oligosaccharides and other glycosaminoglycans have been undergoing clinical trials as anticoagulant/antithrombotic agents for use in a wide variety of disease states ranging from deep vein thrombosis to non-hemorrhagic stroke.[1,9,10-12] These agents include both natural products, chemically or enzymatically derived from heparin, as well as fully synthetic acidic oligosaccharides and polysaccharides. This chapter focuses on the enzymatic preparation of oligosaccharides from heparin and other glycosaminoglycans. Although they have been primarily studied in their soluble form, methods are being developed to incorporate these oligosaccharides into synthetic polymers.

EXPERIMENTAL

1. Materials

Heparin, sodium salt, from porcine intestinal mucosa (approximately 150 U/mg) was obtained from Hepar (Franklin, OH). Other glycosaminoglycans including heparan sulfate, chondroitin sulfates A and C, dermatan sulfate and hyaluronic acid (also sodium salts) were from Sigma (St. Louis, MO). Raw (unbleached) heparin, in which approximately 10% of the polysaccharide chains contain peptide bonded covalently at their reducing end, was either prepared in our laboratory from porcine intestinal mucosa or purchased from Sigma. Lyase enzymes including heparinase, heparinase II, heparitinase, and chondroitinases A, AC and B were prepared in our laboratory.[12-14] These enzymes, as well as chondroitinase ABC and hyaluronidase, are also commercially available from Sigma and Seikagaku Kogyo America (Rockville, MD).[15] Reactigel (HW-65F), a polyvinyl alcohol based beaded resin pre-activated with 1,1'-carbonyliimidizole was from Pierce (Rockford, IL). Sulfopropyl Sephadex C-50 was from Sigma. Controlled pore dialysis bags were from Spectrum Medical (Los Angeles, CA).

2. Methods

2A. Lyase Catalyzed Depolymerization of Glycosaminoglycan

Heparin was prepared in 5 mM sodium phosphate buffer (pH 7.0) containing 150 mM sodium chloride at 10 mg/mL. To 80 μl of heparin, 20 μl of heparin lyase (0.03 IU in the same buffer) was added. The reaction mixture was incubated overnight at 30°C. After the reaction was complete, protein was removed from the reaction mixture by adjusting its pH to 3.0 and passing this solution through a 1 mL column of sulfopropyl-Sephadex adjusted with 30 mM hydrochloric acid to the same pH. The material failing to bind to this support was collected and its pH adjusted to 7.0 with sodium hydroxide solution. The oligosaccharide products could be desalted by exhaustive dialysis using 500 or 1000 molecular weight cut-off bags. The resulting salt-free solution of heparin oligosaccharides were concentrated by freeze-drying and reconstituted in distilled water. Oligosaccharides derived from other glycosaminoglycans were similarly prepared by depolymerization using the appropriate enzyme, reaction time and temperatures.[14,15]

2B. Fractionation of Acidic Oligosaccharides

Two methods were used to fractionate oligosaccharide mixtures, preparative strong anion-exchange high pressure liquid chromatography[16] and preparative gradient polyacrylamide gel electrophoresis.[17] These methods were used to prepare multimilligram quantities of homogeneous oligosaccharides for structural characterization, further chemical or enzymatic modification and ultimately for chemical coupling to polymer surfaces.[18-21]

2C. Covalent Binding of Peptide-Containing Glycosaminoglycans

Raw, unbleached, heparin was prepared at 100 mg/mL in 0.1 M sodium borate buffer at pH 8.5. To this heparin, 2 mL of activated beaded resin (activation level: >50 μmoles/mL gel) was added and the 8 mL of suspension was shaken at 25°C for 24 h. After the reaction was complete, the beads were washed two times with 3 volumes of 6 M urea followed by two 3 volume washes with distilled water, two 3 volume washes with 2M sodium chloride and finally by two 3 volume washes with distilled water. After resuspension in distilled water the content of immobilized heparin was measured by dye binding assay.[22]

RESULTS AND DISCUSSION

1. Heparin's Biological Activities

The biological activities of heparin and other glycosaminoglycans are primarily mediated through their binding to proteins and subsequent regulation of their activities. To be effective, substitute polyanions must bind to these proteins and regulate these same activities. Heparin's binding capability is primarily through electrostatic interactions and depends on its high charge density. Most heparin substitutes that have been proposed are highly sulfated polyanions, prepared enzymatically or chemically, and can take the place of heparin by the positioning of these charged groups. Ideally, these heparin substitutes should exhibit tighter binding to these proteins to have higher potency. Only the antithrombin III (ATIII) binding site (Figure 3) in heparin has been sufficiently studied to develop a well defined structure activity relationship. A synthetic heparin pentasaccharide containing this binding site has an ATIII binding constant that is comparable to heparin's.[23]

In addition to its anticoagulant activity, heparin exhibits a number of side-effects. Not all of heparin's side-effects, however, are undesirable. Some of these side-effects might be exploited and new pharmacologically active agents prepared. For example, heparin releases and activates lipoprotein lipase (LPL) but does so only at concentrations that are fully anticoagulating.[24] If a heparin were prepared devoid of this anti-

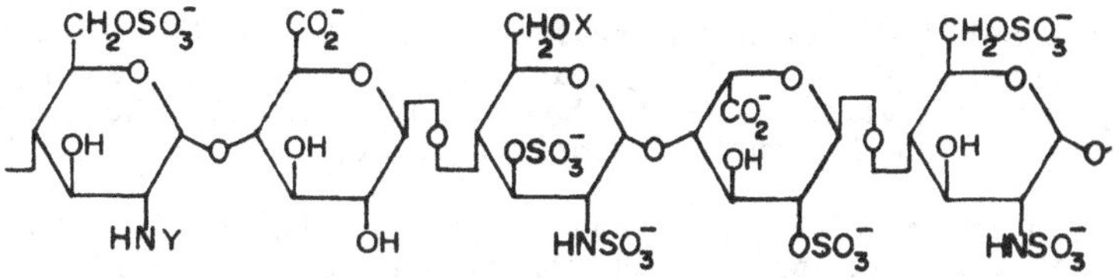

Figure 3. Heparin's antithrombin III-binding site. A pentasaccharide sequence at which antithrombin III can
tightly bind is shown where X = H or SO_3^- and Y =
SO_3^- or $COCH_3$.

coagulant activity, but with high LPL releasing activity, it might represent a useful agent in the treatment of atherosclerosis. Heparin inhibits
complement activation but only at concentrations much greater than those
required for full anticoagulation.[8] Recent results in our laboratory
demonstrate that it is possible to prepare oligosaccharides from heparin
that are equipotent with heparin (on a weight basis) in inhibiting complement activation but without anticoagulant activity.[8,21] Such a drug
might be very useful in preventing complement activation in extracorporeal therapy. There are scores of other heparin side-effects that might
be usefully exploited resulting in the preparation of new classes of
therapeutic agents.

Hemostasis is the "spontaneous arrest of bleeding from ruptured blood
vessels".[25] This broad physiological process includes the blood coagulation system and involves plasma coagulation factors, platelets, mono-
cytes, and endothelial cells that line the blood vessels. The coagulation
cascade consists of a sequence of reactions in which protease precursors
are converted from enzymatically inactive to enzymatically active forms.
In the final stages of the coagulation cascade, fibrinogen is converted
by thrombin (factor IIa) into the spontaneously polymerizable fibrin
monomer. Polymerization and subsequent crosslinking of fibrin monomers
produces gelatinous fibers which enmesh platelets forming a primary hemostatic plug.

Heparin catalyzed anticoagulation is primarily attributable to its
binding to antithrombin III (ATIII), a serine protease inhibitor. On
binding to heparin, ATIII undergoes a conformational change that enhances
its activity as serine protease inhibitor. Thrombin, a serine protease
then binds to the heparin-ATIII complex, acts on ATIII, and is irrevers-
ibly inactivated. Thus the conversion of fibrinogen to fibrin clot is
blocked and the blood is anticoagulated. Heparin also inhibits coagulation and thrombosis through a number of other mechanisms the details of
which are beyond the scope of this chapter.[1]

2. Polymers As Heparin Substitutes

Polymers that are structurally related to heparin and that possess
certain of its biological properties, such as anticoagulant activity, are
commonly called heparinoids. Non-heparin glycosaminoglycans having anti-
coagulant/antithrombotic activities are sometimes classified as heparinoids. Heparinoids are also prepared by modification of naturally occurring polysaccharides, by the total synthesis of heparin-like polymers,
and most recently by the synthesis of small sulfated heparin-like oligomers.

Dermatan sulfate and heparan sulfate have been used in animal studies and in clinical studies in Europe as antithrombotic agents.[10,26,27] Other glycosaminoglycans, such as hyaluronic acid and chondroitin sulfates, may also have low antithrombotic activity.[10]

Chitin, a major organic component of the exoskeleton of insects,[2] can be de-N-acetylated to prepare chitosan. On chemical sulfation and/or carboxymethylation, chitosan affords a polyanion with certain structural and activity similarities to heparin.[28]

Pentosan, extracted from the bark of the beech tree, can be sulfated by chemical methods resulting in an anticoagulant with one-tenth of heparin's activity on a weight basis.[29,30]

Dextran, a branched glycan polymer, can be chemically sulfated to prepare dextran sulfate,[31] having low anticoagulant activity.[2] Dextran sulfate has been used as a heparin replacement in anticoagulation and has recently been immobilized on plastic tubes to prepare non-thrombogenic surfaces.[32]

Synthetic polymers such as poly(vinyl sulfate) and poly (anethole sulfonate) are highly charged heparin-like polyanions that also exhibit anticoagulant activity.[33] However, these agents are not used *in vivo* as they are resistant to metabolism and thus remain in the body for extended periods resulting in toxic side effects.[1,33]

3. Defined Oligosaccharides as Heparin Substitutes

A synthetic 3-O-sulfated pentasaccharide, representing heparin's ATIII binding site (Figure 3) was first prepared by Choay, et.al., in a multi-step synthesis.[34] Clinical studies on this pentasaccharide, as an antithrombotic agent, demonstrated that it was not as effective as heparin itself and its cost probably precludes its use as a therapeutic agent.[35]

Heparin-oligosaccharides of defined structure have been prepared by our laboratory using lyases (Figure 4) as described in the experimental section. These heparin oligosaccharides have important biological activities including complement inhibitory activity with an *in vitro* potency nearly equal to heparin on a weight basis.[21] Methods are being developed

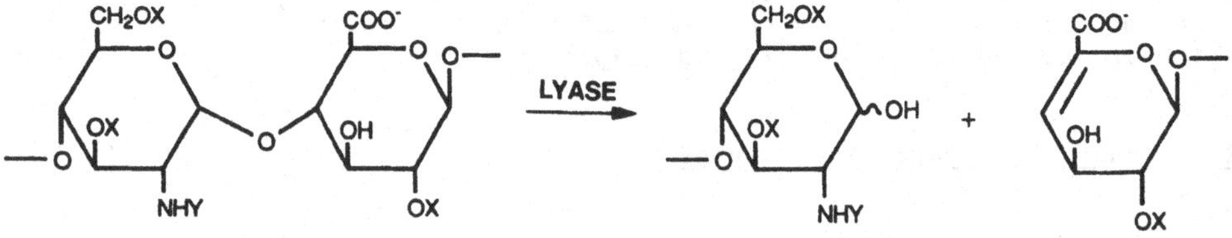

Figure 4. Depolymerization of heparin using polysaccharide lyase. The glycosidic linkage between a substituted glucosamine (X = H or SO_3^-, Y = SO_3^- or CH_3CO) and uronic acid (iduronic or glucuronic, X = SO_3^- or H) is cleaved eliminatively affording oligosaccharide products. The non-reducing end of each product formed contains a Δ-4,5 unsaturated uronic acid residue.

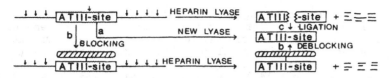

Figure 5. Strategies to prepare homogeneous oligosaccharides
having an intact antithrombin III-binding site. (a)
Heparin is depolymerized using a new highly specific lyase that cuts only glycosidic linkages outside
the ATIII-binding site. (b) The ATIII-binding site
in heparin is blocked while heparinase depolymer-
izes the surrounding heparin chain. The blocking
agent is then removed. (c) Two partial ATIII-
binding sites are ligated by reverse enzymic catalysis using heparinases to form an intact ATIII-
binding site.

for the large-scale and inexpensive preparation of these heparin oligosaccharides.[16,17]

The ATIII-binding site contains an unique arrangement of five different saccharide units. Because this binding site contains such a wide
variety of saccharide types; no method of chemical depolymerization has
yet been devised that is capable of excising this ATIII-binding site from
a heparin chain without breaking it. Enzymatic methods using lyases also
cleave heparin's ATIII-binding site.[14,16] Multistep, low-yield chemical
synthesis remains the only reliable method to prepare heparin oligosaccharides containing intact ATIII-binding sites. Herein lies the problem:
to prepare in a few steps, and in high yield and high purity, oligosaccharides having intact ATIII-binding sites. We are examining three
methods to prepare these ATIII-binding sites (Figure 5).

The first method (Figure 5a) uses a lyase to cleave selectively at
(→4)-β-D-N-sulfated glucosamine 6-sulfate (1→4)-α-L-iduronic acid 2-
sulfate (1→) linkages occurring only outside the ATIII-binding site. An
ATIII-binding octasaccharide from porcine heparin (or hexasaccharide from
bovine heparin) could be prepared using this lyase in a single step in
high yield. Of the lyases that we have characterized, heparin lyase comes
closest to meeting this specificity requirement; but its specificity also
permits it to cut the (→4)-β-D-N-sulfated glucosamine 3,6-disulfate-
(1→4)-α-L-iduronic acid 2-sulfate (1→) linkage, located in the center of
the ATIII-binding site. This linkage, containing an additional 3-sulfate
group, seems to be cut preferentially, probably because of a reduced Km
of the enzyme for this linkage.[16] Reaction conditions are being examined
that decrease the action of heparin lyase at this linkage within the
ATIII-binding site. The specificity of new lyases are being examined to
see if these can cleave heparin while leaving the ATIII-binding site
intact.

The second method (Figure 5b) approaches this problem by selectively
blocking the action of heparin lyase at the ATIII-binding site. ATIII (Kd
~10⁻⁷ M) has been used successfully to compete with heparin lyase (Km
~10⁻⁸ M) for heparin's ATIII-binding site. ATIII non-covalently binds to
heparin, blocking lyase action at this site, and affording oligosaccharides with intact ATIII-binding sites.[36]

The third approach (Figure 5c) involves the enzymatic or chemical
ligation of a broken ATIII-binding site. There is no reported reverse

161

enzymic catalysis utilizing polysaccharide lyases. Such an approach, however, if successful, might be extremely useful in re-assembling oligosaccharides into acidic polysaccharides of defined structure and with a variety of desirable biological properties. Our laboratory prepared a hexasaccharide, using heparin lyase, in a single step (Figure 4) from porcine mucosal heparin.[18,37] Approximately 95% of this hexasaccharide is recovered and it represents about 5-10% of heparin's mass. Based on the principle of microscopic reversibility, heparin lyase should catalytically re-form an appropriate glycosidic linkage between two hexasaccharides (each containing a portion of heparin's ATIII binding site). The result would be a single intact ATIII-binding site.

4. Preparing Stable Blood Compatible Surfaces

The major problem associated with the design and preparation of blood compatible surfaces is the lack of a complete understanding of coagulation and thrombosis. The ideal blood compatible polymer should mimic blood's natural container, the vessel lined with endothelial cells, as closely as possible. The surface should be stable and survive enzymatic and chemical attack from the components present in the circulation. The vascular endothelium is lined with heparan sulfate, a heparin-like molecule containing ATIII-binding sites.[1] The production of stable heparinized, antithrombotic surfaces, however, is very difficult. Simple adsorption of heparin onto a polymer produces a blood compatible surface that only lasts a short period of time until the heparin leaches from the surface.[38,39] Covalent immobilization should offer an alternative, chemically stable linkage. Many of these surfaces, however, also have a short lifetime possibly due to the enzymatic stripping of heparin from the surface. Enzymes that act on heparin, including both exoglycuronidases and endoglycuronidases are present in the circulation.[40-42] The precise

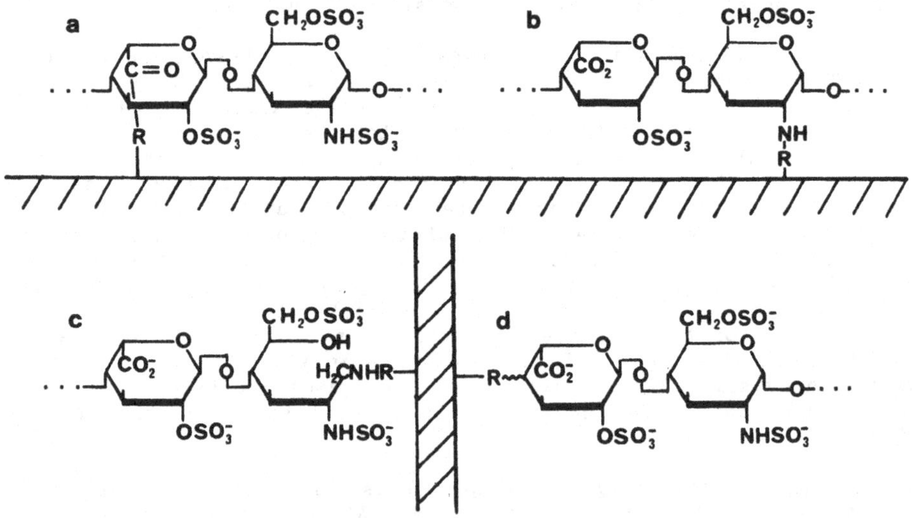

Figure 6. Orientations of immobilized heparin chains on the surface of a polymer. Possible orientations include coupling through: the center of a heparin chain utilizing (a) carboxyl or (b) amino functionality; (c) heparin's reducing-end by reductive amination or through a peptide linkage; and (d) heparin's non-reducing end.

nature of the surface-heparin linkage as well as its orientation (Figure 6) may control enzyme access to immobilized heparin chains resulting in stable heparinized surfaces.

Our laboratory is examining the question of how the orientation of a heparin chain that is immobilized to a surface (i.e., coupled through either its reducing-end or its non-reducing end), affects its linkage stability when exposed to enzymes in the circulation (Figure 6). The first of these defined heparinized surfaces have been prepared by coupling heparin at its reducing end to a synthetic polymer. The coupling chemistry relies on residual peptide present in raw (unbleached) heparin. This heparin can be immobilized with a high loading (1 mg heparin/mL of swelled polymer bead). A peptide linkage is used to assure the chemical stability of this heparinized surface. Studies of the *in vitro* stability in blood as well as the *in vivo* stability of this heparinized surface are currently underway. Surfaces with heparin (or oligosaccharides prepared from heparin using lyases) bound in other orientations (Figure 6) are also being prepared for testing.

Finally, it may be possible to incorporate heparin directly into the backbone of a synthetic polymer. The approach makes use of heparin oligosaccharides prepared using a lyase (Figure 4). These oligosaccharides contain an unusual unsaturated uronic acid at their non-reducing end.[15] This unsaturated uronic acid residue represents an activated site through which the oligosaccharide might be incorporated into a synthetic polymer. Although this unsaturated uronic acid residue resembles an acrylic acid type Michael acceptor, it is extremely unreactive towards nucleophiles. This low reactivity may rule out ionic-based polymerization methods for incorporation into synthetic polymers. An alternative method of incorporating these oligosaccharides into synthetic polymers relies on free-

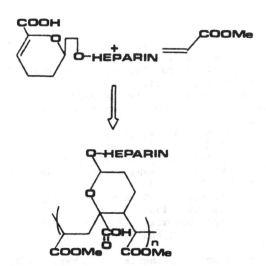

Figure 7. Scheme for the free-radical copolymerization of unsaturated heparin oligosaccharide and an acrylic acid based monomer. A heparin oligosaccharide (R = 1,3,5... sugar residues) prepared using a lyase is copolymerized with methyl acrylate in the presence of a peroxide catalyst and initiator. The polymer prepared contains the non-reducing terminal sugar of the oligosaccharide in the backbone of the polymer.

radical polymerization methods. Here a monomer such as methyl acrylate might be copolymerized with a heparin oligosaccharide to form a heparin-containing synthetic polymer (Figure 7). Such an approach has recently resulted in the successful incorporation of lyase prepared heparin oligosaccharides into polyacrylamides.[43]

5. Future Prospects

Heparin has been in widespread clinical use for over a half a century. Despite heparin's structural complexity, multiple activities and numerous side-effects, its use continues and has even expanded during the past decade. The major factor behind heparin's success is that it remains the most effective anticoagulant/antithrombotic agent available. Thus, the prospects for new and improved heparins or heparinized surfaces are very good. The twenty-first century should bring an array of monodisperse, homogenous, potent heparinoid drugs with high therapeutic indices and increased specificities. In addition, heparinized biomaterials will probably be in great demand for use in both extracorporeal devices and implanted artificial organs.

ACKNOWLEDGMENTS

The authors acknowledge the support of the National Institutes of Health for this research in the form of research grants HL 29797, GM 38060 and AI 22350.

REFERENCES

1. R. J. Linhardt and D. Loganathan, in: "*Biomimetic Polymers,*" C. G. Gebelein, Ed., Plenum Press, 1990, pp.135-173.
2. J. F. Kennedy and C. A. White, "*Bioactive Carbohydrates,*" Ellis Horwood Limited, New York, 1983.
3. W. D. Comper, "*Heparin (and related polysaccharides),*" Polymer Monographs, Volume 7, E. M. B. Huglin, Gordan and Breach Science Publ., New York, 1981.
4. L. B. Jacques, Science, **206**, 528 (1979).
5. B. Casu, Adv. Carbohydr. Chem. Biochem., **43**, 51 (1985).
6. J. Folkman, Biochem. Pharmacol., **34**, 905 (1985).
7. J. Folkman, R. Langer, R. J. Linhardt, C. Haundenschild and S. Taylor, Science, **221**, 719 (1983).
8. R. J. Linhardt, K. G. Rice, Y. S. Kim, J. D. Engelken and J. M. Wieler, J. Biol, Chem, **263**, 13090 (1988).
9. J. Fareed, Sem. Thromb. Hemostas., **11**, 227 (1985).
10. D. L. Gordon, R. J. Linhardt and H. P. Adams, Clin. Neuropharmacol, **13**, 522 (1990).
11. G. Bratt, E. Tornebohm, S. Granquist, W. Aberg and D. Lockney, Thromb. Haemostas., **54**, 813 (1985).
12. J. Biller, E. W. Massay, H. P. Adams, J. N. Davis, J. R. Marler, A. Bruno, R. A. Henriksen, R. J. Linhardt, L. Goldstein, M. Alberts, C. Kisker, G. Toffol, C. Greenberg, K. Banwart, C. Bertels, M. Waller and H. Maganani, Neurology, **39**, 262 (1989).
13. V. C. Yang, R. J. Linhardt, H. Bernstein, C. L. Cooney and R. Langer, J. Biol. Chem. **260**, 1849 (1985).
14. R. J. Linhardt, J. E. Turnbull, H. M. Wang, D. Loganathan and J. T. Gallagher, Biochem., **29**, 2611 (1990).

15. R. J. Linhardt, P. M. Galliher and C. L. Cooney, Appl. Biochem. Biotechnol. **12**, 135 (1986).
16. K. G. Rice and R. J. Linhardt, Carbohydr. Res. **190**, 219, (1989).
17. A. Al-Hakim and R. J. Linhardt, Electrophoresis, **11**, 23 (1990).
18. L. M. Mallis, H. M. Wang, D. K. Loganathan and R. J. Linhardt, Anal. Chem. **61**, 1453 (1989).
19. D. Loganathan, H. M. Wang, L. M. Mallis and R. J. Linhardt, Biochem., **29**, 4362 (1990).
20. R. J. Linhardt, D. Loganathan, A. Al-Hakim, H. M. Wang, J. M. Walenga, D. Hoppensteadt and J. Fareed, J. Med. Chem., **33**, 1639 (1990).
21. R. J. Linhardt and J. M. Weiler, U.S. Patent No. 4,916,219 (1990).
22. P. K. Smith, A. K. Mallia and G. T. Hermanson, Anal. Biochem., **109**, 466 (1980).
23. D. H. Atha, A. W. Stephens, A. Rimon and R. D. Rosenberg, Biochem., **23**, 5901 (1984).
24. Z. M. Merchant, E. E. Erbe, W. P. Eddy, D. Patel and R. J. Linhardt, Atherosclerosis, **62**, 151 (1986).
25. P. N. Walsh, in: *"Hemostasis and Thrombosis,"* R. Colman, J. Hirsh, V. J. Marder and E. Salzman, Eds., Lippincott Co., Philadelphia, 1981, p. 404.
26. R. J. Linhardt, A. Al-Hakim, S. Y. Liu, Y. S. Kim and J. Fareed, Sem. in Thromb. & Haemostas., **17**, 15 (1990).
27. R. J. Linhardt, A. Al-Hakim, S. Ampofo and D. Loganathan, in: *"New Trends in Haemostasis: Coagulation Proteins Endothelium, and Tissue Factors,"* G. Schettler, L. Heene and J. Harenberg, Eds., Springer Verlag, 1990, pp. 12-26.
28. S. Nishimura, N. Nishi and S. Tokura, Carbohydr. Res., **156**, 286 (1986).
29. G. O. Aspinall, Adv. Carbohydr. Chem., **14**, 429 (1983).
30. M. F. Scully, K. M. Weerasinghe, V. Ellis, B. Djazaeri & V. V. Kakkar, Thrombosis Res., **31**, 87 (1983).
31. C. A. Rickets & K. W. Walton, U.S. Patent No. 2715091, (1955).
32. G. Oshima, Thromb. Res., **49**, 353 (1988).
33. R. Langer, R. J. Linhardt, M. Klein, P. M. Galliher, C. L. Cooney & M. M. Flanagan, in *"Biomaterials: Interfacial Phenomenon and Applications,"* S. Cooper, A. Hoffman, N. Pepas & B. Ratner, Eds., American Chemical Society, Washington, DC, 1982, Advances in Chemistry Symposium Series, Chapter 31, p. 493.
34. P. Sinay, J.-C. Jacquinet, M. Petitou, P. Duchaussoy, I. Lederman, J. Choay & G. Torri, Carbohydr. Res., **132**, c5 (1984).
35. H. K. Breddin, J. Fareed & N. Bender, Eds., *"Low-Molecular-Weight Heparins, 4th Congress on Thrombosis and Haemostasis Research,"* in: Haemostasis, **18**, 1-87 (1988).
36. Y. S. Kim, Doctoral Dissertation, The University of Iowa, August, 1988.
37. R. J. Linhardt, K. G. Rice, Y. S. Kim, D. L. Lohse, H. M. Wang and D. Loganathan, Biochem. J. **254**, 781, (1988).
38. N. A. Plate and L. I. Valuev, Adv. Polym. Sci., **79**, 95 (1986).
39. C. G. Gebelein and D. Murphy, Polym. Sci. Technol. (Plenum), **35**, 277 (1987).
40. A. K. Larsen, R. J. Linhardt, K. G. Rice, W. Wogan and R. Langer, J. Biol. Chem., **264**, 1570 (1989).
41. A. K. Larsen, *"Toxicological Aspects of an Enzymatic System for Removing Heparin in Extracorporeal Therapy,"*, Ph.D. Thesis, MIT, 1984.
42. A. K. Larsen, R. J. Linhardt, K. G. Rice, G. Wogan and R. Langer, J. Biol. Chem. **264**, 1570 (1989).
43. M. A. Mazid, E. Moase, E. Scott, R. Hanna and F. M. Unger, Abstracts of the Division of Carbohydrate Chemistry, American Chemical Society Meeting, Boston, April 23, 1990.

STRUCTURE-CONTROLLED SYNTHESIS OF REGIOSPECIFICALLY MODIFIED

POLYSACCHARIDES STARTING FROM A PYROLYSIS PRODUCT OF CELLULOSE

Kazukiyo Kobayashi

Faculty of Agriculture
Nagoya University
Chikusa, Nagoya 464-01, Japan

Starting from a pyrolysis product of cellulose, 1,6-anhydro-β-D-glucopyranose derivatives having regiospecific substituents and protecting groups were prepared. Ring-opening polymerization using these anhydro sugar derivatives as monomers, followed by debenzylation, afforded structurally well-defined (1→6)-α-D-glucopyranan (linear dextran) derivatives: (a) 3-O-Methyl-(1→6)-α-D-glucopyranan, (b) 3-O-octadecyl-(1→6)-α-D-glucopyranan, (c) 3-C-methyl-3-O-methyl-(1→6)-α-D-glucopyranan, (d) 3-deoxy-(1→6)-α-D-ribo-hexopyranan, (e) 3,4-dideoxy-(1→6)-α-D-threo-hexopyranan, and (f) 2,4-dideoxy-(1→6)-α-D-threo-hexopyranan and (g) 2,4-dideoxy-3-O-(β-D-galactopyranosyl)-α-D-threo-hexopyranan. Heteropolysaccharides with the required amount of substituents at relatively uniform intervals were also prepared via copolymerization.

Partial to complete 3-O-octadecylated polysaccharides exhibited characteristic solution and solid properties based on hydrophilic-hydrophobic structures. These polymers are suggested to form micellar conformations in water and in chloroform polysaccharide-coated liposomes, polymeric membranes, and thermotropic liquid-crystalline mesophase, depending on the octadecyl content. Hydrolysis of 3-deoxygenated, 3-O-methylated, and 3-O-octadecylated dextrans by an endo-acting dextrans is compared. The possibility of a comb-shaped branched polysaccharide toward cell-specific biomedical materials is discussed.

INTRODUCTION

Naturally-occurring polysaccharides exhibit a variety of distinct physicochemical, physiological, and pharmaceutical properties.[1] These properties and functions of polysaccharides depend on their chemical structures, which differ slightly in the kinds of monosaccharide components and functional groups, their positions, modes of linkage, and so on. The primary structures influence polysaccharide conformations, which determine highly-ordered solid structures and solution properties.

Biotechnology and Polymers, Edited by C.G. Gebelein
Plenum Press, New York, 1991

Understanding the structure-function relations is important to applying polysaccharides more extensively. Synthesis of structurally well-defined polysaccharides, coupled with elucidation of their properties, is indispensable for these purposes. We have been interested in regiospecifically modified, stereoregular polysaccharides. We believe that new types of polysaccharides with novel properties will be developed through these processes.

Various attempts to modify polysaccharides regioselectively have been reported in the literature.[2] In cellulose and amylose, for example, the reactivity of primary hydroxyl groups in position 6 is higher than those in secondary hydroxyl groups in position 2 and 3. By applying this difference in reactivity, several types of chemical transformations were carried out to obtain regioselectively modified polysaccharide derivatives. Strictly speaking, however, the modifications were sometimes incomplete in both aspects of the regioselectivity and degree of substitution. One of the principal disadvantages of polymeric reactions is that impure components generated in the polymer chain can not be removed by purification procedures. These impure components are often accumulated in the polymer sequences when the reaction is multistep.

In these respects, we have been engaged in chemical synthesis of polysaccharides, especially via ring-opening polymerization of anhydro sugar derivatives.[3-5] Our synthetic strategy leading to regiospecifically modified polysaccharides is shown in Figure 1. Anhydro sugar derivatives having one substituent and two regiospecific benzyl protecting groups were synthesized and polymerized, and then the protection was removed from the resulting polymerizates.[6-10] The present article describes molecular design of polysaccharides and their characteristic functions.

RESULTS AND DISCUSSION

1. Pyrolysis of Microcrystalline Cellulose: Preparation of 1,6-Anhydro-β-D-glucopyranose [1]

The starting material of this strategy is microcrystalline cellulose. As shown in Figure 2, its pyrolysis product, 1,6-anhydro-β-D-glucopyranose, was selectively dibenzylated and then the free hydroxyl group of the resulting 1,6-anhydro-2,4-di-O-benzyl-β-D-glucopyranose was chemically modified.

Pyrolysis of polysaccharides has been a subject for investigation on flame retardants, sources of fuel, and preparation of chemical feedstocks. When powdery cellulose or starch are treated at high temperature under reduced pressure, a complex mixture of tar and gaseous substances

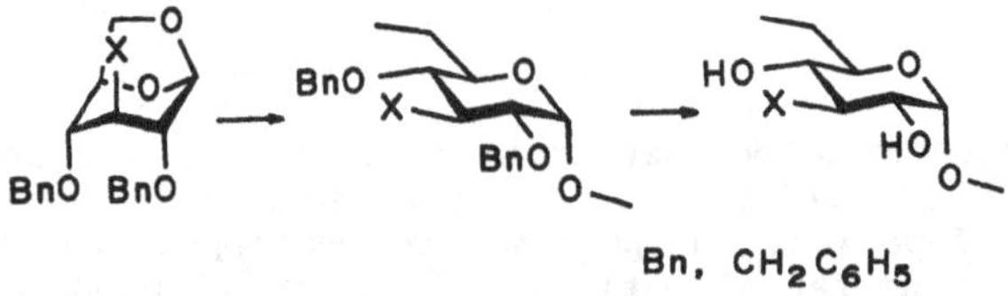

Bn, $CH_2C_6H_5$

Figure 1. Synthetic strategy of regiospecifically modified polysaccharides via ring-opening polymerization of anhydro sugar derivatives.

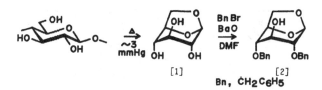

Figure 2. Pyrolysis of cellulose followed by selective protection with benzyl groups.

is distillated. The main product is levoglucosan (1,6-anhydro-β-D-gluco-
pyranose, [1]),[11] but its isolation procedure is tedious and the yield is
not reproducible. Recently, pyrolysis of microcrystalline cellulose is
found to afford levoglucosan in high yield.[12] The procedure was scaled up
(150 g of cellulose) and the supply of levoglucosan (20-39 g) was
facilitated.

2. Chemical Transformation of 1,6-Anhydro-β-D-glucopyranose

The skeleton of the anhydro sugar is a bicyclic acetal which is subject to ring-opening polymerization using Lewis acids as initiator. Lewis
acids are inactivated with hydroxyl groups, hence appropriate protection
of the anhydro sugar is required. Our intention is to introduce different
types of protecting and blocking groups regiospecifically. The hydroxyl
group in position 3 of [1] is less reactive than those in positions 2 and
4, the latter of which can be benzylated selectively.[13,14] Levoglucosan
was treated with benzyl bromide in dimethylformamide in the presence of
activated barium oxide at 75°C and 1,6-anhydro-2,4-di-O-benzyl-β-D-
glucopyranose [2] was isolated in a 92% yield.

The free hydroxyl group in position 3 of [2] was then chemically
transformed to afford some intended compounds. Figure 3 lists the structures of the resulting compounds used as the monomers, together with
those of the intermediates. The methylated compound [3] was obtained by
treatment of [2] with sodium hydride and then with methyl iodine in dimethyl sulfoxide.[6] Similar alkylation using octadecyl bromide afforded
the compound [4].[15] The 3-C-methyl-3-O-methyl compound [5] was synthesized via the following three reaction steps:[8] (1) Oxidation of the secondary hydroxyl group of [2] with pyridinium chlorochromate; (2) Grig-
nard addition to carbonyl group of the ulose; (3) methylation of the
tertiary hydroxyl group with potassium hydride and methyl iodine. A de-
oxygenated derivative [6][9,10] was prepared via (1) treatment with N,N'-
(thiocarbonyl)-diimidazole and (2) radical reduction with tributyl-
stannane.

Figure 4 shows another analogous synthetic scheme. Selective ditosyl-
ation of [2] and subsequent reduction with super hydride yielded dideoxygenated derivatives [7][16] and [8].[17] An anhydro disaccharide derivative [9][18] was also prepared via glycosidation of an intermediate with a
galactose derivative.

It is noteworthy to point out that the reaction sequences of Figures
3 and 4 include a lot of reagents and methods recently developed in the
field of synthetic carbohydrate chemistry.[19-23] Each reaction proceeded
regio- and stereospecifically in a high yield.

169

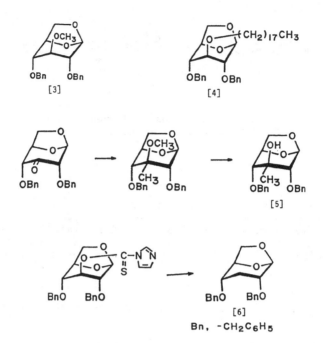

[3]

[4]

[5]

[6]

Bn, -CH₂C₆H₅

Figure 3. Anhydro sugar derivatives obtained from 1,6-
anhydro-2,4-di-O-benzyl-β-D-glucopyranose.

3. Ring-Opening Polymerization of Anhydro Sugar Derivatives

A representative procedure of cationic ring-opening polymerization
using high vacuum techniques (2 x 10⁻⁵ mmHg) is as follows.[24] A monomer
solution in dichloromethane, dried over calcium hydride and degassed

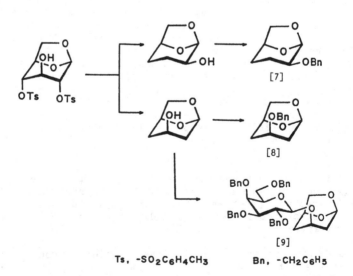

[7]

[8]

[9]

Ts, -SO₂C₆H₄CH₃   Bn, -CH₂C₆H₅

Figure 4. Anhydro sugar derivatives obtained from 1,6-anhy-
dro-2,4-di-O-p-toluenesulfonyl-β-D-glucopyranose.

under vacuum, was cooled in a liquid nitrogen bath. The catalyst break-seal was broken and phosphorus pentafluoride was condensed in the solution by heating the catalyst precursor, p-chlorobenzenediazonium hexafluorophosphate. The tube was sealed off and the polymerization was carried out at -60 and -78°C.

The polymerization was terminated with a large excess of cold methanol (containing a small amount of pyridine, if necessary) to precipitate a white powdery polymer. It was purified by reprecipitation from a chloroform solution into methanol, and freeze-dried from benzene.

The monomers listed in Figures 3 and 4 are highly reactive and their polymerizations proceeded rapidly to give stereoregular polymers in high yields. The molecular weights of the resulting polymers were in the range of 1.6×10^4 to 3.8×10^5.

The ring-opening polymerization proceeded stereospecifically to give the corresponding polysaccharide derivatives of α-anomeric configuration. The propagation mechanism is illustrated in Figure 5, as proposed for 1,6-anhydro-2,3,4-tri-O-benzyl-β-D-glucopyranose [10].[3] The growing terminal unit is assumed to be the cyclic trialkyloxonium ion. A monomer attacks the acetal carbon exclusively from the opposite (σ) side of the C-O+ bond and, as a result, inversion of configuration at C-1 occurs to give an α-anomeric structure.

The polymerization rate, polymer yield, molecular weight, and stereoregularity of each polymer depended on the monomer structure and polymerization conditions. Discussion on the polymerization reactivity of each monomer is an important subject of polymerization chemistry, which is described in the respective original papers.

4. Debenzylation

A conventional method of debenzylation of synthetic polysaccharide derivatives is reduction with sodium in liquid ammonia.[24] A solution of benzylated polysaccharide in a toluene/1,2-dimethoxyethane mixture was stirred with an excess amount of liquid ammonia at -33°C. Small pieces of sodium were added gradually to generate solvated-electrons, a powerful reducing agent, in liquid ammonia layer. As debenzylation proceeded, the resulting polysaccharide became soluble in liquid ammonia and debenzyla-

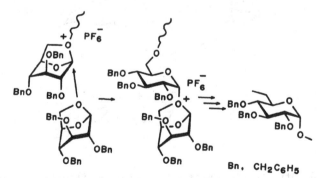

Figure 5. Trialkyloxonium ion mechanism in the ring-opening polymerization of 1,6-anhydro sugar derivatives leading to stereoregular (1→6)-α-D-glucopyranans.

tion was accelerated. A water-soluble product was dialyzed with water and isolated by freeze-drying from water.

The conventional debenzylation was unsuccessful for octadecylated polysaccharide, probably because both the reactant and product were highly lipophilic and hence incompatible with the reducing agent and inorganic solvent. Another debenzylation method via radical bromination and hydrolysis[25] was successfully performed.[7,26] The method was effective when the product was soluble in organic solvents, although a cleavage of the main chain occurred to some extent.

5. Structural Analysis of Regiospecifically Modified Polysaccharides

Structural units of the homopolysaccharide derivatives synthesized are summarized in Figure 6. Such well-defined regiospecific modifications cannot be attained by direct reactions of polysaccharides. Their systematic nomenclatures are as follows. [P3]: 3-O-Methyl-(1→6)-α-D-glucopyranan. [P4]: 3-O-octadecyl-(1→6)-α-D-glucopyranan. [P5]: 3-C-methyl-3-O-methyl-(1→6)-α-D-glucopyranan. [P6]: 3-deoxy-(1→6)-α-D-ribo-hexopyranan. [P7]: 3,4-dideoxy-(1→6)-α-D-threo-hexopyranan. [P8]: 2,4-dideoxy-(1→6)-α-D-threo-hexopyranan. [P9]: 2,4-dideoxy-3-O-(β-D-galactopyranosyl)-(1→6)-α-D-threo-hexopyranan.

The ^{13}C-NMR spectrum of [P5] as an example is shown in Figure 7.[8] The simple pattern is an indication of well-defined stereoregular polysac-

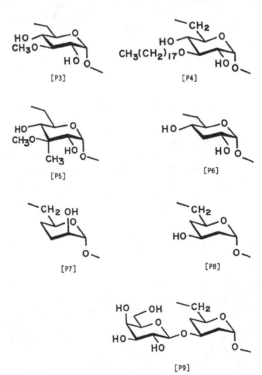

Figure 6. Structural units of regiospecifically modified stereoregular polysaccharides. The numbers correspond to those monomers illustrated in Figures 4 and 5.

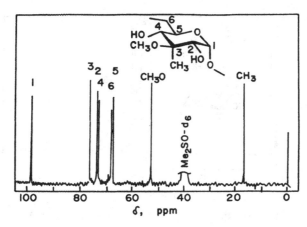

Figure 7. The ^{13}C-NMR spectrum of 3-C-methyl-3-O-methyl-
(1→6)-α-D-glucopyranan [P5]. 50 MHz, in Me$_2$SO-d^6,
80°C.

charides. There appeared only one anomeric signal assignable to the α-C-1
carbon (98.04 ppm), but no β-C-1 carbon signal was detectable. 3-C-
Methyl and 3-O-methyl signals at 16.9 and 53.5 ppm, respectively, were
distinguishable from others. The C-3 and C-6 resonances could be assigned
by off-resonance techniques and those of C-2, C-4, and C-5 were assigned
as indicated.

High stereoregularities are also confirmed from specific rotation of
the polysaccharides. 3-O-Methyl [P3] and 3-deoxy [P6] homopolymers were
soluble in water and dimethyl sulfoxide. 3-C-methyl-3-O-methyl [P5] and
dideoxygenated polymers [P7], [P8], and [P9] were partially soluble in
water and soluble in dimethyl sulfoxide. 3-O-octadecyl polysaccharide
[P4] was insoluble in water and soluble in chloroform and benzene.

6. Copolymerization: a Useful Procedure for Molecular Design

Heteropolysaccharides with various compositions can be designed by a
copolymerization method, which is one of the merits of the present strategy. Required amount of substituents can be introduced into polysaccharide chains regiospecifically and at relatively uniform intervals by
using 1,6-anhydro-2,3,4-tri-O-benzyl-β-D-glucopyranose [10] as comonomer.
The comonomer is also useful as a reference compound in comparison on
polymerization reactivities.

Figure 8 represents one of the ^{13}C-NMR spectra of heteropolysaccharides consisting of 3-deoxygenated and non-deoxygenated (1→6)-α-D-gluco-
pyranan units obtained via copolymerization of [6] and [10].[10] All peaks
were superimposable on peaks of one or the other homopolymers. The α-C-1
resonance of each structural unit was completely separated from the
other, and the copolymer compositions could be estimated from their area
ratios. In the benzylated precursors, some of the resonances were split
into two peaks whose intensity varied with copolymer compositions. The
splittings are due to diad sequences between the two crossover units.
This is evidence of randomly distributed copolymers. These homo-and
heteropolysaccharides will serve as a tool to investigate properties and
functions.

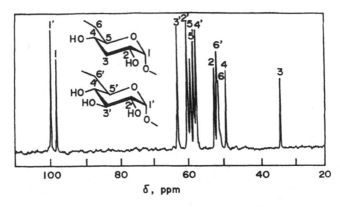

Figure 8. The ^{13}C-NMR spectrum of a partially 3-deoxygenated (1→6)-α-D-glucopyranan. 50 MHz, in D_2O, room temp.

7. Functions of Hydrophillic-Hydrophobic Amphiphilic Polysaccharides

Figure 9 schematically illustrates three different types of octadecyl polysaccharides: A: homopolymer [P4].[7] B: partially 3-O-octadecylated (1→6)-α-D-glucopyranans.[15] C: (1→6)-α-D-glucopyranans consisting of non-substituted and 2,3,4-tri-O-octadecylated glucose units.[26] Polymer C was prepared from 2,3,4-tri-O-octadecyl-β-D-glucopyranose and comonomer [10]. These polysaccharide derivatives can be regarded as a lipid-polysaccharide conjugate or a polymeric model system of glycolipids.

7a. Micellar conformation in water.[15,26]

The copolymers B and C of lower octadecyl content were soluble in water and interacted with some organic solutes in water. The copolymers induced a blue shift in the emission maximum of magnesium 1-anilino-8-naphthalenesulfonate and enhanced its fluorescence intensity. The copolymer B of DS = 0.07 interacted most strongly. We assumed that the polysaccharide formed a polymeric micellar conformation in water and the organic solute was bound to nonpolar regions surrounded by long hydrocarbon chains (Figure 10). It is suggested that both of the copolymers B and C, having a similar octadecyl/hydroxyl ratio, form a conformation with a similar hydrophobic microenvironment.

7b. Reversed micellar conformation in chloroform.[26]

When chloroform-soluble polymer A or C was mixed with an aqueous methyl orange solution, the chloroform layer was colored, as detected with visible spectra. The methyl orange solution was solubilized into chloroform with the help of the amphiphilic polysaccharide, since methyl orange is insoluble in chloroform and an aqueous methyl orange solution is not miscible with chloroform. We assume that the polysaccharides form reversed micellar conformations in chloroform and that the methyl orange solution is solublized to the hydrophilic boundary (Figure 10). As judged from the absorbance, the solubilizing ability of polymer C is higher than that of polymer A by a factor of 10, while the octadecyl/hydroxyl ratio of C (DS = 0.29) is almost the same as that of B (DS = 0.33).

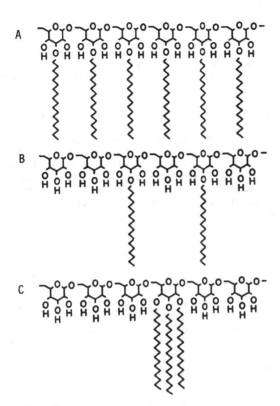

Figure 9. Three types of regiospecifically octadecylated (1→6)-α-D-glucopyranans.

7c. Polysaccharide-coated liposomes.[15,27]

There occurred strong interactions between copolymer B and DL-α-di-palmitoylphosphatidilcholin (DPPC) liposomes. Polysaccharide-coated lipo-

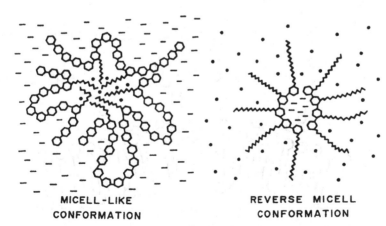

MICELL-LIKE
CONFORMATION

REVERSE MICELL
CONFORMATION

Figure 10. Possible conformations of octadecylated dextrans in solution.

somes were prepared by mixing copolymers B with multilamellar DPPC lipo-somes in aqueous solution and then isolated by chromatography on a Sepha-dex G-50 column. The interaction between liposomes and the dextrans of octadecyl content (DS = 0.01-0.05) was strong enough, and the octadexyl dextran of DS = 0.03 yielded the largest amount of polysaccharide-coated liposomes.

It is important to note that the liposomes obtained were surrounded by a larger amount of the polysaccharides than the amount calculated by assuming that the polysaccharide chains were spread flat in single layers over the outer surface of multilamellar liposomes. We propose a binding mode between liposomes and polysaccharides as illustrated in Figure 11. This conformational model can be regarded as a modification of the "loop-train-tail" model that has been widely accepted from both theoretical and experimental view points as the conformational model for macromolecules adsorbed on an interface.

7d. Formation of polymeric membranes.[15]

Naturally occurring dextrans have no ability to form a stable mem-brane, but copolymer B, with a suitable amount of octadecyl group (DS = 0.07-0.22), was found to form transparent, flexible membranes by casting a polymer solution from Me_2SO on a glass plate.

7e. Thermotropic properties.[28]

Polymer A showed a crystalline X-ray diffraction pattern and an endo-thermic peak at 31-42°C in differential scanning thermograms. The polymer had birefringence below 90°C under a polarizing microscope. It is of interest to speculate as follows. Both of the hydrocarbon side chains and the polysaccharide main chains are crystalline and the polymer A exhibits a unique two-stage melting process of both crystalline regions. The poly-mer possibly forms a liquid-crystalline mesophase between these two tran-sition temperatures.

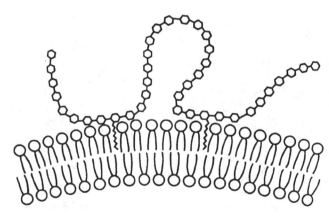

Figure 11. A loop-train-tail conformational model of polysac-charide-coated liposome.

8. Enzymatic Degradation by an Endo-Acting Dextranase[29]

Hydrolysis of synthetic polysaccharides were investigated with an endo-acting dextranase. The percent of glycoside linkage cleaved was determined from the concentration of reducing sugars and expressed as the degree of hydrolysis. The maximum hydrolysis of a nonsubstituted dextran was 53%, and the main product was a disaccharide isomaltose.

The 3-deoxy and 3-O-methyl homopolysaccharides ([P6] and [P3]) were not hydrolyzed, but (1→6)-α-D-glucopyranans containing these structural unit were degraded. The heteropolysaccharide of the [P6] unit composition (DS) of 0.59 gave the maximum hydrolysis of 16%. This value agreed with the diad sequence of {G-G} = 0.17 estimated using the copolymer composi-tions {G} and {R}, where G and R represents "glucose" and "deoxy-ribo-hexose" units, respectively. In addition, kinetic treatments suggested that the hydrolysis of a nonsubstituted dextran was retarded in the pre-sence of [P6]. We assume that the polysaccharide [P6] is a potent compe-titive inhibitor in the hydrolysis of dextran. In other words, the 3-deoxy sequence is bound to the active site of the enzyme.

The 3-O-methylated synthetic dextran of DS = 0.57 gave the maximum hydrolysis of 16%, which is almost the same as that of 3-deoxygenated one. On the other hand, the maximum degree of hydrolysis of the octadecy-lated dextran (polymer B in Figure 9) od DS = 0.22 was only 14%. It sug-gests that the bulkiness of the octadecyl chains interfered with the hydrolysis of several units adjacent to the octadecylated units.

9. A Comb-Shaped Branched Polysaccharide: Cell-Specific Biomedical Materials[18]

The term "comb-shaped branched polysaccharide" is defined as illus-trated in Figure 12. It has a linear polysaccharide backbone and pendant monosaccharide units regularly substituted on each sugar unit in the main chain.

This type of polysaccharide may be prepared from anhydro disacchar-ides. Polymerization reactivity of anhydro disaccharide derivatives, however, is lower than that of the corresponding anhydro monosaccharide derivatives. In order to obtain comb-shaped polysaccharides of moderate molecular weight, highly reactive disaccharide anhydrides are required. Hence we gave attention to a dideoxygenated anhydro-ring structure.

Monomer [9] was polymerized with 20 mol% phosphorus pentafluoride as initiator in dichloromethane at -78°C for 6 min and the corresponding polymer was obtained in 67% yield. Its conventional debenzylation yielded a comb-shaped polysaccharide [P9] of DPn = 40.

Naturally occurring polysaccharides often have branched structures. The length, interval, and density of branches are intrinsic for their biological origins. The nature of branches influences physical and biolo-gical functions of polysaccharides. Comb-shaped polysaccharides have high density of pendant carbohydrates. According to inspection of CPK molecu-lar models, large pendant sugar units in the comb-shaped polysaccharide are very crowded with respect to each other.

It is noted that branched oligosaccharides on glycoproteins and glycolipids in cell surface membranes are key participants in cell-cell recognition during development. β-Galactose is one of the major recogni-

LINEAR POLYSACCHARIDE

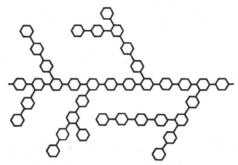

BRANCHED POLYSACCHARIDE

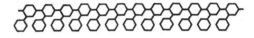

COMB-SHAPED POLYSACCHARIDE

Figure 12. A schematic representation of comb-shaped branched polysaccharides.

tion markers. We have found that a polystyrene derivative having a β-galactose residue on each repeating unit (Figure 13) is useful as substratum materials for culture of liver cells.[38] It is reasonable to assume that β-galactose-specific lectins on the surface of liver cells act as receptors of the substratum. The interaction was quite strong, probably because the density of β-galactose units on the surface of the homopolymer chain was quite high owing to the hydrophilic-hydrophobic structure.

In these respects, we expect that the well-defined comb-shaped polysaccharide [P9] will serve as a model system of naturally occurring branched polysaccharides and also as cell-specific biomedical materials using high-density β-galactose moiety as a recognition marker.

CONCLUSION

The polysaccharide synthesis via ring-opening polymerization method is an attractive method leading to structurally well-defined polysaccharides which are rarely attained by direct chemical modifications of natural polysaccharides. The synthetic polysaccharides obtained are useful tools to elucidate relations between the structure and functionality of polysaccharides and to develop new types of polysaccharides with novel properties.

ACKNOWLEDGMENTS

The author thanks the coworkers for their dedicated efforts in this research: Prof. C. Schuerch, Prof. H. Sumitomo, Prof. M. Okada, Dr. H. Ichikawa, Mr. H. Shiozawa, Mr. M. Takahashi, and Mr. H. Sugiura.

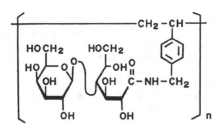

Figure 13. A polystyrene derivative having a β-D-galactose residue on each structural unit: cell-specific biomedical materials suitable for culture substratum of liver cells.

REFERENCES

1. G. O. Aspinall, Ed., "*The Polysaccharides*," Academic Press, New York, vol 1-3, 1982-1985.
2. M. Yalpani, "*Polysaccharides: Synthesis, Modifications and Structure/Property Relations*," Elsevier, Netherlands, 1988, p. 189.
3. C. Schuerch, Adv. Carbohydr. Chem. Biochem., 39. 157 (1981).
4. H. Sumitomo and M. Okada in: "*Ring-Opening Polymerization*," vol 1, K. J. Ivin and T. Saegusa, Eds., Elsevier, London, 1984, p. 299.
5. N. K. Kochetkov, Tetrahedron, 43, 2389 (1987).
6. K. Kobayashi and H. Sumitomo, Macromolecules, 14, 259 (1981).
7. K. Kobayashi, H. Ichikawa and H. Sumitomo, Macromolecules, 23, 3708 (1990).
8. K. Kobayashi, H. Sumitomo and M. Takahashi, Polym. J., 21, 559 (1989).
9. K. Kobayashi and H. Sumitomo, Maromolecules, 16, 710 (1983).
10. K. Kobayashi, H. Sumitomo and H. Shiozawa, Polym. J., 21, 137 (1989).
11. R. B. Ward, Methods Carbohydr. Chem., 2, 394 (1963).
12. F. Shafizadeh, R. H. Furneaux, T. G. Cochran, J. P. Scholl and Y. Sakai, J. Appl. Polym. Sci., 23, 3525 (1979).
13. M. Cerny and I. Stanek, Jr., Adv. Carbohydr. Chem. Biochem., 34, 23 (1977).
14. T. Iversen and D. R. Bundle, Can. J. Chem., 60, 200 (1982).
15. K. Kobayashi, H. Sumitomo and H. Ichikawa, Macromolecules, 19, 529 (1986).
16. H. Ichikawa, K. Kobayashi, M. Okada and H. Sumitomo, Polym. J., 19, 873 (1987).
17. K. Kobayashi, H. Sumitomo, H. Ichikawa and H. Sugiura, Polym. J., 18, 927 (1986).
18. H. Ichikawa, K. Kobayashi and H. Sumitomo, Macromolecules, 23, 1884 (1990).
19. D. H. R. Barton and S. W. McCombie, J. Chem. Soc., Perkins Trans., 1, 1574, (1975).
20. J. R. Rasmussen, J. Org. Chem., 45, 2725 (1980).
21. B. B. Bissember and R. H. Wightman, Carbohydr. Res., 81, 187 (1980).
22. E. C. Ashby and J. T. Laemmle, Chem. Rev., 75, 521 (1975).
23. A. G. Kelly and J. S. Roberts, Carbohydr. Res., 77, 231 (1979).
24. C. Schuerch and T. Uryu, Macromol. Synth., 4, 151 (1972).
25. J. N. BeMiller, R. E. Wing and V. Y. Meyers, J. Org. Chem., 33, 4292 (1968).
26. H. Ichikawa, K. Kobayashi and H. Sumitomo, Macromol. Chem., 189, 1019 (1988).
27. J. Sunamoto, K. Iwamoto, M. Takada, T. Yuzuriha and K. Katayama in: "*Polymers in Medicine*," E. Chiellini and P. Giusti, Eds., Plenum, New

York, 1984, p. 157.

28. K. Kobayashi, H. Ichikawa and H. Sumitomo in: "*Cellulose: Structural and Functional Aspects*," J. F. Kennedy, G. O. Phillips and P. A. Williams, Eds., Ellis Horwood, Chichester, 1989, p. 207.

29. K. Kobayashi, H. Sumitomo and H. Shiozawa, Japan-U.S. Symposium, Kyoto, 1985, Prepr., p. 151.

30. K. Kobayashi, H. Sumitomo, A. Kobayashi, and T. Akaike, J. Macromol. Sci.-Chem., **A25**, 655 (1988).

BIOLOGICAL GELS: THE GELATION OF CHITOSAN AND CHITIN

Shigehiro Hirano, Ryuji Yamaguchi, Nobuaki
Fukui, and Mamoru Iwata

Department of Agricultural Biochemistry and
Biotechnology
Tottori University
Tottori 680 Japan

A chitosan oxalate gel, a chitosan gel and a chitin gel were prepared from chitosan, and a partially O-acetylated chitin gel and a chitin gel were prepared from chitin. Their intramolecular gel conversion was demonstrated. A transparent chitosan oxalate gel was formed from chitosan in aqueous oxalic acid (1.5-2.0 mol./GlcN). The gel melted on heating at 80-90°C and solidified on cooling. It was composed of one mol. equiv. of oxalic acid per GlcN, indicating the absence of the crosslinking of oxalic acid and chitosan chains. An opaque chitosan gel was formed after the complete removal of oxalic acid by treating the chitosan oxalate gel with an aqueous NaOH solution. The chitosan gel was treated with acetic anhydride at room temperature to give a transparent chitin gel. Chitin solution in dimethylacetamide - 5%LiCl was treated with acetic anhydride-pyridine at 70-100°C for 6 hrs. to give a partially O-acetylated chitin gel, and its O-deacetylation in aqueous NaOH solution gave a chitin gel. These biological gels are usable for preparation of films and sheets and for immobilization of enzymes and cells.

INTRODUCTION

Polysaccharide hydrogels are generally known to be composed of fibrous solid and aqueous liquid phases, in which a porous framework in the solid phase is filled with liquid.[1] A number of neutral [e. g., curdlan[2] and agarose[3]] and acidic [e.g., carrageenans[4,5] and alginic acid[6]] polysaccharide gels are known. However, little is known about basic (amino) polysaccharide gels. Chitin, a linear (1→4)-linked 2-acetamido-2-deoxy-β-D-glucan, and its N-deacetylated product (chitosan) are the main structural elements in crab and shrimp shells which are abandoned as biomass from marine products-processing. Natural chitin has strong hydrogen bonds in the molecule,[7,8] and the destruction of the hydrogen bonds and their reorganization into a specific structure give rise to gels. N-Acyl,[9] N-arylidene[10] and N-alkylidene[11] chitosan gels have been reported, and these gels are thermally irreversible.

Biotechnology and Polymers, Edited by C.G. Gebelein
Plenum Press, New York, 1991

This paper describes a novel thermally reversible chitosan oxalate gel, a thermally irreversible chitosan gel, and a thermally irreversible chitin gel (N-acetylchitosan gel or regenerated chitin). An intramolecular conversion of the chitosan oxalate gel into the chitin gel via the chitosan gel is discussed. The chitin gel may play a role of the structural processing of crab and shrimp shells and insect cuticle. A sheet of transparent film has been prepared from the gel slice by air-drying,[12] and a sheet of non-transparent porous by freeze-drying. The gel may be usable as a digestible material for drug delivery systems and as media for the immobilization of enzymes and cells.

EXPERIMENTAL

1. General Methods

IR spectra (KBr) were recorded with a Hitachi 215 grating spectrometer, specific rotation with a JASCO Dip-181 digital polarimeter, and ^{13}C CP/MAS (cross polarization/magic angle spinning)-NMR spectra with a Jeol-Fx200FT NMR spectrometer. The elastic hardness ($dyne/cm^2$) and elastic breaking point ($dyne/cm^2$) of gels were measured with an IIo M-302 curdmeter. A Hitachi S-500 scanning electron microscope was used by operating at an accelerating voltage of 20 kV.

2. Materials

Crab shell chitin (d.s. 0.95 for NAc) was obtained from Katakurachikkarin Co., Ltd., Tokyo. Chitosan ($[\alpha]_D^{25}$ -9.6 (c 0.8, methanesulfonic acid), d.s. 0.10 for NAc) was prepared in our laboratory. The d.s. for the N-acetyl group was calculated from the C/N value in the elemental analysis. Lysozyme (hen egg-white, grade III, Sigma) was a commercial product.

3. Chitosan Oxalate Gel From Chitosan

Chitosan (0.97 g. d.s. 0.10 for NAc) was dissolved in aqueous oxalic acid solution (30 mL, 0.97 g, 1.8 mol./GlcN) on a boiling water bath, and a transparent gel was obtained by cooling for 18 hrs. The gel was dialyzed in distilled water at room temperature for 3 days, and a transparent chitosan oxalate gel was obtained. The gel was composed of about 5% chitosan oxalate (1.0 for oxalic acid/GlcN) and about 95% water. A portion of the gel was homogenized, washed with distilled water, and dried to give the xerogel (1.3 g). ν^{max}_{KBr} 1740 (COOH), and 2600, 1620 and 1400 cm^{-1} ($NH_3^+COO^-$); ^{13}C CP/MAS-NMR data: ^{13}C, δ 167 and 164 (C=O for COOH), 106 (C_1), 84 (C_4), 74 (C_3), 76 (C_5), 61 (C_6), and 54 (C_2).

Anal. Calc for $[C_6H_{10.9}NO_4(C_2H_3O)_{0.10} \cdot 0.90C_2H_2O_4 \cdot 0.56H_2O]n$
        C, 37.46; H, 5.50; N, 5.46.
Found: C, 37.61; H, 5.51; N, 5.48.

4. Chitosan Gel From Chitosan Oxalate Gel

Chitosan oxalate gel (20 g by wet weight) was treated with 1N NaOH solution (500 mL) at room temperature for five days, and then dialyzed in distilled water (500 mL) for 3 days by changing with fresh distilled water several times to give an opaque chitosan gel (18 g wet weight). A portion of the gel was homogenized, washed with distilled water and dried to give amorphous powders. ν^{max}_{KBr} 1600 cm^{-1} (NH_2); ^{13}C CP/MAS-NMR data: ^{13}C, δ 107 and 105 (C_1), 86 and 82 (C_4), 76 (C_3), 76 (C_5), 60 (C_6), and 58 (C_2). No acetyl or oxalate carbon signals were detected.

Anal. Calc. for $[C_6H_{10}NO_4(C_2H_3O)_{0.12} \cdot (H)_{0.88}H_2O_4 \cdot 0.43H_2O]n$:

C, 37.46; H, 5.50; N, 5.46.

Found: C, 37.61; H, 5.51; N, 5.48.

5. Chitin Gel From Chitosan Gel

The chitosan gel obtained above was treated with acetic anhydride (about 10 mols/GlcN) in 70% aqueous methanol solution at room temperature for 18 h, and a transparent chitin gel was obtained. A portion of the gel was homogenized, washed with distilled water, and dried to give amorphous powders. ν^{max}_{KBr} 1650 (C=O for NAc), 1560 cm^{-1} (C=O for NAc); ^{13}C CP/MAS-NMR data: ^{13}C, δ 174 (C=O), 104 (C_1), 84 (C_4), 75 (C_5), 75 (C_3), 56 (C_2), 61 (C_6), and 61 (Me).

Anal. Calc. for $[C_6H_{10}NO_4(C_2H_3O)_{0.96} \cdot (H)_{0.04}H_2O_4 \cdot 0.47H_2O]n$:

C, 37.46; H, 5.50; N, 5.46.

Found: C, 37.61; H, 5.51; N, 5.48.

6. Partially O-Acetylated Chitin Gel From Chitin

Chitin (0.2 g) was dissolved by stirring at room temperature in 40 mL of N,N-dimethylacetamide (DMA) in the presence of 5% LiCl. To the solution were added acetic anhydride (1.5 mL) and pyridine (50 mL), and the mixture solution was allowed to stand at 100°C for 6 hrs. to give a rigid transparent gel. The gel was dialyzed against running water for 2 days to give a gel. A portion of the gel was homogenized, washed well with distilled water, and dried to give amorphous powders. $[\alpha]_D^{18}$ +27°(ca. 1, methanesulfonic acid). ν^{max}_{KBr} 1760 and 1240 (C=O and C-O for OAc), 1650 and 1560 (C=O and NH of NAc) cm^{-1}.

Anal. Calc. for $[C_8H_{12}NO_5(C_2H_3O)_{0.85} \cdot (H)_{0.15} \cdot 0.63H_2O]n$:

C, 46.55; H, 6.38; N, 5.60.

Found: C, 46.50; H, 6.38; N, 5.56.

7. Chitin Gel From Partially O-Acetylated Chitin Gel

The partially O-acetylated chitin gel obtained above was suspended in aqueous 0.1% NaOH solution at room temperature overnight to give a chitin gel. The absence of O-acetyl group in the product was confirmed by the disappearance of IR absorptions at 1760 and 1240 cm^{-1}, indicating that the O-deacetylation effected little on the gel properties.

8. Hydrolysis Rate By Lysozyme

Each (15 mg, >80 mesh) of chitin xerogel prepared above was mixed with a solution of hen egg-white lysozyme (1 mL, 10 mg/mL, Sigma, 57,200 U/mg) in 0.05M citric acid - 0.1M Na$_2$HPO$_4$ (McIlvain) buffer solution (2 mL, pH 6.2). The suspension mixture was incubated for 1 hr. at 37°C with mechanical shaking. Aliquots were withdrawn from the supernatant solution, and the increases in the reducing value was analyzed by a modified method of Schales and Schales,[13] and calculated in umol of N-acetyl-D-glucosamine.

RESULTS AND DISCUSSION

1. Gelation of Chitosan: Chitosan Oxalate Gel, Chitosan Gel and Chitin Gel

The thermally reversible gel was formed from >2% chitosan solution (<0.2 NAc/GlcN) in 3% aqueous oxalic acid (1.0-2.0 mols/GlcN). The chitosan oxalate gel melted on heating at 80-90°C and solidified again on cooling under these conditions. Furthermore, the heating and cooling cycle was able to be repeated without affecting the gel properties. The elastic hardness of the gel increased almost linearly with chitosan concentration (3.0-4.5%) at a molar ratio of 1:1.5 for GlcN:oxalic acid (Figure 1).

Maximum values for its elastic hardness and breaking point at 3% chitosan concentration were in the range of 1.0-1.5 mol. equiv. of oxalic acid to GlcN (Figure 2).

The gel did not undergo syneresis, and it was stable in water and in 2% aqueous acetic acid solution at room temperature for several months. The excess of oxalic acid present in the gel was removed by dialysis in distilled water to give the chitosan oxalate gel. Its elemental analysis data proved that the chitosan oxalate xerogel is composed of one mol. equiv. of oxalic acid per D-glucosaminyl residue, indicating the absence of the cross-linking of oxalic acid in chitosan chains. The structure was

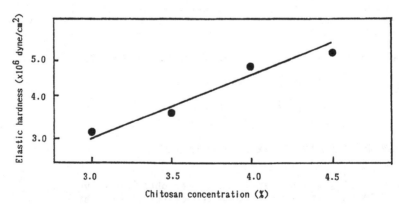

Figure 1. Effect of chitosan concentration on the elastic hardness of the chitosan oxalate gel at a molar ratio of GlcN:oxalic acid = 1.0:1.5

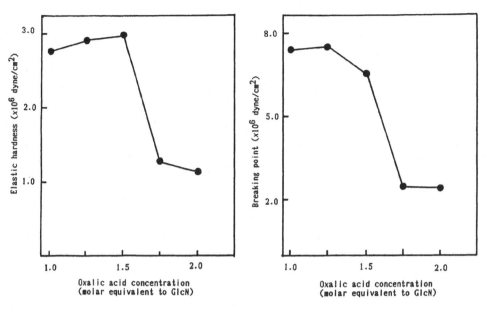

Figure 2. Effects of oxalic acid concentration on the elastic hardness and the breaking point of the chitosan oxalate gel at 3% chitosan concentration.

confirmed by the IR absorptions at 1700 (COOH), and 2600, 1620, and 1400 cm^{-1} (ionic salt) and by ^{13}C-resonances at 167 and 164 (δ) (C=O of oxalic) in the ^{13}C CP/MAS-NMR spectrum (Figure 3).

The gel formation was inhibited by the presence of N-acetyl group (d.s. >0.2/GlcN) in chitosan. The corresponding gel did not form from chitosan when it was dissolved in either an aqueous solution of malonic acid (pK$_1$, 2.86) or in an aqueous solution of succinic acid (pK$_1$, 4.16), both of which are weaker acids than oxalic acid. The gelation is initiated by formation of the ionic salt with oxalic acid (pK$_1$, 1.27) at d.s. >0.8/GlcN. The gelation mechanism of the chitosan oxalate gel seems to differ from the crosslinking of chitosan chains, because the minimum d.s. value required for gel formation by crosslinking was about 0.002, as previously reported in the reaction of chitosan with m- or p-phthalade-hyde.[14] Oxalic acid forms a salt with the amino groups of chitosan, and its gelation mechanism is similar to those of chitin gel[15] and N-aryli-dene- and N-alkylidene-chitosan gels (monoaldehydes),[10,11] in which the d.s. for N-substitution should be higher then 0.7/GlcN.

The gel shape was little changed, even after the complete removal of oxalic acid by treating the gel with aqueous NaOH solution; however, the gel turned opaque (Figure 4). The opaque xerogel was identified as chito-san on the basis of its IR and ^{13}C CP/MAS-NMR spectra, and on the elemen-tal analysis data. The elastic hardness and the breaking point of the chitosan gel were about two times higher than that of the chitosan oxalate gel (Table 1).

The opaque chitosan gel was treated with acetic anhydride in an aque-ous methanol solution at room temperature overnight, and a transparent gel was obtained. The gel was indistinguishable from chitin gel prepared from chitosan by N-acetylation in aqueous acetic methanol[14] on the basis of its IR and ^{13}C CP/MAS-NMR spectra, on the hydrolysis rate by lysozyme

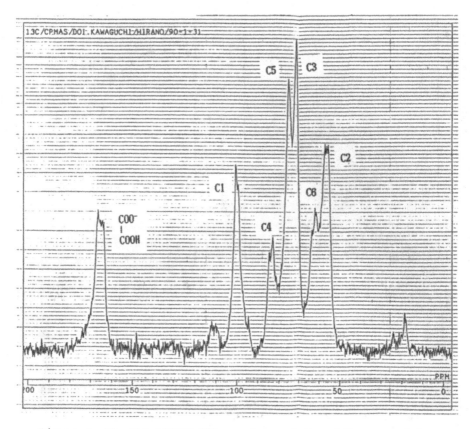

Figure 3. The ^{13}C CP/MAS-NMR spectrum of the chitosan oxalate xerogel.

(Table 2), and on a scanning electron microscopic observation of the xerogel.[16] The present study strongly indicates the presence of a common solid phase among the chitosan oxalate, chitosan, and N-acetylchitosan gels.

Figure 4. A view of the chitosan oxalate gel (right) and the chitosan gel (left).

Table 1. Some properties of the gels derived from chitin and chitosan.

| | Gels[a] | | |
	Chitosan oxalate	Chitosan	Chitin
Color	colorless	white	colorless
N-substituent	oxalate	free	acetamido
transparency	transparent	opaque	transparent
thermal property	reversible	irreversible	irreversible
stability in aqueous acids	stable	unstable	stable
stability in aqueous alkalis	unstable	stable	stable
elastic hardness ($\times 10^6$ dyne/cm^2)	2.8	3.0	6.4
breaking point ($\times 10^6$ dyne/cm^2)	1.9	14.0	1.5

(a) The chitosan oxalate gel was prepared from chitosan (0.9 g), oxalic acid (0.63 g, 1.25 molar equivalent to GlcN) and water (30 mL) as described in the experimental section. The chitosan and chitin gels were prepared from the chitosan oxalate gel.

2. Gelation of Chitin: Partially O-Acetylated Chitin Gel and Chitin Gel

Chitin dissolved in DMA in the presence of 5% LiCl was treated with acetic anhydride-pyridine at 70-100°C for 6 hrs. to give a partially O-acetylated chitin gel. The presence of O-acetyl group in the product was detected by IR absorption at 1760 and 1240 cm^{-1}, and its d.s. was 0.85/GlcNAc. The location of the O-acetyl group may be on C_6 because of its high activity. The O-acetylation did not effect on the gel properties. The gel was not distinguishable from the chitin gel produced by N-acetylation of chitosan in aqueous acetic acid-methanol.[15]

3. Hydrolysis Rate by Hen Egg-White Lysozyme

As shown in Table 2, the xerogels prepared by these methods were hydrolyzed at almost the same rate by hen egg-white lysozyme, and the rate was about 6 times higher than that of crab shell chitin. The hydrolysis rate by lysozyme was slightly inhibited by the presence of O-acetyl group in chitin xerogel, probably because of the steric hindrance of the O-acetyl group at C_6 for forming an enzyme substrate complex. The increase in the rate of enzymic hydrolysis also proves the formation of the same chitin gel by three independent methods.

In conclusion, this is the first report dealing with a basic polysaccharide gel that is thermally reversible and dealing with an intramolecular conversion of one gel to another gel in the solid phase. The chitin gels, which were produced by N-acetylation of the chitosan gel via the chitosan oxalate gel, and by O-deacetylation of the partial O-acetylated

Table 2. Hydrolysis rate by hen egg-white lysozyme.	
Xerogels	Reducing sugar value (μmol GlcN/mL)
Chitin gel from chitosan gel	0.31
Partially O-acetylated chitin gel	0.20
Chitin gel from partially O-acetylated chitin	0.35
Chitin gel from chitosan[14]	0.32
Native chitin (crab shell)	0.06

chitin gel prepared from chitin in DMA, were not distinguishable from the chitin gel prepared from chitosan by N-acetylation.[15]

REFERENCES

1. D. A. Rees, Chem. & Ind., (London), 630 (1972).
2. T. Harada, A. Misaki, and H. Saito, Arch. Biochem. Biophys., 24, 292 (1968).
3. M. Tako and S. Nakamura, Carbohydr. Res., 180, 277 (1988).
4. M. Tako and S. Nakamura, Carbohydr. Res., 155, 200 (1986).
5. M. Tako, S. Nakamura, and Y. Kohda, Carbohydr. Res., 161, 2147 (1987).
6. O. Smidsrod, Faraday Disc. Chem. Soc., 57, 263 (1974).
7. R. Mink and J. Blackwell, J. Mol. Biol., 120, 1167 (1978).
8. K. H. Gardener and J. Blackwell, Biopolymers, 14, 1481 (1975).
9. S. Hirano, Y. Ohe, and H. Ono, Carbohydr. Res., 47, 315 (1976).
10. S. Hirano, N. Matsuda, O. Miura, and H. Iwaki, Carbohydr. Res., 71, 339 (1979).
11. S. Hirano, N. Matsuda, O. Miura, and T. Tanaka, Carbohydr. Res., 71, 344 (1979).
12. S. Hirano, Agric. Biol. Chem., 42, 1939 (1978).
13. T. Imoto and K. Yagishita, Agric. Biol. Chem., 35, 115 (1971).
14. S. Hirano and M. Takeuji, Int. J. Biol. Macromol., 5, 373 (1983).
15. S. Hirano and R. Yamaguchi, Biopolymers, 15, 1685 (1976).
16. S. Hirano, R. Yamaguchi, and N. Matsuda, Biopolymers, 16, 1987 (1977).

TRANSPORT PROPERTIES OF MEMBRANES CONTAINING CHITOSAN DERIVATIVES

Yasuo Kikuchi and Naoji Kubota

Marine Science Laboratory
Faculty of Engineering
Oita University
Dannoharu, Oita 870-11, Japan

Water-insoluble polyelectrolyte complexes (PEC) containing chitosan derivatives, glycol chitosan (GC)-poly(vinyl
sulfate) (PVSK) and methyl glycol chitosan (MGC)-GC-PVSK
systems, were prepared in aqueous solution at various hydrogen ion concentrations. The molecular composition of each PEC
was dependent on pH, since the degree of dissociation and
conformation of GC were affected by pH. PEC membranes were
prepared, and their transport properties were investigated.
Measurements of the transport ratio of Na^+ and K^+ and electric potential difference between both sides of the membrane
suggested that the driving force of the uphill transport is
due to the membrane potential, that is, the sum of the Donnan
potential and diffusion potential. The transport selectivity
of K^+ was higher than that of Na^+. The permeabilities of KCl,
urea, and sucrose were also determined at various pH in the
case of the GC-PVSK membrane, and were found to be control-
lable by the solution pH. GC was modified with 3,3'-dithiodi-
propionic acid (DTPA) which causes the thiol$\leftrightarrow$disulfide transition through a redox reaction. The permeabilitites of KCl
and sucrose through the GC membrane modified with DTPA are
discussed based on the view that the thiol$\leftrightarrow$disulfide transition is responsible for the water content of the membrane.

INTRODUCTION

Chitin, obtained mainly from the cuticle of the marine crustacean, is
presently of considerable interest in industrial and medical fields,[1,2]
since it is a mucopolysaccharide which resembles cellulose structurally
and is present in nature as much as cellulose.[3] Deacetylation of the
acetamide group at the 2-position in the acetylglucosamine unit of chitin
by alkaline hydrolysis yields chitosan, a cationic polyelectrolyte. Chitosan appears more useful than chitin, having both hydroxyl and amino
groups which can be modified easily.[4] Therefore, these chitin and chitosan derivatives are finding a broad range of applications.

Polyelectrolyte complexes (PEC), prepared from cationic and anionic polyelectrolytes, should have many uses owing to the diversity of their structures and properties.[3,6] In particular, their applications as membranes have been widely proposed, and studies on the permeability of ions and low-molecular-weight solutes through PEC membranes have already been conducted.[6,7] The PEC membrane, however, has been studied primarily in regard to synthetic polymers, while only limited research has been on PEC membranes made from polysaccharides.[8,9] PEC can be prepared in cationic or anionic form as well as neutral polymer complexes through the proper choice of components, mixing ratio, and solvent pH, and can be cast into membranes of various ion exchange capacities (0-2 meq/g dry membrane).[10] Consequently, it is possible to control the transport of alkali metal and halide ions by such PEC membranes. The specific transport of ions is major common function of a biomembrane. Stimulus-sensitive membranes whose permeability changes according to the stimulus such as pH, redox reagent, or light have also been reported recently.[11-13]

The present article describes the characterization of glycol chitosan (GC), chemical reactions of GC and/or methyl glycol chitosan (MGC) with poly(vinyl sulfate) (PVSK), and general features of the resulting PEC in GC-PVSK and MGC-GC-PVSK systems. The transport of alkali metal ions through PEC membranes is discussed, with attention to the relation between the transport ratio and membrane potential. Assessment is also made of the feasibility of the permeability control of KCl, urea, and sucrose on the GC-PVSK membrane, whose compactness changes with pH, and the GC membrane modified with 3,3'-dithiodipropionic acid (DTPA), which causes the thiol↔disulfide transition with redox reagents. The results obtained provide support for the underlying concepts of the drug delivery system, release control capsule, and other matters.

EXPERIMENTAL

1. Materials

GC, MGC, and PVSK were the reagents for colloidal titration (Wako Pure Chemical Industry, Ltd., Osaka). DTPA was purchased from Aldrich Chemical Company, Inc., Milwaukee. The other chemicals were of reagent grade. For the potentiometric titration, deionized water (0.3 μS/cm) was used.

2. Characterization of Polyelectrolytes

Deionization of GC solution (0.02 base mol/L) was performed by use of a mixture of cationic and anionic exchangers. After adding NaCl up to 0.1 mol/L, 50 mL of GC solution was neutralized with HCl, followed by back titration with 0.5 mol/L NaOH at 25°C in nitrogen atmosphere. Titrations were carried out with a Hiranuma UCB-7 autoburet and a Horiba M-8 pH meter. Intrinsic viscosity of GC, MGC, and PVSK was measured with an Ostwald viscometer in the presence of 1 mol/L NaCl at 25°C.

3. Preparation of PEC

Reactions were run in 7%, 4%, 1% HCl, pH 2.0, 6.5, 11.0, 13.0, and 5% NaOH solutions. In the system of MGC-GC-PVSK, the molar ratio of the reactive groups of MGC to those of GC in the polycation solution was

unity. Polycation solution (2 g/L) was added dropwise to polyanion solution (2 g/L) and vice versa, where both solutions were adjusted to the
same [H$^+$], at a rate of 50 mL/30 min with stirring at 22 ± 2°C. After
being left for 30 min, the precipitate was separated by centrifugation.
It was washed with methanol, and dried *in vacuo* at room temperature.
Determination of nitrogen was carried out by the Kjeldahl method, and
quantitative analysis of sulfur was performed at the Institute of Physical and Chemical Research, Japan. IR spectra of the products were taken
with a Hitachi 270-50 spectrophotometer. The miscibilities of GC, MGC,
PVSK, and PEC with a ternary solvent system (NaBr/acetone/H$_2$O) were
examined as described previously.[14]

4. Modification of GC

A mixture of DTPA (0.32 g) and 1-ethyl-3(3-dimethylaminopropyl) car-
bondiimide hydrochloride (EDC) (0.58 g) was stirred in N,N-dimethylform-
amide (10 mL) at room temperature. The resulting clear solution was added
dropwise to a vigorously stirred aqueous solution of GC (0.21 g/10 mL)
which was cooled with ice-cold water. After standing overnight at room
temperature, the gel was precipitated with acetone, washed with ethanol,
and centrifuged. Disulfide content of the GC modified with DTPA was
determined by titration with KBrO$_3$ solution.

5. Membrane Application

PEC membranes were obtained by casting. The method and experimental
conditions were described in a previous paper.[15] PEC membranes were fixed
tightly with silicone rubber between two chambers of the diapragm-type
cell having an effective area of 4.0 cm^2. For the uphill transport, 25 mL
of 0.1 mol/L NaOH solution was introduced into the right-hand chamber,
and 25 mL of HCl or poly(styrene sulfonic acid) (HSS) solution of one of
various concentrations including 0.1 mol/L NaCl or 0.1 base mol/L sodium
poly(styrene sulfonate) (NaSS) into the left-hand chamber simultaneously.
The cell was then placed in a thermostat controlled at 30°C. At proper
intervals 0.1 mL samples from both chambers were withdrawn and measured
for the Na$^+$ concentration with a Shimadzu AA-640 atomic absorption spectrophotometer. At the same time, the concentration of Cl$^-$ was determined
by the volumetric titration with Hg(NO$_3$)$_2$, and the concentration of SS$^-$
from the absorbance at 260 nm. The electric potential difference between
the left- and right-hand sides of the membrane was obtained by measurement with a Horiba F-7 potentiometer and Ag-AgCl electrodes. The ion-
exchange capacity of the membrane was determined by titration.

For the selective transport, 25 mL of 0.05 mol/L NaOH and KOH mixture
solution was introduced into the right-hand chamber, and 25 mL of HCl
solution of one of various concentrations into the left-hand chamber.
Other procedures were the same as for the uphill transport experiment.

In the GC-PVSK system, the resulting precipitate of PEC was also
dried at 30°C on a Petri dish to obtain the membrane. For measurement of
the permeabilities of KCl, urea, and sucrose, 25 mL of pH 3.0 solution
was introduced into the left-hand chamber, and 25 mL of 0.05 mol/L KCl,
0.01 mol/L urea, or 0.02 mol/L sucrose into the right-hand chamber, where
pH value was adjusted as the same as that in the left-hand chamber. Similarly, the measurement was performed under proper pH from 3.5 to 7.0.
Other procedures were the same as for the uphill transport experiment.
Urea and sucrose were determined with a Hitachi U-2000 spectrophotometer

by the biacetyl monoxime method and the phenol-sulfuric acid method, respectively. For the GC modified with DTPA, we prepared the membrane as follows: The GC gel modified with DTPA, obtained in the reaction mixture, was washed well with an acetone-water (1:1) mixture, then washed with water, and centrifuged. It was held between silicone rubbers (50 x 50 x 0.5 mm), pressed with PMMA plates (50 x 50 x 5 mm) and dried at 30°C. The reduction and the oxidation of the membrane were achieved by immersing it in a 156 mmol/L methanol-water solution of Bu_3P and a 10 mmol/L aqueous solution of iodine, respectively, overnight. The measurement of the permeabilities of KCl and sucrose was carried out in the same manner as for the GC-PVSK membrane. The degree of swelling and water content of each membrane were calculated from the difference in the weights of the water-swollen and dried membranes.

RESULTS AND DISCUSSION

1. Characterization of Polyelectrolytes

GC is a chitosan derivative which is 2-hydroxyethylated at the 6-position. The degree of dissociation (protonation) varies with the pH of the solution, since GC has a primary amino group at the 2-position. It is water-soluble at all pH, in contrast to chitosan which is soluble in acidic medium but insoluble in basic medium. On account of this property, GC is quite usable as a reagent for colloidal titration. In addition, GC seems more suitable for studying the solution properties of chitosan derivatives. For example, the potentiometric titration of glycol chitosan hydrochloride (GCH) in the presence of 0.1 mol/L NaCl gave the pK-α curve shown in Figure 1. This curve indicates the dependence of the dissociation constant, K, on the degree of dissociation, α, of GCH.

The intrinsic and apparent pK of GCH in 0.1 mol/L NaCl solution are found to be 6.35 and 6.63, respectively. It is due to the good solubility of GC that such a pK-α curve was obtained over a wide range of α. The slope of the curve demonstrates the intramolecular electrostatic interaction between neighboring dissociable functional groups on the GC chain. A break is observed in the pK-α curve when the polyelectrolyte brings about a reversible conformational change such as the helix↔ coil transition of poly(L-glutamate)[16] or poly(L-lysine).[17] The linear pK-α curve obtained for GC, however, suggests that such a conformational change in the backbone of GC did not occur. That is, GC is a rather rigid polyelectrolyte whose β-(1,4) linkage is identical to that of cellulose. The flexibility of some polysaccharides was investigated in terms of "stiffness", i.e., the dependence of the intrinsic viscosity on the ionic strength. Terbojevich et al. obtained values of stiffness from 0.043 to 0.091 for chitosan with degrees of acetylation ranging from 52.2% to 12.1%.[18] These values are essentially the same as those for carboxymethylcellulose (0.065) and hyaluronate (0.080), greater than DNA (0.0055), and less than poly(acrylate) (0.23). GC also seems to have the same stiffness as these polysaccharides. Hwang et al. found the conformational change to occur in the chitosan backbone based on measurement of the intrinsic viscosity, [η].[19] They noted the intrinsic viscosity or hydrodynamic volume of chitosan to decrease with increase in the solution pH. In our study, the intrinsic viscosity of GC at high pH was found to slightly exceed that at low pH, as shown in Figure 2. This may be due to the 2-hydroxyethyl groups.

Figure 2 shows also the relationship between [η] and pH of MGC and PVSK, a strong polybase and strong polyacid, respectively. Linear curves

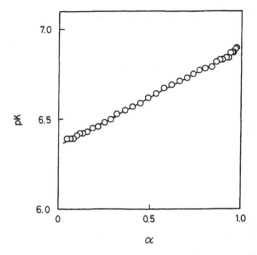

Figure 1. Dependence of the dissociation constant on the
degree of dissociation of GCH in the presence of
0.1 mol/L NaCl at 25°C.

were obtained from [η]-pH plots of MGC and PVSK. [η] of these two polyelectrolytes is not influenced by pH, because the degree of dissociation
of strong polyelectrolytes is constant over wide range of pH.

2. Structures and Properties of PEC

Figure 3 shows the dependence on the hydrogen ion concentration of

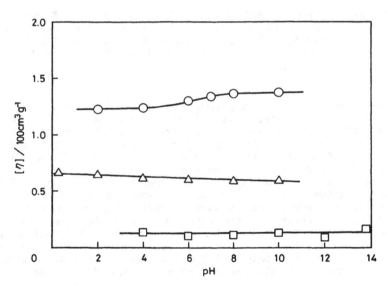

Figure 2. The pH dependence of the intrinsic viscosity of
polyelectrolytes in the presence of 1 mol/L NaCl at
25°C. o = GC, △ = PVSK, □ = MGC.

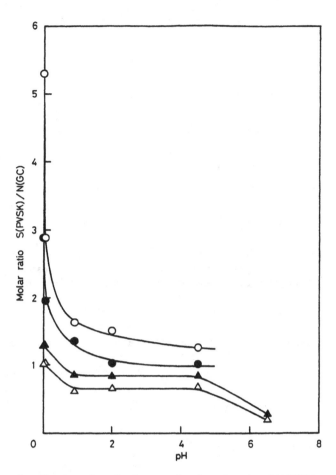

Figure 3. Start and end of coagulation in the GC-PVSK system.
o△ = start of coagulation, ●▲ = end of coagulation,
o● = GC solution was added dropwise to PVSK solution, △▲ = PVSK solution was added dropwise to GC
solution.

the molar ratio of the reacting groups in PVSK and GC in a reaction mixture solution at the start and end of coagulation of PEC. The molar
ratio, S(PVSK)/N(GC), increased with increase in [H+] and no coagulation
occurred in solutions at [H+] less than pH 6.5. This means that larger
amounts of PVSK solution are necessary to coagulate GC solution with
increase in the hydrogen ion concentration. The reason for this is that
the degree or dissociation of GC decreases with decrease in [H+] and
eventually reaches zero, whereas that of PVSK remains unchanged. In the
pH 6.5 solution, coagulation occurred on adding PVSK solution to GC solution but not when reversing this addition. A probable reason for this is
that the concentration of PEC and ionic strength in the reaction mixture
in the former situation are initially higher than those in the reaction
mixture in the latter situation.[6] Too, the degree of dissociation of GC
may easily change, with the reaction pH being near the apparent pK.

Experimental conditions and results of elemental analyses for each
PEC prepared following coagulation are given in Table 1. The amounts of
sulfur and nitrogen in PEC which reflect the molar ratio of PVSK to GC

Table 1. Reaction conditions and elemental analyses of PEC consisting of GC and PVSK.

		Reaction Conditions			
Sample[a]	[H$^+$]	Molar ratio in mixture (S/N)	Sulfur (%)	Nitrogen (%)	Molar ratio in PEC (S/N)
1-A	7% HCl	2.00	7.03	2.67	1.15
1-B	4% HCl	1.60	7.40	2.59	1.25
1-C	1% HCl	1.10	7.73	2.29	1.48
1-D	pH 2.0	1.00	7.61	2.18	1.53
1-E	pH 4.5	0.90	7.41	2.10	1.54
1-F	pH 6.5	0.50	4.64	5.19	0.39
2-A	7% HCl	2.00	7.04	2.15	1.43
2-B	4% HCl	1.60	6.51	2.02	1.41
2-C	1% HCl	1.10	7.01	2.07	1.48
2-D	pH 2.0	1.00	7.85	2.43	1.41
2-E	pH 4.5	0.90	8.13	2.60	1.37

(a) Series 1: PVSK solution was added dropwise to GC solution.
Series 2: GC solution was added dropwise to PVSK solution.

are roughly the same, except in the case of PEC prepared in the pH 6.5 solution. The reactions appear to proceed in the same way when PEC was prepared following coagulation in the solutions of [H$^+$] > pH 6.5. Thus, the hydrogen ion concentration in this system is an essential factor in determining the ratio of PVSK/GC in PEC. Since the IR spectra of PEC prepared when PVSK solution was added to GC solution are identical with those when GC solution was added to PVSK solution, only the former are shown in Figure 4. IR spectra of PEC are almost the same as those of a mixture of GC and PVSK. The absorption band at 1540 cm^{-1} assigned to NH$_3$' in GC is present in all PEC, whereas that assigned to -NH$_2$ which should appear at 1600 cm^{-1} is absent. The absorption band at 1230 cm^{-1}, assigned to -OSO$_3$$^-$ in PVSK, is also present in each PEC. The intensity of this absorption is nearly equal in all PEC except in the case of the pH 6.5 solution, whose absorbance at 1230 cm^{-1} is weak. Thus, PEC may possibly be produced in the same way, and this view is consistent with the results of elemental analyses in Table 1. The absorption bands around 3500 cm^{-1} assigned to -OH groups are present in all PEC, indicating the possible formation of inter- and intramolecular hydrogen bonding. These IR spectra indicate PEC prepared in 7% HCl-pH 4.5 solutions to contain a greater fraction of PVSK, whereas PVSK content in PEC prepared in the pH 6.5 solution is low. Consequently, PEC consisting of GC and PVSK seems to possess cation-exchange capacity.

A schematic model of the GC-PVSK system is shown in Figure 5, where GC is a rigid molecule and the density of the cation site of GC is less than that of the anion site of PVSK.[8] PEC is insoluble in common organic solvents, but soluble in specific ternary solvent mixtures (e.g., NaBr/acetone/H$_2$O and CaCl$_2$/1,4-dioxane/H$_2$O) without heating.[6] Phase diagrams were established for all PEC in the NaBr/acetone/H$_2$O system, as shown in Figure 6. There is a small region in the solvent composition field where PEC remains in solution to form a homogeneous, viscous syrup. The miscibility limit of each PEC is essentially constant except for that of PEC

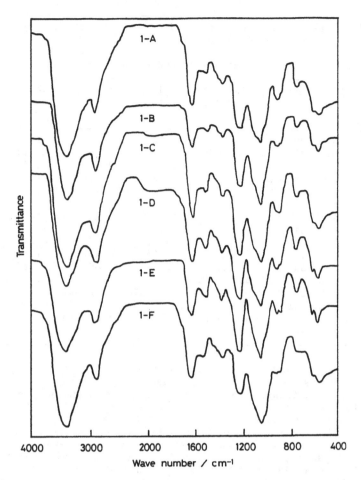

Figure 4. IR spectra of PEC in the GC-PVSK system. Sample codes correspond to those in Table 1.

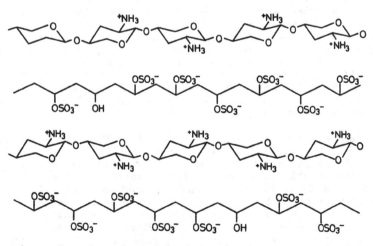

Figure 5. Schematic model of PEC in the GC-PVSK system.

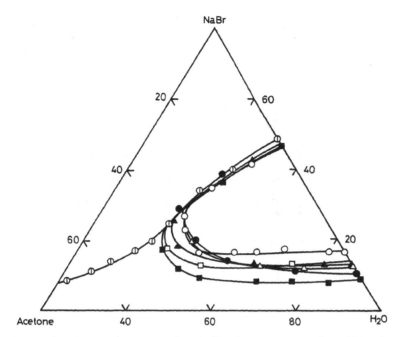

Figure 6. Phase diagrams of PEC in the ternary solvent system
at 30°C. Φ = NaBr/acetone/H₂O, o = 7% HCl, ● = 4%
HCl, Δ = 1% HCl, ▲ = pH 2.0, □ = pH 4.5, ■ = pH
6.5.

prepared in the solution of pH 6.5. These experimental results support
the consideration that PEC prepared in 7% HCl-pH 4.5 solutions have constituents in common and similar compositions, while PEC prepared in the
pH 6.5 solution differs from these, due to change in the degree of
dissociation of GC with $[H^+]$.

Figure 7 shows the dependence of the molar ratio of the reacting
groups in PVSK and MGC+GC in the reaction mixture solution at the start
and end of coagulation of PEC on the hydrogen ion concentration in the
MGC-GC-PVSK system. The molar ratio, S(PVSK)/N(MGC+GC), increased with
increase in $[H^+]$. Experimental conditions and results of elemental analyses for each PEC prepared following coagulation are given in Table 2.
Sulfur and nitrogen is present in about the same amounts in PEC. Coagulation did not occur in the solution of lower $[H^+]$ than pH 6.5 in GC-PVSK
system, and thus only MGC may react with PVSK in the solution of $[H^+]$ <
pH 6.5.

IR spectra of PEC prepared when PVSK solution was added to MGC+GC
solution are shown in Figure 8 and are almost the same as those of a
mixture of MGC, GC, and PVSK. The absorption band at 1540 cm⁻¹ assigned
to -NH₃⁺ in GC is present in all PEC except for PEC prepared in the pH
13.0 solution, and that assigned to -NH₂ which should appear at 1600 cm⁻¹
is absent. The absorption band at 1230 cm⁻¹, also present in each PEC, is
assigned to -OSO₃⁻ in PVSK. The intensities of these absorptions are
nearly equal in all PEC. Thus, PEC is apparently produced in the same way
except for that prepared in the pH 13.0 solution, when PEC was prepared
following coagulation.

Phase diagrams obtained for all PEC in the system of NaBr/acetone/H₂O

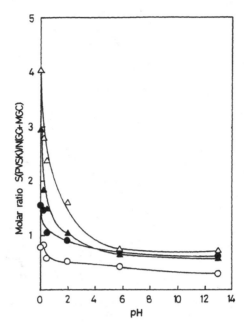

Figure 7. Start and end of coagulation in the MGC-GC-PVSK
system. o△ = start of coagulation, ●▲ = end of
coagulation, o●= PVSK solution was added dropwise
in MGC+GC solution, △▲= MGC+GC solution was added
dropwise in PVSK solution.

Table 2. Reaction conditions and elemental analyses of PEC
consisting of MGC, GC, and PVSK.

Reacting Conditions					
Sample[a]	[H+]	Molar ratio in mixture (S/N)	Sulfur (%)	Nitrogen (%)	Molar ratio in PEC (S/N)
1-A	7% HCl	2.00	6.70	2.63	1.11
1-B	4% HCl	1.60	7.02	2.98	1.03
1-C	1% HCl	1.20	8.15	3.13	1.14
1-D	pH 2.0	1.00	8.12	3.18	1.12
1-E	pH 6.5	0.80	8.10	3.04	1.16
1-F	pH 13.0	0.70	6.82	2.88	1.03
2-A	7% HCl	2.00	6.83	2.08	1.43
2-B	4% HCl	1.60	6.93	2.75	1.10
2-C	1% HCl	1.20	7.68	3.10	1.08
2-D	pH 2.0	1.00	8.17	3.22	1.11
2-E	pH 6.5	0.50	6.51	3.19	0.89
2-F	pH 13.0	0.50	8.19	3.13	1.14

(a) Series 1: PVSK solution was added dropwise to MGC+GC
solution.
Series 2: MGC+GC solution was added dropwise to PVSK
solution.

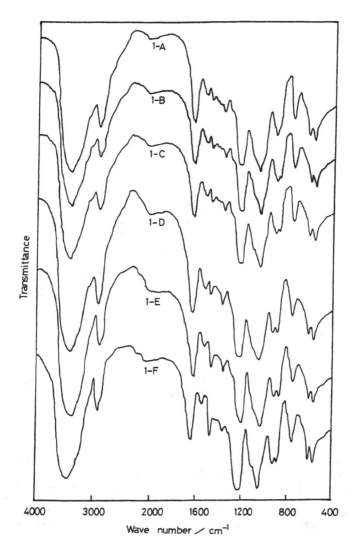

Figure 8. IR spectra of PEC in the MGC-GC-PVSK system. Sample
codes correspond to those in Table 2.

are shown in Figure 9. The miscibility limit of PEC prepared in acidic
solution is different from that of PEC prepared in alkaline solution.
This is consistent with the idea that PEC prepared in acidic solution
have constituents in common and similar compositions, while PEC prepared
in alkaline solution differs from these. This concept is attributed to
the degree of dissociation of MGC, GC, and PVSK which is in accordance
with $[H^+]$.

3. PEC membranes and Uphill Transport

The properties of PEC membranes in acidic and basic solution are
given in Table 3. The membranes formed at 50% and 90% relative humidity
(R.H.) except for 1-D, which was formed at pH 2.0, resisted acidic solu-

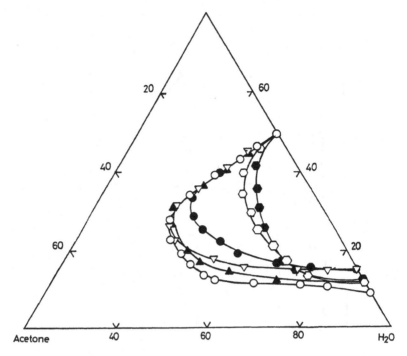

Figure 9. Phase diagrams of PEC in the ternary solvent system at 30°C. o = 7% HCl, ▲ = 4% HCl, ▽ = 1% HCl, ● = pH 2.0, ⬡ = pH 6.5, ⬢ = pH 13.0.

tion. The membranes except for 2-B, which was formed in a 4% HCl solution at 50% humidity, did not resist basic solution. The 2-B membrane was not damaged in either a 0.1 mol/L NaOH or 0.1 mol/L HCl solution, even after 5 days. The membrane obtained by casting PEC prepared in 4% HCl solution was used to study the transport of alkali metal ion, Na⁺, this membrane being the most stable in acidic and alkaline solution. The concentrations of Na⁺ in the left- and right-hand chambers of the cell were set the same

Sample[a]	30°C, 50% R.H.[b]		30°C, 90% R.H.[b]	
	0.1 mol/L HCl	0.1 mol/L NaOH	0.1 mol/L HCl	0.1 mol/L NaOH
1-B	o	Δ	o	Δ
1-D	–	–	–	–
2-B	o	o	o	Δ
2-D	o	–	Δ	–

Table 3. Stability of membranes in acidic and alkaline solution.

(a) Sample codes corresponding to those in Table 1.
(b) Relative humidity.

o = stable, Δ = slightly unstable, – = soluble.

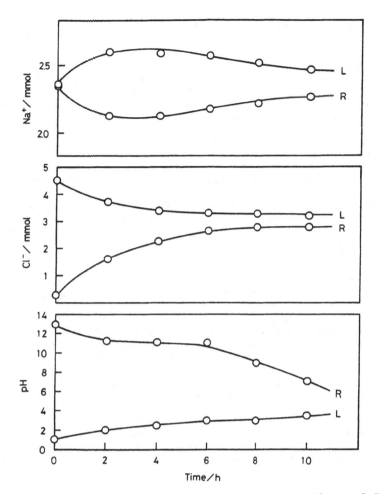

Figure 10. Changes in the concentration of Na⁺, Cl⁻, and H⁺
with time at the uphill transport.

initially. The left-hand chamber also contained HCl, while the right-hand chamber NaOH.

Figure 10 shows concentration changes in Na⁺, Cl⁻, and H⁺ with time in both chambers, with the left-hand chamber containing 0.1 mol/L HCl. Though both sides of the membrane were originally equal in the concentration of Na⁺, an increase in Na⁺ in the left-hand chamber was observed and change in the concentration of Na⁺ in the right-hand one was quite reverse. The uphill transport of Na⁺ may thus possibly occur from the right-hand chamber, alkaline solution, to the left-hand chamber, acidic solution, through the membrane. The concentration of Na⁺ in the left-hand chamber increased to a maximum, and then decreased with time. This back transport of concentrated Na⁺ is due to decrease in the hydrogen ion concentration difference between both sides of the membrane and permeation of Cl⁻. The ratios of transported ions were calculated by Equation 1, where $[Na^+]_{max}$ is the concentration of Na⁺ in the left-hand chamber at maximum, and $[Na^+]_o$ is the mean of the initial concentration of Na⁺ in both chambers.

$$\text{Transport ratio (\%)} = 100 \ ([Na^+]_{max}-[Na^+]o)/[Na^+]_o \qquad \text{(Equation 1)}$$

The dependence of the transport ratio of Na^+ on the hydrogen ion concentration in the left-hand chamber is shown in Figure 11. The transport ratio corresponds to $[H^+]$, and is high at $[H^+]$ above 0.1 mol/L, whereas low at $[H^+]$ below 0.01 mol/L. Hence, the driving force for the transport of Na^+ may be the hydrogen ion concentration difference between both side chambers, causing an electric potential gradient. Accordingly, the membrane potential difference should be measured. Figure 12 indicates the membrane potential difference to increase with the concentration of HCl in the left-hand chamber, this being consistent with the results in Figure 11. The membrane potential difference, however, decreased with time, due not only to the transport of Na^+ from the right- to left-hand chamber but also to the permeation of Cl^- from the left- to right-hand chamber. As shown in Figure 11, this Cl^- permeation is large, and thus an investigation was made of the transport for cases that negative charges in the left-hand side of the membrane are fixed. The results obtained were analyzed based on the above conclusion.

The results when using HSS and NaSS instead of HCl and NaCl, respectively, are shown in Figures 13 and 14. The transport ratios were determined by Equation 1, taking into consideration the transfer of water based on the Donnan osmosis. The concentration of Na^+ in the left-hand chamber increased and that in the right-hand one decreased with time as in the case of HCl. The concentration of SS^- did not change, in contrast to Cl^- in the case of HCl. Moreover, the transport ratios of Na^+ considerably exceeded those in the case of HCl, as shown in Figure 14. This is compatible with the assumption that the membrane potential difference is maintained for a long time because SS^- is fixed well in the left-hand side of the membrane.

This assumption is supported by the results in Figure 15. The membrane potential difference was maintained for more than 24 h. The transport of the alkali metal or halide ions was formerly explained as active

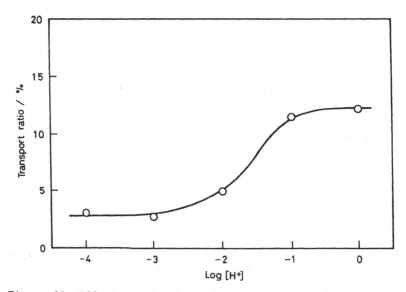

Figure 11. Effect of the hydrogen ion concentration on the transport ratio of Na^+ in the case when using HCl.

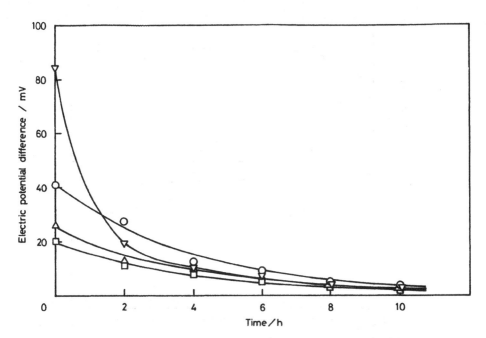

Figure 12. Changes in the electric potential difference with time in the case when using HCl. ∇ = 1 mol/L HCl, o = 0.1 mol/L HCl, $\triangle$ = 0.01 mol/L HCl, $\square$ = 0.001 and 0.0001 mol/L HCl.

transport by a proton-pump mechanism and ring opening-closing reaction.[20,21] Based on the present results the transport of Na^+ should occur as follows: PEC membrane in this study is considered a sort of charged membrane due to the functional groups of $-OSO_3^-$ in PVSK, since the molar ratio of S/N is larger than unity. The transport in a case where a charged membrane is used or a certain ion is fixed in one side of the membrane is explained by the Donnan membrane equilibrium. It is difficult to analyze such a system comprising many kinds of ions. However, assuming the membrane potential to be the concentration membrane potential raised by HCl or HSS, the membrane potential $\Delta\phi$ can be regarded as the sum of the Donnan potentials $\Delta\phi_{don,L}$ and $\Delta\phi_{don,R}$ at the two solution-membrane interfaces and diffusion potential $\Delta\phi_{dif}$ in the membrane.[22]

$$\Delta\phi = \Delta\phi_{don,L} + \Delta\phi_{dif} + \Delta\phi_{don,R} \qquad \text{(Equation 2)}$$

A tentative membrane potential and transport of Na^+ deduced from the above conclusions are shown in Figure 16. In this system, the membrane surface potential is low because the PEC membrane is anionic, and the potential in the membrane increases to the right owing to the large mobility of H^+. Thus, the right-hand chamber is higher than the left-hand chamber in electric potential. Na^+ is transported from the right- to left-hand chamber, i.e., from the alkaline to acidic side, through the membrane in accordance with the gradient of electric potential. However, the Donnan exclusion for Cl^- is small due to the small cation-exchange capacity (0.14 meq./g dry membrane) and there are $-NH_3^+$ groups in the membrane capable of attracting Cl^-, so that the membrane potential difference can not be maintained constant very long by Cl^- permeation. Then,

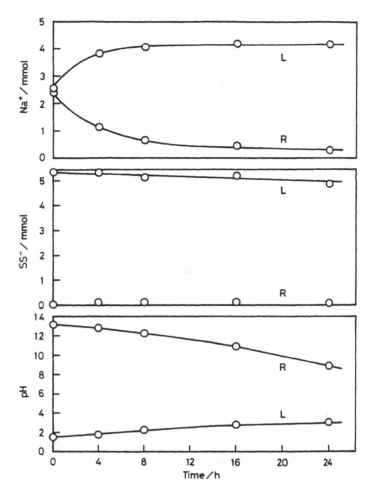

Figure 13. Changes in the concentration of Na⁺, SS⁻, and H⁺
with time at the uphill transport.

the transport ratio of Na^+ seems low in the case of HCl. In the case of
HSS, $\Delta\phi_{don,L}$ and $\Delta\phi_{dif}$ are larger than those in the case of HCl for fixed
SS^-, although $\Delta\phi_{don,R}$ is equal to that in the case of HCl. The retention
time of constant membrane potential difference is long. Therefore, the
transport of Na^+ in accordance with the membrane potential gradient is
appreciable. Recently, quantitative analysis of the uphill transport was
made by Higa et al.[23] In the MGC-GC-PVSK membrane, similar transport was
observed to occur by the same mechanism.[24] Changes in the concentration
of K^+ in the KCl-KOH system were somewhat larger than those in Na^+ in the
NaCl-NaOH system. The difference in the transport ratios of K^+ and Na^+
against the hydrogen ion concentration of the two systems were hardly
discernible at $[H^+]$ below 0.01 mol/L, whereas that of K^+ was larger than
that of Na^+ at $[H^+]$ above 0.1 mol/L. The transport rate was proportional
to the initial hydrogen ion concentration. It is of significance in sav-
ing energy and resources that the PEC membrane carrying a fixed functional group is capable of the uphill transport of alkali metal ions.

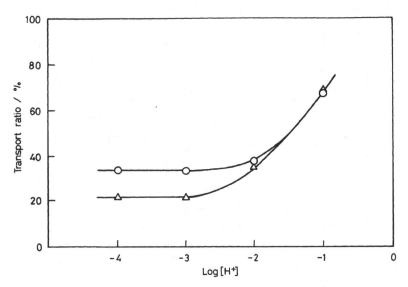

Figure 14. Effect of the hydrogen ion concentration on the transport ratio of Na$^+$ in the case when using HSS. o = Mol. Wt. 5.0 x 10^5, △ = Mol. Wt. 1.0 x 10^4.

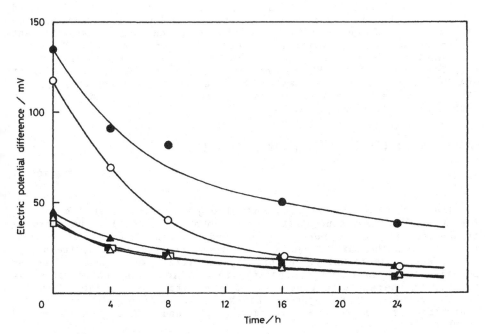

Figure 15. Changes in the electric potential difference with time in the case when using HSS. ● = 0.1 base mol/L HSS (Mol. Wt. 5.0 x 10^5), ▲ = 0.01 base mol/L HSS (Mol. Wt. 5.0 x 10^5), ■ = 0.001 and 0.0001 base mol/L HSS (Mol. Wt. 5.0 x 10^5), o = 0.1 base mol/L HSS (Mol. Wt. 1.0 x 10^4), △ = 0.01 base mol/L HSS (mol. wt. 1.0 x 10^4), □ = 0.001 and 0.0001 base mol/L HSS (Mol. Wt. 1.0 x 10^4).

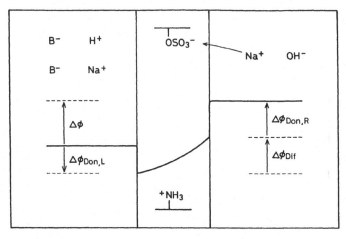

Figure 16. Schematic representation of the membrane
potential.

4. Selective Transport

Selective transport from the right-hand chamber to the left-hand
chamber through the membrane was observed at a 0.05 mol/L concentration
of both NaOH and KOH solution in the right-hand chamber and at different
hydrogen ion concentrations in the left-hand chamber at 30°C. The ratio
of transported ions, i.e., the selectivity, was calculated as follows
from Equation 3, where $[K^+]_{L,t}$ and $[Na^+]_{L,t}$ are the concentrations of K^+
and Na^+ in the left-hand chamber at a certain time, respectively, and
$[K^+]_{R,o}$ and $[Na^+]_{R,o}$ are the initial concentrations of K^+ and Na^+ in the
right-hand chamber, respectively.

$$\text{Selectivity} = ([K^+]_{L,t} / [K^+]_{R,o}) / ([Na^+]_{L,t} / [Na^+]_{R,o})$$

(Equation 3)

The selectivity of the transport through membrane after 4 h at different hydrogen ion concentrations is shown in Figure 17. The ratio of
transported K^+ to transported Na^+ changed with the hydrogen ion concentration. When the initial hydrogen ion concentration was low, the selectivity was close to unity. In this region, alkali metal ions are transported without any great or specific interaction with the membrane. In
the region where the hydrogen ion concentration was higher than 0.1
mol/L, more K^+ were transported than Na^+. The optimum hydrogen ion concentration is 0.1 mol/L at 0.05 mol/L NaOH and KOH in the right-hand
chamber.

These results suggest that the main contributing factor in the selectivity of the transport of alkali metal ions is the hydrogen ion concentration. The conclusion that PEC moiety extends on the alkaline side and
shrinks on the acidic side is well confirmed by the following experimental results of the swelling or shrinking of the membrane immersed in basic
or acidic solution. The horizontal and vertical dimensions of the membrane immersed in the pH 1.0 and 13.0 solutions were 25 mm x 28 mm and 33

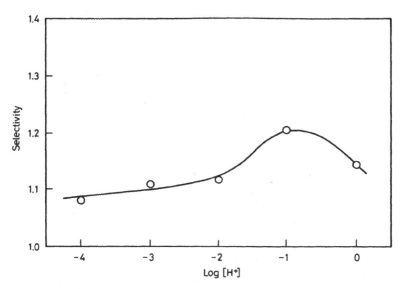

Figure 17. Effect of the hydrogen ion concentration on the selectivity in the transport of Na⁺ and K⁺ after 4 h.

mm x 34 mm, respectively. When the above membranes, pretreated with the pH 1.0 and pH 13.0 solution, were further immersed in the pH 13.0 and pH 1.0 solution, respectively, the above two dimensions were about 31 mm x 34 mm and 30 mm x 30 mm, indicating the membrane to swell in basic solution and shrink in acidic solution. In the region where the hydrogen ion concentration was larger than 0.1 mol/L in the left-hand chamber, K⁺ was transported faster than Na⁺ since the radius of the hydrated K⁺ is smaller than that of the hydrated Na⁺, due to dense polymer network in the membrane. In the region where hydrogen ion concentration was smaller than 0.01 mol/L, no appreciable selectivity of the transport could be found owing to the indiscriminate transport of alkali metal ions, as a result of loose polymer network formation in the membrane. The degree of swelling and shrinking of the PEC membrane may thus possibly control the selective transport through the membrane.

5. Permeability Properties of PEC membrane

Based on the above, the GC-PVSK membrane was used for a permeability control experiment. GC solution was added dropwise to PVSK solution at pH 3.0. PEC consisting of GC and PVSK has been considered a ladder-like structure of ionic bond between $-NH_3^+$ groups in GC and $-OSO_3^-$ groups in PVSK as shown in Figure 5. Further, inter- and intramolecular hydrogen bonding seems to take place due to the fact that GC is a polysaccharide having many hydroxyl groups.

The GC-PVSK membrane was quite sturdy and resembled a cellophane membrane. It has been reported that if PEC contains the strong polyelectrolyte PVSK, its ionic bond is maintained throughout a wide pH range,[25] and then the GC-PVSK membrane is also stable over a wide range of pH. The stability of GC-PVSK membrane seems to be due not only to the above reason but also to the packing of GC. PEC consisting of GC and PVSK is thus favorable as membrane material. The permeability of KCl, urea, and

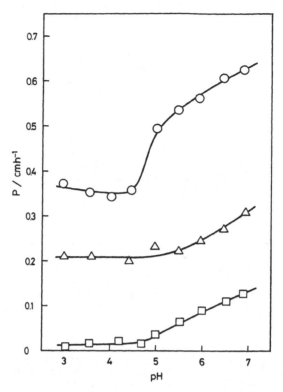

Figure 18. The pH dependence of the permeabilities of solutes through PEC membrane consisting of GC and PVSK at 30°C. o = KCl, △ = urea, ☐ = sucrose.

sucrose through this membrane was investigated, and the results are shown in Figure 18.

The permeability was estimated by the following Equation 4, derived from the Fick's law of diffusion,[26] where V is the volume of each chamber of the cell, A is the effective area of membrane, t is the time, ΔC is the difference in the concentrations of the right- and left-hand chambers, and C_0 is the initial concentration of solutes in the right-hand chamber.

$$\text{Permeability} = -V/2At \ \ln(\Delta C/C_0) \qquad \text{(Equation 4)}$$

Figure 18 shows the permeability of KCl, urea, and sucrose to depend on molecular weight and pH. Each value for the permeability decreased with molecular weight in the order, KCl > urea > sucrose. Since the GC-PVSK membrane is a cation-exchange membrane as described above, the electric potential in the left-hand chamber must be higher than that in the right-hand chamber when salt such as KCl is permeated. The electric potential difference between both sides of the membrane was actually 7-8 mV at pH 3.0 even after 10 h. In the IR spectra of the GC-PVSK membranes treated for 4 h with the buffer solutions of pH 3.0, 4.0, 5.0, 6.0, and 7.0, the absorption band at 1530 cm[-1] assigned to $-NH_3^+$ groups was present in all membranes, whereas 1600 cm[-1] assigned to $-NH_2$ groups of GC was absent. That is, the dissociation of GC in the membrane was not observed at pH from 3.0 to 7.0. It is clear that the breaking of ionic bonds does not lead to an increase in the permeability. Since GC is extended with increase in pH, as mentioned above, and PVSK has a flexible polymer chain, the GC-PVSK membrane should become somewhat swollen so that the permeability increases.[27] Hwang et al. prepared micro capsules whose diffusion rate for γ-globulin and BSA decreased by decreasing in the solution pH, at which chitosan was gelled with alginate.[19] The PEC membrane in the GC-PVSK system can control the permeability of KCl, urea, and sucrose on pH, even if PEC is prepared at a constant pH value. These investigations suggest a possibility of permeation control by the PEC membrane consisting of GC, an amino polysaccharide, and PVSK, having a flexible polymer chain, because the permeability can be controlled by the solution pH which controls the compactness and packing of PEC.

6. Permeability Properties of Modified GC Membrane

GC was used as the membrane matrix and functional groups, which change through redox reactions, were introduced into the side chains in the manner shown in Figure 19. Figure 20 shows IR spectra of GC, GC modified with DTPA, and N-(3-mercaptopropionyl)GC obtained in reduction of GC modified with DTPA. GC has the absorption band assigned to free amino groups at 1600 cm^{-1} whereas GC modified with DTPA does not, but rather the absorptions assigned to amide groups at 3090, 1660, and 1550 cm^{-1}. The shoulder at 1730 cm^{-1} of GC modified with DTPA indicates the presence of the ester carbonyl resulting from combination of hydroxyl groups of GC with carbonyl groups of DTPA. This shoulder seems to include the absorption of carboxyl groups of DTPA condensed with GC on one side, since its intensity decreases in the next reduction step, as will be discussed later. The absorption at 2560 cm^{-1} assigned to thiol groups is observed in the spectrum of N-(3-mercaptopropionyl)GC. Thiol groups have been found generally unstable in the monomer form, but appreciably stable in the polymer matrix.[28] Their stability in GC was thus studied by examining change in the absorption at 2560 cm^{-1}. The absorption was reduced to about half the initial value after standing for 30 days in the atmosphere, as shown in Figure 20.

An attempt was made to cast a polymer membrane from GC modified with DTPA, but without success because the polymer obtained was completely insoluble in common organic solvents and water. In consideration of the fact that chitin derivatives reproduced the hydrogen bonding to yield films when dried after being washed with water,[29] we tried to prepare the membrane as follows: The GC gel modified with DTPA, which was obtained in the reaction mixture, was washed well with an acetone-water mixture to keep the precipitate highly swollen, washed again with water, and dried while being held between silicone rubbers and PMMA plates. By this procedure, the GC membrane modified with DTPA was obtained. The thiol$\leftrightarrow$disulfide transition was found to occur in the membrane matrix as evident from the finding that the IR absorption at 2560 cm^{-1} disappeared by iodine oxidation and reappeared by reduction with Bu$_3$P.

The modified GC membrane was then placed in a diaphragm type cell and the permeabilities of KCl and sucrose were measured. The values, determined from Equation 4, are summarized in Table 4 along with the sulfur content, degree of swelling, and water content of each membrane. The permeability of KCl through the GC membrane modified with DTPA (M_{DTPA}) increased by reduction (M_{SH}), and that through the M_{SH} membrane decreased by oxidation (M_{SS}). Takizawa et al. found the polypeptide membrane containing thiol groups to show a 90% decrease in the permeability of KCl by the oxidative reaction,[12] while the decrease in the permeability of our

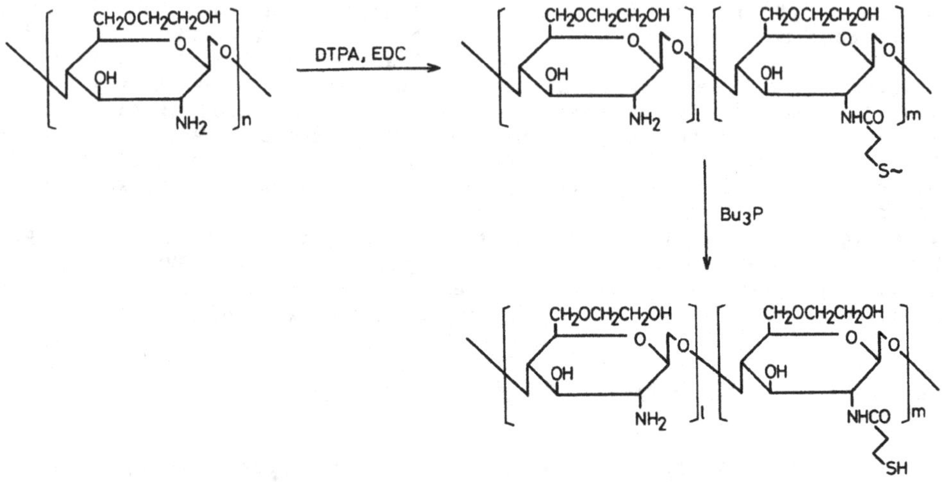

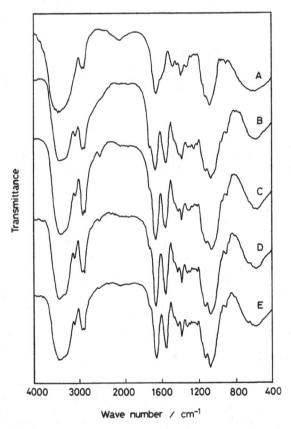

Figure 19. Synthetic route of GC modified with thiol groups.

Figure 20. IR spectra of GC and each product. A = GC, B = GC modified with DTPA, C = N-(3-mercaptopropionyl)GC, D = N-(3-mercaptopropionyl)GC after standing for 30 days, E = N-(3-mercaptopropionyl)GC after standing for 90 days.

system was about 60% the permeability which had increased by the previous reduction. The permeability of sucrose had the same tendency as shown in Table 4. In the case of sucrose, all of the permeabilities are lower than those of KCl, possibly due to the molecular and ionic size of each solute.

The sulfur contents of the M_{DTPA} and M_{SH} membranes were 9.32 and 5.99 wt%, respectively. The decrease in this parameter may have resulted from the elimination of 3-mercaptopropionic acid produced in reduction of DTPA condensed with GC on one side. Based on the difference in the sulfur contents of the M_{DTPA} and M_{SH} membranes, the amount of 3-mercaptopropionic acid eliminated was estimated as 44%. The resulting 3-mercaptopropionyl groups in the M_{SH} membrane may be the cause for the reversible thiol↔disulfide transition through redox reaction. Figure 21 presents a schematic structure of each membrane. In the case of the M_{DTPA} membrane, 1.45 mmol/g of disulfide groups, as measured by elemental analysis and titration with $KBrO_3$, were introduced into the side chains. In the case of the M_{SH} membrane, 44% of 3-mercaptopropionic acids were eliminated by reduction. In the M_{SS} membrane, the remaining thiol groups, 1.87 mmol/g, combined again to form disulfide crosslinkings by the oxidative reaction. Actually, the M_{SH} membrane shrank immediately when immersed in a solution of iodine. The backbone of GC, however, is somewhat stiff so that the vacancy due to the elimination of 3-mercaptopropionic acid cannot be filled completely. Consequently, the structure of the M_{SS} membrane may be somewhat more dilated than that of the M_{DTPA} membrane. To control the permeability by introducing thiol groups into the side chains, a more flexible macromolecule for the membrane matrix than GC should perhaps be used though assessment should be made of the ability to form a membrane.

The degree of swelling and water content, which reflect the degree of crosslinking and compactness of the membrane, support the hypotheses indicated by the schematic models in Figure 21. The degree of swelling and water content were in the order of M_{DTPA} < M_{SS} < M_{SH}. In parallel, the permeabilities of KCl and sucrose through each membrane were in the same order, M_{DTPA} < M_{SS} < M_{SH}. The relation between the permeability coefficient in cm^2/S and water content was studied previously, and the permeability through the hydrophilic and hydrophobic membranes was found to depend on the water content.[30] The relation between the permeability coefficients of KCl and sucrose and water content is shown in Figure 22. Although the water content of our membranes is only in the high region, the logarithmic permeability coefficients and water contents are shown to be linearly related and the slope in the case of sucrose considerably exceeds that for KCl. This suggests that the permeability of sucrose can be controlled more easily than that of KCl in our system.

The GC membrane modified with thiol groups, whose functional groups change reversibly by redox reaction, was found capable of controlling the permeabilities of KCl and sucrose to some extent. Change in the permeabi-

Table 4. Properties and permeabilities of modified GC membranes.

Membrane	Sulfur (%)	Degree of swelling	Water (%)	Permeability	
				KCl (cm/h)	sucrose (cm/h)
M_{SS}	9.32	2.24	55.4	0.564	0.114
M_{SH}	5.99	5.90	83.1	0.836	0.237
M_{DTPA}	–	3.76	73.4	0.628	0.179

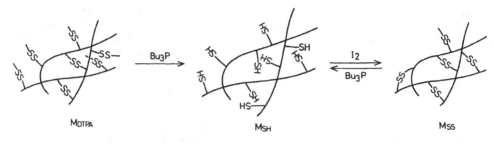

Figure 21. Schematic representation of the structure of modified GC membranes.

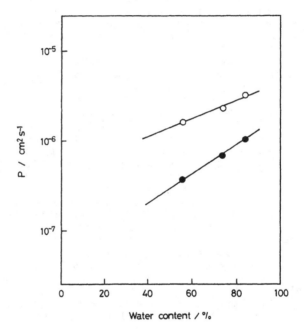

Figure 22. Relationship between the permeability coefficient
and water content. o = KCl, ● = sucrose.

lity can be explained on the basis of the extent of disulfide crosslinking and water content of the membrane. To achieve the ON/OFF permeation
control, the water content should perhaps be changed more by increasing
the number of thiol groups.

REFERENCES

1. R. A. A. Muzzarelli, C. Jeuniaux & G. W. Gooday, Eds., "*Chitin in Nature and Technology*," Plenum Press, New York, 1986.
2. E. R. Pariser & D. P. Lombardi, "*Chitin Sourcebook*," John Wiley & Sons, Inc., New York, 1989.
3. M. V. Tracey, Rev. Pure Appl. Chem., 7, 1 (1957).
4. S. Hirano, Kobunshi, 37, 608 (1988).
5. H. J. Bixler & A. S. Michaels, Encycl. Polym. Sci. Technol., 10, 765 (1969).
6. E. Tsuchida & K. Abe, Adv. Polym. Sci., 45, 18 (1982).
7. S. M. Ryu & I. F. Miller, J. Biomed. Mater. Res., 5, 287 (1971).
8. A. Nakajima & K. Shinoda, J. Appl. Polym. Sci., 21, 1249 (1977).
9. C. M. Berke, U.S. Patent 4501835 (1985); Chem. Abstr., 102, 169601v (1985).
10. L. L. Markley, H. J. Bixler & R. A. Cross, J. Biomed. Mater. Res., 2, 145 (1968).
11. M. Maeda, M. Kimura, Y. Hareyama & S. Inoue, J. Am. Chem. Soc., 106, 250 (1984).
12. A. Takizawa, T. Kinoshita, A. Ohtani & Y. Tsujita, J. Polym. Sci., Polym. Chem. Ed., 24, 665 (1986).
13. Y. Okahata, Acc. Chem. Res., 19, 59 (1986).
14. Y. Kikuchi, K. Hori & Y. Ohnishi, Nippon Kagaku Kaishi, 1980, 1157.
15. Y. Kikuchi & N. Kubota, Bull. Chem. Soc. Jpn., 58, 2121 (1985).
16. M. Nagasawa & A. Holtzer, J. Am. Chem. Soc., 86, 538 (1964).
17. J. Hermans, Jr., J. Phys. Chem., 70, 510 (1966).
18. M. Terbojevich, A. Casani, M. Scandola & A. Fornasa in: "*Chitin in Nature and Technology*," R. A. A. Muzzarelli, C. Jeuniaux & G. W. Gooday, Eds., Plenum Press, New York, 1986, p. 349.
19. C. Hwang, C. K. Rha & A. J. Sinskey in: "*Chitin in Nature and Technology*," R. A. A. Muzzarelli, C. Jeuiaux & G. W. Gooday, Eds., Plenum Press, New York, 1986, p. 389.
20. T. Shimidzu, M. Yoshikawa, M. Hasegawa & K. Kawakatsu, Macromolecules, 14, 170 (1981).
21. N. Ogata, K. Sanui & H. Fujimura, J. Appl. Polym. Sci., 25, 1419 (1980).
22. Y. Kikuchi & N. Kubota, Bull. Chem. Soc. Jpn., 60, 375 (1987).
23. M. Higa, A. Tanioka & K. Miyasaka, J. Membr. Sci., 37, 251 (1988).
24. Y. Kikuchi & N. Kubota, J. Appl. Polym. Sci., 35, 259 (1988).
25. H. Sato, M. Maeda & A. Nakajima, J. Appl. Polym. Sci., 23, 1759 (1979).
26. N. Kubota, J. Appl. Polym. Sci., in press.
27. Y. Kikuchi & N. Kubota, Bull. Chem. Soc. Jpn., 61, 2943 (1988).
28. T. Saegusa, S. Kobayashi, K. Hayashi & A. Yamada, Polymer J., 10, 403 (1978).
29. Y. Kobayashi, M. Nishimura, R. Matsuo, S. Tokura & N. Nishi, "*Proc. Second Internat. Confer. on Chitin and Chitosan*," Sapporo, 1982, p. 239.
30. H. Yasuda, L. D. Ikenberry & C. E. Lamaze, Makromol. Chem., 125, 108 (1969).

SORPTION BEHAVIOR OF CHEMICALLY MODIFIED CHITOSAN GELS

Toshihiro Seo* and Toshiro Iijima†

Department of Polymer Science
Tokyo Institute of Technology
Ookayama, Meguro-ku, Tokyo 152, Japan

The chemical modification of chitosan by acylation of the amino group with various carboxylic anhydrides was performed to introduce a new functionality. The state of water in N-acylated chitosan gels was studied by DSC technique. The amounts of two states of water, freezable and nonfreezable water, were found to depend on the chemical and physical structure of the gels. The interaction between chitosan and its acyl (octanoyl, dodecanoyl, octadecanoyl and benzoyl) derivatives and the dyes carrying the ionic, hydrophilic and hydrophobic nature was discussed in detail. The sorption isotherms were interpreted by means of a dual mechanism which comprises partition and Langmuir sorption modes. The dyes having $-OH$, $-N(C_4H_9)_2$, and a naphthalene nucleus show a remarkable increase of the equilibrium sorption compared with that of Methyl Orange. The contribution of the electrostatic, hydrophobic and hydrogen bonding interaction in the sorption was discussed. On the other hand, the sorption of D, L-amino acids by chitosan gels with hydrophobic group was investigated under various conditions. The L-amino acids are sorbed to a greater extent than their isomers, which suggests that chemically modified chitosan gels are able to separate D,L-amino acids.

INTRODUCTION

Chitosan, [1], obtained by the deacetylation of chitin, is an aminosaccharide which constitutes a useful natural polymer resource. Recently, some interesting results have been presented in which the utility of chitosan as a source of value-added polymers with selective sorption functions has been explored.[1-5] On the other hand, the basic researches on the interaction between those materials and small molecules are rather few.[6]

---

* To whom correspondence should be addressed.
† Present address: Dept. of Textiles, Jissen Women's University, Oosakaue, Hino 191, Japan.

In exhibiting the sorptivity of both the natural and synthetic polymers, water in a polymer is known to have an important role.[7] The properties of polymer are strongly affected by the state of water in the polymer matrices.[8] In this context, the differential scanning calorimetry (DSC) was performed on the sorbed water/chitosan gel systems.

In the present work, anionic dyes were used as the small sorbate molecules. The various azo dyes can be prepared by designing the ionic group, the hydrophobic group, and the groups which have a specific function as, for example, hydrogen bond formation. To elucidate the mechanism of the sorption of small molecules by chitosan gels having hydrophobic groups [2], the sorption isotherms of these dyes were investigated.

It is well known that the natural polymer cellulose and its derivatives have the ability to separate a racemate into enantiomers.[9,10] Consequently, further work was carried out on the selective sorption of various D, L-amino acids by chemically modified chitosan gels.

EXPERIMENTAL

1. Materials

Chitosan which have the degree of the substitution of amino group, 0.85 (sample I) and 0.77 (sample II) (Kyowa Yushi Co. Ltd.) were used. The chitosan flakes were dissolved in 10% acetic acid and methanol and acylated by the carboxylic anhydride. The gel formed was washed with water and neutralized with 0.1N KOH. Finally the products were washed with water and dried. A gel particles obtained were of diameter 50 to 300 μm.[5,6] The carboxylic anhydrides used were octanoic, dodecanoic and octadecanoic derivatives. The chitosan gels obtained are shown in Figure 1. Figure 2 shows the variety of chemical structures for the dyes and the amino acids used, respectively. All of the dyes were prepared by the conventional methods, or purchased from Tokyo Kasei Co. D,L-amino acids were obtained from Nakarai Chemicals Co. The figures in the parentheses indicate Rekker's hydrophobic fragmental constant, Σf.[11]

2. DSC Measurement

Freezing and melting of water-swollen gels were investigated by DSC (DSC-10, Seiko I. and E. Ltd.). Adhering external water was carefully removed from the membrane with filter paper, and the gels were sealed in

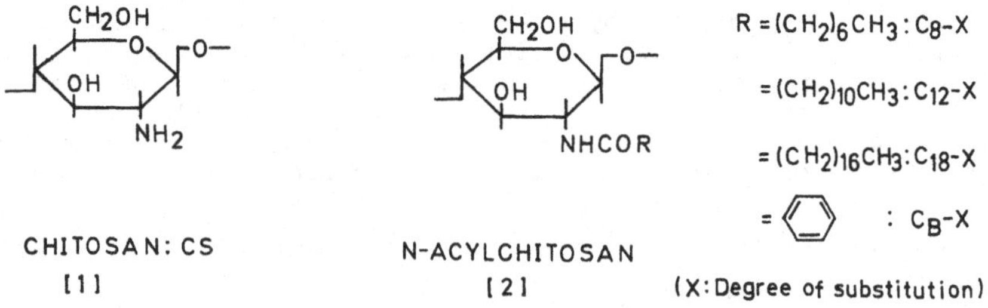

Figure 1. Structural formulas of chitosan and N-acylchitosan gels.

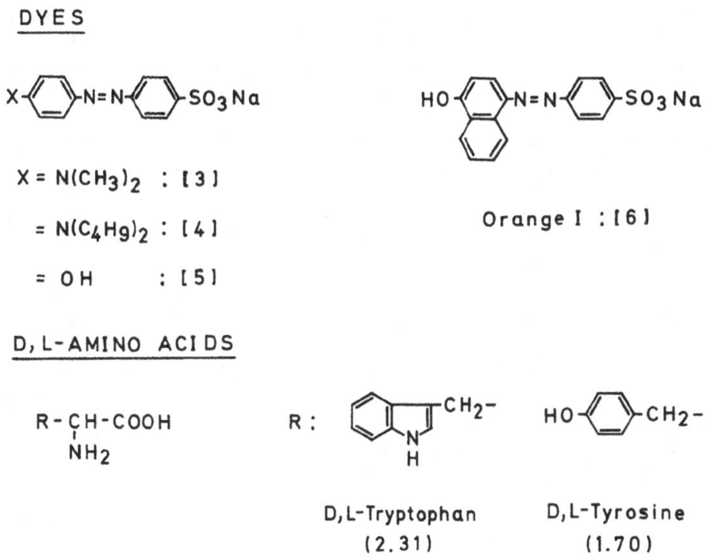

<u>DYES</u>

X—⟨◯⟩—N=N—⟨◯⟩—SO₃Na

$X = N(CH_3)_2$: [3]

$= N(C_4H_9)_2$: [4]

$= OH$: [5]

HO—⟨◯◯⟩—N=N—⟨◯⟩—SO₃Na

Orange I : [6]

<u>D,L-AMINO ACIDS</u>

R—CH—COOH
⎮
NH₂

R : ⟨indole⟩—CH₂—

D,L-Tryptophan
(2.31)

HO—⟨◯⟩—CH₂—

D,L-Tyrosine
(1.70)

Figure 2. Types of azo dyes and amino acids used.

an aluminum pan. The sample pan was cooled from 300 to 170°K at a speed of 2.5°K min⁻¹ and then heated to 300°K at the same speed. The amounts of the freezable water were estimated from the endothermic peak in the thermogram obtained. The rest of the total water content was assigned to the nonfreezable water.

3. Equilibrium Sorption

Weighted amounts of gels were soaked in an aqueous solution of dyes or amino acids of the required concentration and shaken 4 days at 30°C. The attainment of equilibrium sorption was ascertained in preliminary experiments. The amounts of equilibrium sorption was determined by measuring the initial and final concentrations spectrophotometrically.

RESULTS AND DISCUSSION

1. Properties of N-Acylchitosan Gels

The characteristics of the gels are summarized in Table 1. Octanoyl chitosan with various degree of substitution up to approximately 1.0 was prepared by reaction at 25°C for 20 h, as described in the previous paper.[6] Dodecanoyl and benzoyl chitosan was obtained at 50°C for 2 h and further standing at 25°C for 20 h. Octadecanoyl chitosan was obtained at higher temperature. A D.S. lower than 0.3 makes the gels highly soluble in acetic acid and D.S. higher 0.5 increases their solubility in a LiCl (1 g)/N-methylpyrrolidone (10 mL)/N,N-dimethylacetamide (10 mL) mixture. In the case of the longer octadecanoyl and benzoyl gels, the affinity to the organic solvents becomes larger.

The water content (g/g of dried polymer) of CS-I and its octanoyl

Table 1. Characteristics of N-acylchitosan gels.				
			Solubility in[e] LiCl/NMP/DMAc	
Sample No.	D.S.[a]	10% AcOH	(1/10/10)	Wc[f]
CS-I	(0.85)[b]	s	i	1.72
CS-II	(0.77)[b]	s	i	3.18
C_8-0.3[c]	0.29	s	i	1.85
0.4	0.41	sw	sw	1.54
0.6	0.58	i	sw	1.39
0.8	0.79	i	s	0.26
1.0	1.02	i	s	0.21
C_{12}-0.3[c]	0.32	s	i	1.24
0.5	0.54	sw	sw	0.58
0.6	0.58	i	sw	0.45
0.9	0.94	i	s	0.11
C_{18}-0.3[c]	0.30	sw	sw	0.98
0.4	0.42	sw	sw	0.89
C_B-0.3[d]	0.32	sw	i	3.33
0.5	0.51	i	sw	1.78
0.7	0.68	i	s	1.75

(a) Degree of substitution of acyl group.
(b) Degree of substitution of amino group.
(c) Prepared from CS-I.
(d) Prepared from CS-II.
(e) i = insoluble, s = soluble, sw = swelling.
(f) Water content (g/g of dried polymer).

derivatives gives a maximum at D.S. = 0.3 and drastically decreases at D.S.> 0.8. In the case of dodecanoyl derivatives, the water content decreases monotonously with increase in D.S. On the other hand, benzoylated CS-II has increasing water content at D.S. = 0.3 and decreasing at D.S. > 0.4. Further increase to D.S. = 0.7 was found to keep the same level of the water content.

Figure 3 is an example of the DSC thermogram at the elevating temperature process. A clear distinction of the thermogram of the gels from that of pure water is the appearance of a shoulder at the lower temperature range in the endothermic peak. The shoulder suggests the existence of different kind of water species, which would be assigned as water weakly interacted with the polymer. The phase transition peak from ice to liquid water is extremely small for the gel of D.S. = 1.0.

In Table 2, the amounts of the freezable and nonfreezable water are given. For the chitosan gel itself, the higher the amino content, the lower the amounts of the freezable water. The introduction of an octanoyl group (D.S. = 0.3-0.4) leads to the increase of the freezable water. Further increase of the degree of substitution over 0.6 gives a decrease and finally reaches zero at D.S. = 1.0. On the other hand, in the case of dodecanoyl chitosan, the amounts of freezable water decrease linearly with increasing D.S. In the introduction of a benzoyl group, the amounts of the freezable water increases at D.S. = 0.3 and then decreases to the

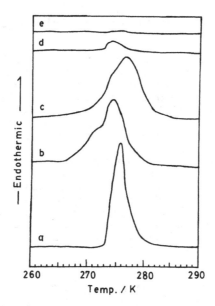

Figure 3. DSC heating curves of water sorbed by chitosan and octanoylchitosan gels. Heating rate = 2.5°K min⁻¹. (a) pure water, (b) chitosan (I), (c) C_8-0.6, (d) C_8-0.8, (e) C_8-1.0.

half of the original chitosan at D.S. = 0.5, and further increase of the D.S. does not change the amounts of freezable water.

Concerning the nonfreezable water, the amounts in chitosan gels with different degrees of amino content are approximately the same as 5 mol/basemol. The nonfreezable water content are 4.5-5 mol/basemol and approx-

	Freezable water W_f	Nonfreezable water W_{nf}
Sample No.	(mol H_2O/base mol)	(mol H_2O/base mol)
CS-I	11.2	5.2
CS-II	25.6	4.7
C_8-0.3	14.4	5.4
0.6	13.6	5.2
0.8	0.6	4.2
1.0	0	4.0
C_{12}-0.3	10.2	4.6
0.5	5.0	2.8
0.6	4.4	2.4
0.9	0.7	1.2
C_B-0.3	31.5	5.9
0.5	14.5	7.1
0.7	15.9	7.2

Table 2. Freezable water and nonfreezable water of gels.

imately 6.5 mol/basemol, for the octanoyl and benzoyl compounds, respectively.

The results suggest that the introduction of a long acyl group makes the crystal structure of the chitosan gels loose in the beginning. With increasing substitution, on the other hand, the more compact structure of the gels, due to the stronger interaction between the long acyl groups as well as the increase of the hydrophobicity, is formed to reduce the amount of free water. The aromatic acyl chitosan has a relatively coarse structure because of a steric effect and a decrease in crystallinity.

On the other hand, as the amounts of the nonfreezable water were scarcely affected by the amino group contents and the degree of acyl group substitution, the nonfreezable water is thought to be bound strongly with the hydroxyl group of glucosamine residue in the structural unit of the gels. A remarkable decrease of the nonfreezable water, approximately 1 mol water/basemol, was found in the dodecanoyl compound with the degree of substitution of 0.9, presumably due to the extremely compact and hydrophobic nature of the gel.

2. Sorptivity of Azo Dyes by N-Acylchitosan Gels

2A. Sorptivity isotherm of dyes by the gels having long acyl groups

The sorption behavior of the chitosan [I] and octanoyl chitosan gels for the azo dyes having the hydrogen bond forming groups is discussed in comparison with that for methyl orange [3]. In Figure 4, [5], having an OH group, shows higher sorption than [3]. The isotherms for [5] and [3] are given in Figure 5. For both of the dyes, the introduction of an octanoyl group in chitosan increases the amounts of the sorption. In the high concentration for chitosan and C_8-0.6, the slope of the isotherm for [5] is larger than [3]; however, the difference is scarcely found for C_8-1.0. Figure 4 shows the sorptivity of various azo dyes for octanoyl chitosan gels. The amounts of sorption give a maximum at D.S. = 0.3 and then decrease with rise of the degree of substitution. Inspecting the chemical constitution of the dyes, it was clearly found that the larger the hydro-

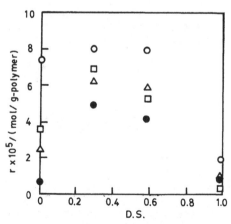

Figure 4. Effects of D.S. on dye sorption by octanoylchitosan gels. Initial concentration of dye = 1 x 10⁻⁴ mol dm⁻³. ● = [3], o = [4], △ = [5], □ = [6].

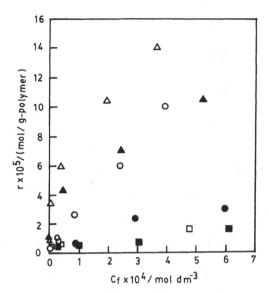

Figure 5. Sorption isotherms of dyes by octanoylchitosan
gels. Open symbols = [5]; filled symbols = [3]. o =
chitosan (I), △ = C_8-0.6, ◻ = C_8-1.0.

phobicity of the dyes, the greater the sorptivity. The larger hydrophobic
nature of butyl orange [4] than [3]; orange I [6] than [5] yield a higher
degree of the sorption. As can be seen from Figure 6, [4] has larger
amounts of sorption than the other dyes, especially for C_{18}-0.3, which
has a longer hydrophobic chain.

2B. Sorption isotherm of dyes by the gels having aromatic acyl groups

Figure 7 shows that the sequence of the amounts of sorption by benzoyl chitosan gels for various dyes is almost the same as with the octanoyl chitosan gels. At low degree of substitution of the benzoyl chitosan

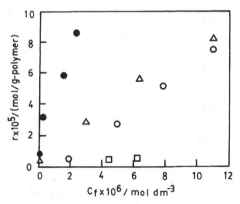

Figure 6. Sorption isotherms of butyl orange [4] by N-acylchitosan gels. o = chitosan (I), △ = C_8-0.6, ◻ =
C_8-1.0, ● = C_{18}-0.3.

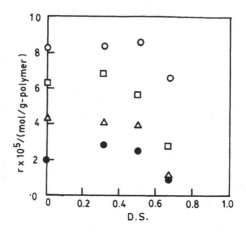

Figure 7. Effect of D.S. on dye by benzoylchitosan gels.
Initial concentration of dye 1 x 10⁻⁴ mol dm⁻³. ● =
[3], o = [4], △ = [5], □ = [6].

gel, the sorption of [3] and [5] is a little lower than the C_8-gel; however, the hydrophobic dyes, [4] and [6], have higher sorptions than the
C_8-gel. In Figure 8, the sorption isotherms of [3], [4], and [6] by a
benzoyl chitosan gel of D.S. = 0.5 are shown. In the low concentration
region, the slopes of the isotherms increase in a sequence of [3] < [6] <
[4], and in the higher concentration region the slope of [6] increases
significantly. The sorption of [4] at low concentration is remarkably
large and increases linearly with increasing concentration.

3. The Analysis of the Interaction by a Dual Sorption Model

In the analysis of the sorption isotherms, Equation 1, derived by a
dual sorption model, was used.[6] The model consists of two modes of the
sorption, i.e., electrostatic (Langmuir) and nonelectrostatic (partition)

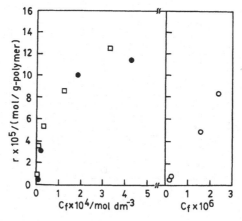

Figure 8. Sorption isotherms of dyes by benzoylchitosan gel
(D.S. = 0.5). ● = [3], o = [4], □ = [6].

Table 3. Sorption parameters.

Sample No.	dye	pH	$K_1 \times 10^2$ (dm^{3}g^{-1})	$K_2 \times 10^{-5}$ (dm^{3}mol^{-1})	$S \times 10^5$ (mol g^{-1})
CS-I	[3]	6.4	4.0	0.6	0.5
	[4]	6.5	550	0.5	0.5
	[5]	6.7	22.6	0.4	1.0
	[6]	6.7	18.8	0.4	6.6
C_8-0.3	[3]	6.3	9.8	3.8	5.2
	[5]	6.7	28.0	2.1	5.2
C_8-0.6	[3]	6.3	13.2	3.9	3.7
	[4]	6.3	500	2.1	3.0
	[5]	6.7	37.3	3.7	4.5
	[6]	6.7	15.2	1.9	5.4
C_8-1.0	[3]	6.0	2.2		
	[4]	6.4	18.8		
	[5]	6.3	1.1		
C_{18}-0.3	[3]	6.4	11.3	5.7	4.1
	[4]	6.6	2380	76.4	2.9
	[5]	6.4	20.7	2.8	4.1
C_B-0.5	[3]	6.4	21.3	3.4	2.6
	[4]	6.5	3480	---	---
	[6]	6.5	18.6	1.1	6.5

modes,[12-14] where r is the amounts of sorption of dye (mol g^{-1} polymer), C_f is the external dye concentration (mol dm^{-3}), K_1, is the partition coefficient, K_2 is the equilibrium constant of the Langmuir sorption, respectively. With use of the nonlinear least-squares method for analysis of the sorption data, K_1, K_2, and S were determined. The results are compiled in Table 3.

$$r = r_1 + r_2 = K_1 C_f + \frac{S K_2 C_f}{1 + K_2 C_f} \qquad \text{(Equation 1)}$$

Sample [5] has a larger value of K_1 than methyl orange [3] for all the gels except C_8-1.0. For CS-I, the values is *ca.* 6 times. Considering this value together with the results of the sorption isotherm, the hydrogen bonding between -NH$_2$ groups of gels and the -OH group of [5] is supposed to contribute to the partition mode sorption.

On the other hand, [4] shows a remarkable increase of K_1 compared with other dyes, especially for C_{18}-0.3 having a long alkyl chain. The significant contribution of the hydrophobic interaction is thought to result in the K_1 value 200 times larger than [3]. In the case of C_8-1.0, due to the lack of an amino group, the sorption is solely governed by the partition mode. In Table 3, K_1 for the C_8-1.0 gel is generally found to be quite small compared with other gels. This is attributed to the compact structure with an extremely limited amount of free water caused by the mutual interaction of long acyl groups. In the limited water sorption, the more hydrophobic [4] is sorbed more than [3] and [5].

The fact that K_2 and S are quite small for chitosan is explained by a compact structure due to the high crystallinity of approximately 40% (density *ca.* 1.4).[15] It is worthwhile to note the difference of the effects of benzoyl and octanoyl groups on the sorptivity of dyes. In comparing the values of K_1 for C_B-0.5 and C_8-0.6, it is concluded that benzoyl groups cannot interact as firmly as the octanoyl group due to the bulky structure of the phenyl ring. Thus the K_1 value of [4] is found to be seven times larger for C_B-0.5 than C_8-0.6.

In Table 4, the temperature dependence of the partition coefficient (K_1) is given for the [4]/C_8-1.0 system in which the partition mode is predominant. The thermodynamic parameters were obtained from the van't Hoff plot. It is clear that the sorption is an endothermic process and is governed by the hydrophobic interaction.

4. Selective Sorption of D,L-Amino Acids by N-Acylchitosan Gels

4A. Effect of Substituents on the Sorption of Amino Acids

Figure 9 shows the sorption of the hydrophobic amino acids, D,L-tryptophan and D,L-tyrosine, by octanoyl chitosan gels. The sorption of L-tryptophan is higher than L-tyrosine and reaches a maximum at D.S. = 0.3. The sequence of the increase of the amounts sorbed, tyrosine < tryptophan, is in agreement with the order of the increase in Rekker's hydrophobic fragmental constant, given in Figure 2. The role of the hydrophobic effect on sorption now becomes clear. If we bear in mind that at pH 5-6, the amino acids are near their isoelectric point, the interaction with chitosan gels can be attributed to hydrogen bonding, in addition to hydrophobic interaction with the acyl chain. This argument is consistent with that resulting from the sorptivity of dyes.

On the other hand, further increase over D.S. = 0.6 decreases sorption. Thus, the decrease of sorption at the higher D.S. is due to a more compact gel structure, which is confirmed by DSC results discussed in the previous section.

The sorption of the D-isomer is almost independent of the D.S. The L-amino acids are sorbed to a greater extent than their isomers. In our previous work,[5] it was observed that the N-octadecanoyl chitosan gel (D.S. = 0.45) selectively sorbs L-tryptophan from an aqueous 1:1 D,L-tryptophan solution giving a ratio of L/D ≈ 60. These results suggest that chemically modified chitosan gels are able to separate D,L-amino acids.

Table 4. Thermodynamic parameters for the interaction between butyl orange [4] and octanoylchitosan of D.S. = 1.0.

$K_1 \times 10^2$ (dm^3 g^{-1})			ΔG	ΔH	ΔS
303°K	313°K	323°K	(kJ mol^{-1})	(kJ mol^{-1})	(J mol^{-1} K^{-1})
20.6	38.2	56.9	3.98[a]	42.4	127
(a) At 303°K					

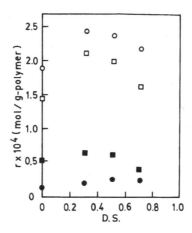

Figure 9. Effect of D.S. on the sorption of amino acids by octanoylchitosan gels. Initial concentration of sorption bath: 1×10^{-3} mol dm^{-3}. o = L-tryptophan, ● = D-tryptophan, □ = L-tyrosine, ■ = D-tyrosine.

4B. Effect of Inorganic Ions on the Sorption of Amino Acids

The effect of inorganic ions that are known to affect the structure of water has been examined. Figure 10 shows the effect of these on the sorption of L-tryptophan. Ammonium sulfate and sodium chloride, which are known for their salting out properties, produce a distinct increase in sorption at ionic strength $\mu = 0.1$. Sodium thiocyanate, a caotropic salt, reduces sorption of L-tryptophan. On the other hand, the sorption of D-isomer becomes almost zero for all the salts used ($\mu = 0.1$). The effects of the added salts on the sorption of D,L-tyrosine are shown in Figure 11. The sequence of sorption agrees with the salting out effect of these

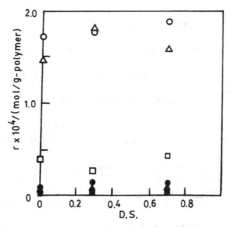

Figure 10. Effect of salts on the sorption of D,L-tryptophan. Initial concentration of sorption bath: 8×10^{-4} mol dm^{-3}. Open symbols = L-tryptophan; filled symbols = D-tryptophan. $(NH_4)_2SO_4$ = o, NaCl = △, NaSCN = □.

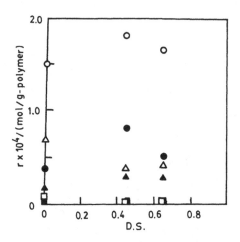

Figure 11. Effect of salts on the sorption of D,L-tryosine.
Initial concentration of sorption bath: 8x10⁻⁴ mol
dm⁻³. Open symbols = L-tyrosine; filled symbols =
D-tyrosine. $(NH_4)_2SO_4$ = o, NaCl = △, NaSCN = □ .

salts, NaSCN < NaCl < $(NH_4)_2SO_4$. The selective sorption of L-tyrosine is
also clearly shown in Figure 11.

We conclude from these results that the chiral structure of chitosan
gels containing hydrophobic groups is effective in the hydrophobic separation and the resolution of optical isomers of D,L-amino acids. Further
work will be applied to the separation of D,L-amino acids and peptides by
liquid column chromatography, using chemically modified chitosan as stationary phase. In addition, the information about equilibrium sorption
obtained from the present investigation is to be expected for supports of
immobilized enzymes and separation membranes.

REFERENCES

1. R. A. A. Muzzarelli, Eds., "*Chitin*," Pergamon Press Ltd., Oxford,
   1977.
2. T. Baba, R. Yamaguchi, Y. Arai & T. Itoh, Carbohydr. Res., **86**, 161
   (1980).
3. R. A. A. Muzzarelli, F. Tanfani, M. Emanuelli & Mariotti, J. Membr.
   Sci., **16**, 295 (1983).
4. T. Uragami, F. Yoshida & M. Sugihara, Makromol. Chem., Rapid Commun.,
   4, 99 (1983).
5. T. Seo, T. Kanbara & T. Iijima, Sen-i-Gakkaishi, **42**, T-123 (1986).
6. T, Seo, T. Kanbara & T. Iijima, J. Appl. Polym. Sci., 36, 1443
   (1988).
7. M. F. Refojo & F. L. Leong, J. Mem. Sci., **4**, 415 (1979).
8. A. Higuchi & T. Iijima, Polymer, **26**, 1207 (1985).
9. G. Hesse & R. Hagel, Justus Liebigs Ann. Chim., **1976**, 996.
10. T. Shibata, I. Okamoto, & K. Ishii, J. Liquid Chromatogr., **9**, 313
    (1986).
11. R. P. Rekker, Eds., "*The Hydrophobic Fragmental Constant*," Elsevier
    Sci. Publ. Co., Amsterdam, 1977, p. 301.
12. H. Zollinger & G. Back, Melliand Textilber., **42**, 73 (1961).
13. T. Tak, J. Komiyama & T. Iijima, Sen-i-Gakkaishi, **35**, T-486 (1979).
14. R. T. Harwood, R. McGregor & R. H. Peters, J. Soc. Dyers Color., **88**,
    216 (1972).

15. K. Sakurai, M. Takagi & T. Takahashi, Sen-i-Gakkaishi, **40**, T-246 (1984).

QUINONE CHEMISTRY: APPLICATIONS IN BIOADHESION

Leszek M. Rzepecki and J. Herbert Waite

College of Marine Studies
University of Delaware
Lewes, DE 19958

Proteins containing the catecholic amino acid L-3,4-dihydroxyphenylalanine (DOPA) as an integral part of their sequence have bioadhesive and sclerotizing functions in the extracorporeal structures of many invertebrates, such as the attachment tendon (byssus) of the marine mussel and the egg-case of parasitic trematodes. The DOPA protein is one part of a natural composite thermoset which also includes stoichiometric quantities of catecholoxidase, an enzymic oxidizing agent, to induce the curing of the resin. An understanding of the molecular mechanism of these functions would enhance the technological and medical exploitation of materials which have been specifically engineered by the course of evolution to possess strong adhesive and cohesive properties. To date, however, we have only a rudimentary appreciation of this mechanism owing to the intractable nature of sclerotized structures and the low solubility of their purified precursors. We have investigated the possibility of using low molecular weight peptidyl DOPA analogues, specifically N-acetyl-DOPA ethyl ester, to explore the potential oxidation chemistry of DOPA proteins. We have found that provided catalytic amounts of a Lewis base are present, that competing reactions such as Michael additions are minimized, and the DOPA quinone moiety formed by oxidation will spontaneously undergo a rearrangement to an α,β-dehydroDOPA residue. This rearrangement occurs at pH 6.0 and above, and has an apparent first order rate constant of the same order of magnitude as those obtained for Michael additions to related quinones. Examination of initial rates of reaction has shown that this rearrangement can account, at pH 6.0, for 60-80% of the reactivity of the quinone. Thus the fate of a DOPA quinone in a natural composite should be exquisitely sensitive to the local environment and may form a variety of previously unexpected reaction products.

INTRODUCTION

Amongst the traditional technological applications of biopolymers has

Biotechnology and Polymers, Edited by C.G. Gebelein
Plenum Press, New York, 1991

been the use of animal or plant products as glues or varnishes.[1] A glue may be defined as a substance that induces adhesion between like or unlike material surfaces, whilst a varnish is a coating applied to a material surface as a liquid which then solidifies to form a protective film. Intriguingly, virtually none of the products historically used for these purposes, e.g. animal gelatin (collagen), starches and celluloses, or various plant phenolics, serve an analogous function in nature. Nevertheless, multicellular organisms are living testaments to the efficacy of adhesion in building large, complex and flexible yet mechanically resilient structures.[2,3] Although none of nature's deliberate adhesive or varnish strategies has been wholly understood at the molecular level, such an analysis could only enhance our exploitation of the adhesive arts by introducing new concepts and materials that may help solve some of the extant problems (e.g. adhesion between wet, uneven, or dirty surfaces; toxicity in medical and dental applications; weakening or erosion of varnishes and adhesives in mechanically or chemically stressful environments)[4] which limit current applications.

Although the chemistries of natural varnishes and adhesives are undoubtedly varied, the strategies involving quinone tanning appear to be a common solution in the animal kingdom, where various implementations of quinone chemistry are found among the invertebrates, especially, but by no means exclusively, the arthropods.[5,6] The arthropods, which provided the first understanding of the quinone tanning mechanisms, utilize low molecular weight catechols to crosslink the protein and carbohydrate components of their exoskeletons to form hard and impervious protective coats. The very abundance of arthropods is in part a tribute to the success of this strategy. An evolutionarily earlier quinone tanning strategy appears to be employed by certain (if not all) invertebrates that live in an aqueous milieu rather than exposed to the open air.[7] This strategy is exemplified by organisms such as the marine mussels, which synthesize a non-living, quinone tanned extracorporeal tendon called the byssus to anchor themselves adhesively to various substrates,[8] and by parasitic trematodes such as liver and blood flukes, which protect their eggs from the natural defenses of their vertebrate, often human, host by a quinone tanned coat or varnish.[9] These invertebrates use rather insoluble high molecular weight proteins, which incorporate catechol into their structure as the post-translationally synthesized amino acid L-3,4-dihydroxyphenylalanine (DOPA), as the direct agents of quinone tanning.[10] These two strategies may have evolved to take advantage of the different solubility and diffusivity properties of low and high molecular weight quinone tanning agents in aqueous or terrestrial environments.

The structures of a number of DOPA protein families have been elucidated by both protein and DNA sequencing techniques, and have been recently reviewed.[7,10] Briefly, DOPA protein have molecular weights in the range 17-130 kDa and each consists of a tandem series of short (2-13 amino acid) oligopeptide repeats. The precise sequence of the repeating unit depends on the source of the DOPA proteins and some currently available consensus tandem repeats, together with the protein molecular weight and pI, are provided in Table 1. These proteins are highly hydroxylated, highly basic, and contain high proportions of the structure-breaking amino acids glycine and/or proline, in addition to considerable amounts of DOPA (100-200 residues per thousand). In addition to the DOPA proteins, the function of the adhesive or varnish depends on the presence of a catecholoxidase (also known as tryosinase, phenolase, polyphenoloxidase, catecholase, etc.) which is required to oxidize the DOPA residues to DOPA quinones as the first step of the quinone tanning process. Preliminary results in our laboratory (Samulewicz, et al., unpublished) suggests that the ratio of DOPA protein to catecholoxidase in the byssal threads of the ribbed mussel may be as low as 1:1. This, and the unusual amino acid composition of this catecholoxidase (60% Gly, Ser, Glx, Asx), suggests the possibility that the catecholoxidase may have more than a catalytic role, and that DOPA protein and catecholoxidase may function as copolymers in a composite thermoset. Such an eventuality has clear significance for the understanding of these adhesive systems and is a focus of current research in our laboratory. The effectiveness of each DOPA protein, either as an adhesive or as a varnish, will presumably reflect (i) its specific sequence and amino acid composition, (ii) its degree of oxidation by catecholoxidases (or other oxidants), and (iii) its interactions with other structural components (e.g. proteins, carbohydrates, metals and lipids) of the encompassing matrix. The specific contributions of these interactions to adhesive and varnish functions have not yet been elucidated, but the ability of catechols such as DOPA to form strong H-bonds and to chelate metals with very high binding constants is undoubtedly a significant factor.[11,10] Oxidation of DOPA proteins by tyrosinase greatly enhances their adsorption to surfaces.[10]

The occurrence of quinone tanning in DOPA protein systems and the high apparent catecholoxidase concentration suggests that exploration of the oxidation chemistry of peptidyl DOPA should be an important research

---

Table 1. Phyletic Distribution and Sequences of Tandem Repeats from some DOPA Protein Precursors.

Organism	$M_r \times 10^{-3}$ (pI)		Tandem Repeat Sequence
Phylum Platyhelminthes (eggcase varnish)			
Fasciola hepatica[39,40]	31	(7.2)	G-G-G-Y̲-X₁X₂Y̲-G-K
(liver fluke)	17	(6.8)	G-X₃
Phylum Annelida (cement)	20	(8.1)	V-G-G-Y̲-G-Y̲-G-A-K
Phragmatopoma californica[a]			
(reef building worm)			
Phylum Mollusca (adhesive and varnish)			
Mytilus edulis[7]	130	(10.0)	A-K-P-S-Y̲-P̲-P̲-T-Y̲-K
(blue Mussel)			A-K-P-T-Y̲-K
Mytilus californianus[7]	85	(10.0)	R-K-P-S-Y̲-P̲-P̲-T-Y̲-K
(pacific mussel)			I-T-Y̲-P̲-P̲-T-Y̲-K-P̲-K
Geukensia demissa[41]	130	(8.2)	Q-T-G-Y̲-D-P-G-Y̲-K
(ribbed mussel)			Q-T-G-Y̲-V-P-G-Y̲-K
			Q-T-G-Y̲-V-L-G-Y̲-K
			T-G-Y̲-S-A-G-Y̲-K
Aulacomya ater[b]	130	(?)	A-G-Y̲-G-G-X₄K
(chilean mussel)			
Septifer bifurcatus[c]	130	(10.0)	Y-P̲-A-K-P-T-S-Y̲-G-T-G-Y̲-K
(scorched mussel)			
Mytella guyanensis[c]	130	(10.0)	S-H-K-P-Y̲-T-G-Y̲-K
(Ecuadorean mussel)			

Y̲ = invariably DOPA, Y = DOPA or Tyr, P̲ = Hydroxyproline, X₁ = Asp, Ala, Glu, or Gly; X₂ = Ser, Asp, or Gly; X₃ = DOPA, Ser or His; X₄ = Leu or Val; other letters are standard amino acid one-letter codes.

(a) Jensen, et al., unpublished.
(b) Burzio & Waite, unpublished.
(c) Waite, et al., unpublished.

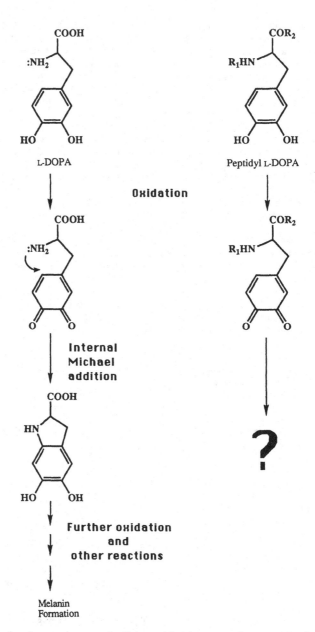

Figure 1. Comparison of the oxidation products of the free amino acid L-DOPA and peptidyl DOPA. R_1 and R_2 represent peptide chains attached by amide linkages on either side of the DOPA residue.

goal in the elucidation of these adhesive strategies. The oxidation chemistry of the free amino acid L-DOPA and its analogue dopamine is well known and proceeds, following the initial oxidation to DOPA quinone, by a relatively rapid ($k = 0.147$-$0.263s^{-1}$ for dopamine) internal Michael addition of the amine moiety.[12,13] Further oxidation and polymerization lead to the formation of an insoluble black precipitate of melanin (Figure 1). The oxidation of peptidyl DOPA is likely to result in radically different products since the reactive amino group is blocked by incorporation in a peptide bond. Unfortunately, none of these potential products has been isolated or characterized to date, owing to the intractability of purified DOPA proteins in physiologically relevant media, and the presumably extensive crosslinking within natural quinone tanned structures. In an effort to circumvent this problem and to assess potential analytical strategies applicable to the more complex protein system, we have focused on the oxidation chemistry of low molecular weight peptidyl DOPA analogues in the anticipation that they may provide clues to the potential reactivities of their higher molecular weight cousins. In this paper, we present a preliminary report of our findings.

EXPERIMENTAL AND DISCUSSION

1. Synthesis of the Peptidyl DOPA Analogue

N-acetylDOPA ethyl ester (NAcDEE) was synthesized by enzymic hydroxylation of N-acetyltyrosine ethyl ester (NAcTEE)(1mM) using mushroom tyrosinase (monophenol dihydroxyphenylalanine:oxygen oxidoreductase, EC 1.14.18.1, Sigma Chemical Co., St. Louis, MO) at 0.1-0.3 mg/mL in 100 mL 0.1 M sodium phosphate, pH 6.7, at room temperature for 60 min with vigorous aeration and stirring.[14] In some cases, the hydroxylation reaction mixture included 25 mM sodium ascorbate to maximize the yield of NAcDEE (Figure 2).[14] NAcDEE was purified from the mixture, following acidification of the reaction by addition of 30 mL glacial acetic acid and concentration by rotary evaporation, by chromatography on a 1.5 x 90 cm column of LH-20 (Pharmacia LKB Biotechnology Inc., Piscataway, NJ) in 0.2 M acetic acid, then by preparative HPLC (Rabbit-HPLX, Rainin Instrument Co. Inc., Woburn, MA) on a 1 x 25 cm Microsorb C 18 column (Rainin Instrument Co. Inc.) using a 15-30% gradient of acetonitrile in water, incorporating 0.1% trifluoroacetic acid. The eluate was monitored at 280 nm using a Gilson Model 116 dual wavelength detector (Gilson Medical Electronics Inc., Middleton, WI). Analytical rechromatography on the same HPLC column showed a single UV absorbing peak, and L-DOPA was the only amino acid detected (System 6300 High Performance Amino Acid Analyzer, Beckman Instruments Inc., Fullerton, CA) after hydrolysis for 22 min at 158°C in 5 N HCL incorporating 8% trifluoroacetic acid and 8% phenol.[15] Ultraviolet difference spectroscopy of NAcDEE in HCl and alkaline borate buffers, a technique which affords spectra characteristic of individual catechols, revealed absorbance maxima coinciding with those found for L-DOPA itself.[16]

2. Oxidation of N-AcetylDOPA Ethyl Ester

Oxidation of pure NAcDEE was effected in most of the following studies by sodium periodate, which is a 2 electron oxidizer of catechols,[17] in order to avoid complications arising from enzymic oxidation by mushroom tyrosinase or byssal catecholoxidase, since these enzymes have relatively low substrate specificity and will generally continue to oxidize any reaction products subsequent to quinone formation. Spectroscopic analyses of NAcDEE oxidation by a sub-stoichiometric aliquot of NaIO4 showed that the formation of NAcDEE quinone, identified by its characteristic absorbance maximum at 392 nm, was complete within the mixing time (Figure 3). Immediately following the oxidation, the absorbance at 392 nm began to decay, indicating further reaction of the quinone, and a new absorbance maximum at about 320-322 nm simultaneously

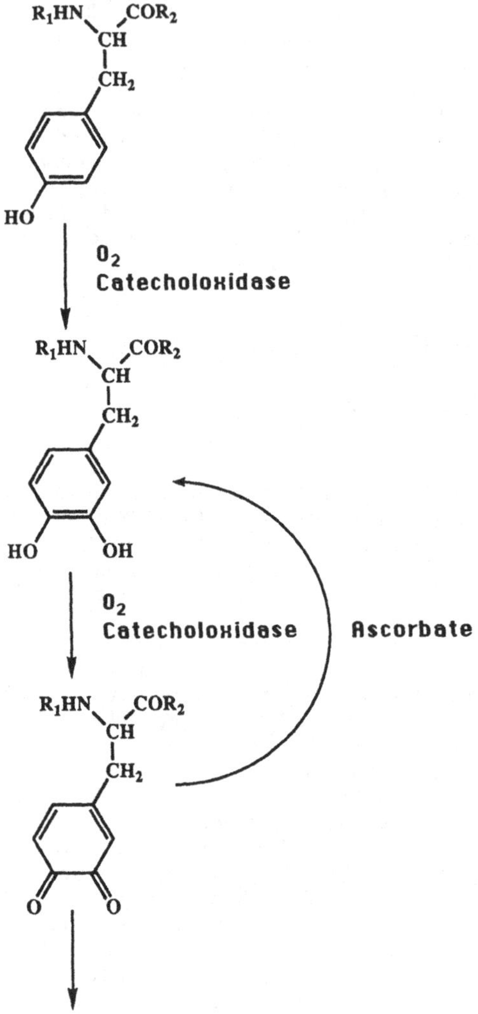

Figure 2. Scheme for the enzymic synthesis of peptidyl DOPA
analogs and other oxidation derivatives. R_1 and R_2
represent the CH_3CO- and $-OCH_2CH_3$ moieties respectively of the NAcDEE used for the studies reported
here.

increased in intensity. Although the particular spectral series shown in
Figure 3 was recorded from a reaction in a highly saline medium, virtually identical spectra were obtained in the absence of NaCl.

3. Characterization of N-Acetyl-α,β-DehydroDOPA Ethyl Ester

The compound absorbing at 320 nm was then isolated by chromatography
on a 1.5 x 90 cm Bio-Gel P-2 column (Bio-Rad Laboratories, Richmond, CA)
in 0.2 M acetic acid, and could be further purified, when desired, by
HPLC using the same protocol as for NAcDEE purification. The UV spectra
and the HPLC profile of the purified compound are shown in Figure 4. The

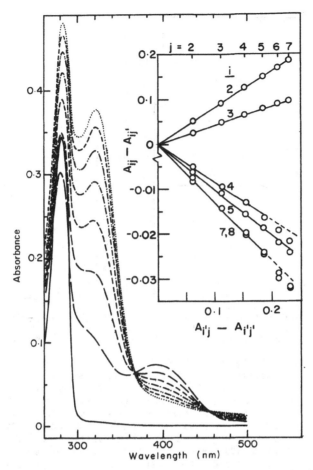

Figure 3. Series of spectra recorded at 2 nm/s following the
oxidation of 113 µM NAcDEE in 1 M NaCl, 0.1 M
sodium phosphate, pH 6.0, by a substoichiometric
aliquot of NaIO₄, which was also added to the
reference curvette. The solid line represents the
spectrum before oxidation; the increasingly finely-
dashed lines are spectra recorded at various times
up to 20 min following oxidation. **Inset:** Matrix
analysis[26] of the spectral series following oxidation using a two-species test assuming restricted
stoichiometry between contributors to the absorbance. A_{ij} is the absorbance at the ith wavelength
of the jth spectrum, where i' is set at 322 nm,
i((2-8) = 336, 348, 376, 388, 400, 412, and 424 nm,
and j' is the first spectrum recorded after
oxidation.

UV spectra, with a maximal absorbance at 322 nm in 0.2 N HCl and 344 nm
in 0.2 M sodium borate, pH 8.5, were characteristic of catechols bearing
an unsaturated carbon substituent on the benzene ring <u>para</u> to the first
hydroxyl, and this oxidation product was tentatively identified as <u>N</u>-
acetyl-α,β-dehydroDOPA ethyl ester (NAcDEE).[16,17] In general, bulk preparations of NAΔDEE used the same enzyme protocol as for NAcDEE synthesis,

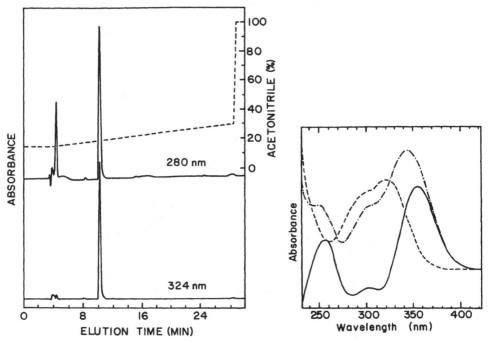

Figure 4. **Left**: HPLC profile (25 cm Vydac C18 column, Vydac, Hesperia, CA) of enzymically synthesized NAcΔDEE isolated by Bio-Gel P-2 chromatography. Solvents were acetonitrile in water and incorporated 0.1% trifluoracetic acid. Elute was monitored at 280 and 324 nm in a dual wavelength UV detector. **Right**: UV spectra of non-enzymically synthesized NAcΔDEE in 0.2 N HCl (----), 0.2 M sodium borate, pH 8.5 (_._.), and the resulting borate-HCl difference spectrum (____).

except that absorbic acid was omitted. Under these circumstances, hydroxylation of the NAcTEE precursor was followed by the oxidation of the initial NAcDEE product to the quinone (Figure 2), with subsequent formation of NAcΔDEE. Overall yields by this method ranged between 3-6% of the initial NAcTEE used. The NAcΔDEE products synthesized by enzymic or non-enzymic oxidation were indistinguishable by HPLC or by UV spectroscopy).

The purified NACΔDEE product reacted positively in the Arnow test, which has been shown to be specific for catechols with at least two unsubstituted vicinal carbons in the benzene ring.[18,19,20] Elemental analysis (185B Carbon Hydrogen Nitrogen Analyzer, Hewlett-Packard, Palo Alto, CA) for carbon and nitrogen (hydrogen was omitted) of two independent NAcΔDEE preparations yielded C/N ratios of 9.71 ± 1.57 and 7.09 ± 0.96 (n = 3 for both), close to the ratio of 8.97 expected for NAcDEE and NAcΔDEE, suggesting that the desaturation was a result of C=C double bond formation rather than oxidation to an aldehyde with concomitant loss of the N-acetyl and ester functions. Amino acid analysis of NAcΔDEE after hydrolysis as outlined above yielded only ammonia, a result consistent with the enamine character of the putative product which would be deaminated upon acid hydrolysis.[21] Catalytic hydrogenation of NAcΔDEE using powdered Pt and H_2 gas in 20 mM acetic acid was confounded by the strong affinity of catechols for metals.[10] Although some regeneration of NAcDEE from hydrogenated NAc△DEE was apparent by HPLC analysis, and L-DOPA was detected (by amino acid analysis after hydrolysis) at higher levels for hydrogenated NAc△DEE than for non-hydrogenated samples, binding of NAc△DEE (and presumably NAcDEE) to platinum interfered with catalysis at low Pt/NAc△DEE ratios, and greatly reduced total yields of catechol at higher ratios. NMR studies (QE-300 NMR Spectrometer, General Electric Co.) of NAc△DEE which had been purified by both Bio-Gel P-2 chromatography and HPLC showed the expected proton signals for an enamine moiety in [^{2}H$_6$]-dimethyl sulphoxide with a singlet signal at 7.05 ppm for the vinyl proton and another for the amide proton at 9.38 ppm.[22] The extinction coefficients (ε) of NAc△DEE are given in Table 2 and were of the same order of magnitude as those of other α,β-unsaturated catechols such as caffeic and chlorogenic acids.[16] Preliminary redox estimates for NAc△DEE (68 μM in 0.1 M sodium phosphate, pH 7.0) by square wave voltammetry (Model 384B Polarographic Analyzer, EG & G Princeton Applied Research, Princeton, NJ) indicated an oxidation potential for NAc△DEE of about 22 mV (vs. SCE) compared with a calculated potential of 145 mV for L-DOPA itself, a result consistent with the expected decrease in oxidation potential arising from the α,β-desaturation of the catechol substituent.[23,24] A similar experiment with NAcDEE (770 μM) gave two apparent oxidation peaks, one at about 144 mV as expected, and the other at 64 mV. This latter anomalous result is as yet unexplained, but may be an artificial result of the higher NAcDEE concentration used.

4. Kinetic Analysis of NAc△DEE Formation

Although the data above established that NAcDEE could undergo an α,β-desaturation on the side chain following oxidation to the quinone, we felt it desirable to see whether further information could be gleaned about the mechanism and rate of synthesis, since we were interested in this reaction as a model for the reactivity of peptidyl DOPA. The chemistry of catechols and their quinones is notoriously complex, and it was not obvious that a realistic extrapolation from this low molecular

Table 2. Extinction coefficients of NAcDEE quinone and NAc△DEE in 0.1M sodium phosphate.

	ε, M^{-1}cm^{-1} (pH)	
	322 nm	392 nm
N-Acetyldopamine quinone	N.D.	1130 (7.5)
NAcDEE quinone	854 (6.0-8.0)	1188 (6.0-8.0)
NAc△DEE	14481 (6.0)	195 (6.0)

Extinction coefficients of NAc△DEE were measured in 0.2 N HCl and data were combined from three separate preparations (ε = 14327 M^{-1}, 322 nm, r^2 = 0.998). NAc△DEE extinctions in 0.1 M sodium phosphate, pH 6.0 were calculated by comparison of spectra in phosphate buffer with spectra in 0.2 N HCl (n = 6). Values of ε for NAcDEE quinone (and N-acetydopamine quinone for comparison) were obtained directly in phosphate buffer. Oxidation was effected by the addition of excess NaIO$_4$. Concentrations of NAcDEE in buffers was determined by the absorbance at 279 nm, assuming the same ε as for L-DOPA (2713 M^{-1}cm^{-1}, ref. 16), since NAcDEE was unexpectedly hygroscopic and weight estimates were unreliable.

weight, soluble compound to a solid state system could be made without a deeper understanding of the underlying molecular mechanism.[25] That NAc△DEE is formed by a rearrangement of NAcDEE quinone is supported by spectroscopic evidence (Figure 3), by HPLC analysis showing sequential formation first of NAcDEE quinone, then of NAc△DEE during oxidation by either NaIO4 or tyrosinase (not shown), and by the inhibition of NAc△DEE formation by ascorbate. However, there are at least two mechanisms by which such a rearrangement might be effected. First, NAcDEE quinone might spontaneously rearrange to form the desaturated isomer. Second, a charge transfer complex between NAcDEE and its quinone might be required to facilitate the rearrangement. Clearly, the nature of the rearrangement mechanism will have profound implications for its likelihood (or otherwise) in the solid state polymer systems of interest.

Matrix analysis of the spectra in Figure 3 suggested that in sodium phosphate buffers at pH 6.0 there were only two species, NAcDEE quinone and NAc△DEE, with significant absorbance in the range 320-400 nm, at least for the first few minutes following oxidation of NAcDEE by NaIO4.[26] A two-species test assuming restricted stoichiometry between contributors to the absorbance suggested, by an initial apparent linearity of the plots (inset to Figure 3), that there was a brief, nearly stoichiometric, rearrangement of NAcDEE quinone to NAc△DEE. However, this ideal relationship rapidly broke down, presumably as competing reactions removed NAcDEE quinone from the pool available for rearrangement, and/or the NAc△DEE decayed in its turn. If the data were replotted using a two-species test with no restrictions on stoichiometry, the results were linear (not shown), confirming that only two species were absorbing in this wavelength region but that conversion of quinone to NAc△DEE was not stoichiometric. These observations suggested, however, that a kinetic analysis of the system should be feasible, provided the precaution was taken of analyzing only the initial rates of quinone decay and NAc△DEE formation, i.e. at a time when NAcDEE and NAc△DEE quinone would be the only species involved in reactions, rather than considering more complete rate curves, which might contain contributions from further reactions involving the initial product, NAc△DEE. Such an analysis in phosphate buffers at pH 6.0 is presented in Table 3 and has several features of interest. Since the reaction kinetics were measured following partial oxidation of NAcDEE by stoichiometric aliquots of NaIO4, reaction conditions could be readily established in which both NAcDEE quinone and residual (unoxidized) NAcDEE were varied by changing the original NAcDEE concentration and/or the amount of NaIO4 added. The two reaction mechanisms briefly outlined above should have rather different kinetic properties. In the case of a simple rearrangement of the quinone, the rate of NAc△DEE formation should depend solely on the NAcDEE quinone concentration, and the observed rate constant, k_{obsd}, should be first order (i.e. rate=k_{obsd}.[NAcDEE quinone], thus k_{obsd} = rate/[NAcDEE quinone], s^{-1}). If the rearrangement were to depend on the formation of a charge transfer complex between the quinone and parent catechol, then the rate should depend on the concentration of both and k_{obsd} should be second order (i.e. k_{obsd} = rate/([NAcDEE][NAcDEE quinone]), M^{-1}s^{-1}.) The results in Table 3 unequivocally show that, at pH 6.0, k_{obsd} for the formation of NAc△DEE is quite independent of the residual NAcDEE concentration and is indeed constant, within experimental error, over a range of quinone concentrations. Thus we can exclude any requirement for charge transfer complex formation in the rearrangement at pH 6.0. When initial rates for NAc△DEE formation were compared with those for quinone decay, they accounted for 60-80% of that decay at low residual NAcDEE concentrations, but this proportion dropped to 30-40% at higher concentrations, an observation explained by the apparent acceleration of quinone decay to other undefined products by high residual NAcDEE levels. Preliminary results at higher pH values indicated that the independence of k_{obsd} from the residual NAcDEE concentration no longer held

Table 3. Kinetics of α,β-dehydroDOPA ethyl ester formation in sodium phosphate buffers at pH 6.0.

Initial [NAcDEEQ] $M \times 10^4$	Initial [NAcDEE] $M \times 10^4$	[Phosphate] M	$k_{obsd} s^{-1} \times 10^4 \pm SE$ (= rate/[NAcDEE quinone)
0.26	13.0	0.1	2.86 ± 0.05
0.51	12.7	0.1	2.75 ± 0.02
1.33	11.7	0.1	2.56 ± 0.04
3.01	10.2	0.1	2.46 ± 0.02
0.25	3.1	0.1	2.59 ± 0.06
0.51	2.8	0.1	2.57 ± 0.04
1.33	1.9	0.1	2.34 ± 0.03
0.21	25.7	0.5	18.6 ± 0.08
0.47	25.5	0.5	16.0 ± 0.03
1.27	24.5	0.5	15.4 ± 0.03
0.24	13.0	0.5	15.9 ± 0.05
0.50	12.7	0.5	15.0 ± 0.03
1.31	11.7	0.5	14.7 ± 0.05
0.26	3.0	0.5	15.6 ± 0.08
0.50	2.8	0.5	15.7 ± 0.04
1.31	2.0	0.5	15.4 ± 0.03

Reaction kinetics were measured at 25°C in sodium phosphate buffers comprising 0.1 or 0.5 M NaH_2PO_4 brought to pH 6.0 with 5 N NaOH. An aliquot of 133.6 mM NAcDEE in 20 mM acetic acid was added to 1 mL of buffer and the reaction was then initiated by the addition of a substoichiometric aliquot of 10.3 mM $NaIO_4$ to this solution and to a reference solution of buffer alone. Oxidation of NAcDEE to NAcDEE quinone was complete within the mixing time. Three separate rate curves were recorded at both 322 and 392 nm for each set of reactant concentrations. Initial rates of absorbance change at each wavelength were calculated from the first derivative at time zero (i.e. dA/dt, t = 0, corresponding to the beginning of the rate curve recorded after mixing) of polynomials of the lowest statistically significant order ($\beta \neq 0$, P < 0.05) fitted to points (n = 20 to 30) obtained from the three separate rate curves over reaction times of 40-100s. The standard errors of the apparent rate constants (k_{obsd} = (initial rate)/[NAcDEE quinone], see text) were calculated from the reported standard error of the first derivative at time zero. The initial NAcDEE quinone concentration was calculated from the average time zero absorbance (see Table 2) at 392 nm (extrapolated from the fitted polynomial), whilst the initial NAcDEE concentration was taken to be the residual NAcDEE following the partial oxidation of the original NAcDEE to the quinone by addition of $NaIO_4$. Initial rates of NAcΔDEE production were calculated by solving the simultaneous equations:

$$dA_{\lambda 1}/dt = (\varepsilon_1\lambda_1 \times (dc_1/dt)) + (\varepsilon_2\lambda_1 \times (dc_2/dt)), \text{ and}$$
$$dA_{\lambda 2}/dt = (\varepsilon_1\lambda_2 \times (dc_1/dt)) + (\varepsilon_2\lambda_2 \times (dc_2/dt)),$$

where ε_1, ε_2 and c_1, c_2 are extinction coefficients and concentrations of reactant (1) and product (2) (see Table 2), and $dA\lambda_1/dt$ and $dA_{\lambda 2}/dt$ are the initial rates of absorbance change at 322 and 392 nm.

(not shown), and that different, or subsidiary, mechanisms might be required to account for the data at higher pH. Table 3 also illustrates the
catalytic effect of phosphate anion on $\underline{k}_{obsd}$. This effect is seen with
other buffer anions such as HEPES, bicarbonate, acetate, etc., but not
with salts such as NaCl or Na_2SO_4. This observation suggests that the
rearrangement is catalyzed by a Lewis base.

CONCLUSIONS

The data above demonstrate that under appropriate conditions, a low
molecular weight analogue of peptidyl DOPA may undergo a spontaneous
rearrangement, following oxidation to the quinone, to an α,β-dehydroDOPA
derivative. This rearrangement, illustrated in Figure 5, does not involve
charge transfer complex formation (at least at pH 6.0), but does require
the participation of a Lewis base, presumably to facilitate the abstrac-
tion of the acidic protons on the α,β carbons of the side chain. The
reaction may proceed through a quinone methide intermediate as shown in
Figure 5, but we have as yet no evidence for that particular pathway.
Quinone methides have been suggested as intermediates in related reactions.[17,27,28] That this rearrangement is critically dependent on a
Lewis base, at least at pH 6.0, creates obvious restrictions on the circumstances in which the reaction might occur in solid state systems such
as mussel byssus or trematode eggcase, where diffusion will be an important factor. Although the eggcase matures in an environment where biological anions might be readily available, the mussel byssus exists in no
such milieu, except perhaps in the case of species such as the ribbed
mussel, which inhabits the relatively acidic and anoxic mud flats of east
coast marshes and estuaries. In this latter case, various organic acids,
amines or other salts derived from decaying vegetation and animal matter
might contribute some catalytic function.

Extensive attempts to identify catechols other than DOPA in the quinone tanned structures under consideration have thus far borne little
fruit. However, it would be surprising if α,β-dehydroDOPA could be readily isolated from tanned structures since: (a) as an enamine, it is destroyed by acid hydrolysis, (b) the option of alkaline hydrolysis is not
open, since resultant β-elimination reactions might effect conversion of
DOPA to α,β-dehydroDOPA, and (c) the lower redox potential of α,β-dehy-
droDOPA would make it more unstable than DOPA under the oxidizing conditions prevailing in nature, and thus significant accumulations of this
derivative would be unlikely.[21,6] Moreover, the novel enamine functions
and resonance forms of α,β-dehydroDOPA (Figure 5) might result in unexpected reaction products, including crosslinks directly involving the
peptide chain or even hydrolytic scission of the chain under relatively
mild conditions.[29,21] These considerations lead to expectations of
intractable difficulty in the analysis of *in situ* oxidation products in
sclerotized DOPA protein matrices, and indeed may be partly responsible
for the frustrating lack of progress in this field.

There does, however, appear to be a family (or families) of low molecular weight DOPA containing oligopeptides, sometimes called peptide
alkaloids, which consists of 2-4 amino acid residues. In several of these
peptides, notably the tunichromes from the sea squirt *Ascidia nigra*, and
the clionamides and celenamides from the sponge *Cliona celata*, one or
more DOPA residues have been desaturated to form α,β-dehydroDOPA, although this is often a trishydroxy derivative.[30-33] Analogous peptides,
the halocyamines, have been found in another sea squirt *Halocynthia
roretzi*.[34] All these peptides bear, in addition (usually) to an internal
DOPA or α,β-dehydroDOPA residue, a decarboxylated, desaturated tyrosine,

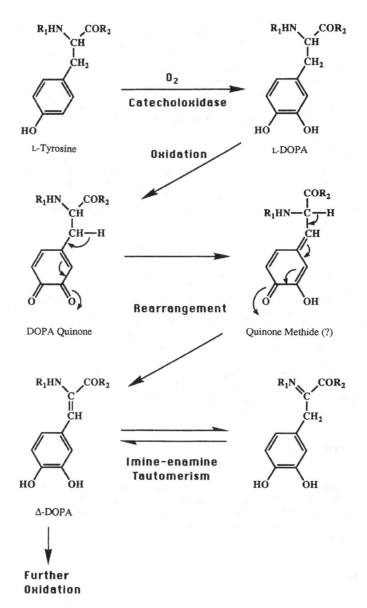

Figure 5. Postulated mechanism for the non-enzymic rearrangement of peptidyl quinone to its α,β-dehydroDOPA
derivative.

DOPA or 6-bromotryptophan residue at the C-terminus. The biosynthetic
route for the formation of these peptides is totally unknown, but their
similar and unusual features strongly suggest homologous origins. Indeed,
it is intriguing to speculate (in the current absence of evidence) that
they may be derived hydrolytically from higher molecular weight DOPA
protein precursors similar to the ferreascidin isolated from the sea
squirt *Pyura stolonifera*.[35] Formation of a α,β-dehydroDOPA residues in
these oligopeptides may well proceed by peptidyl DOPA oxidation chemistry
such as that discussed here, though it is too early to guess whether

these chemistries or their products might be directly involved in any hydrolytic events required for the hypothetical release of these oligopeptides from larger precursors. The function of these peptides is uncertain, but it is probable that in the sea squirt they are involved in iron or vanadium chemistry, and perhaps serve as intermediates in sclerotization.[30,35] Antimicrobial functions have been ascribed to some, while antiviral and antitumor properties have been ascribed to related compounds with pendant caffeoyl moieties.[34,36,37]

It is still premature to assert that α,β-dehydroDOPA is a factor in the quinone tanning of DOPA protein structures, although an analogous catechol, N-acetyldehydrodopamine, has been implicated in the sclerotization of insect cuticles.[38] This study, however, shows that provided the opportunity exists, in the form of adequate catalysis and prevention of competing reactions, desaturation of peptidyl DOPA quinone is a probable event. The major competitors are likely to be various addition reactions of amines, water or hydroxyls to the quinone. The pseudo first order rate constants for such reactions are, however, quite slow and of a similar order of magnitude ($0.0033s^{-1}$ for cystine and $0.0065s^{-1}$ for lysine) to those found for the rearrangement.[13] Proteins contain many residues, such as hydroxylated or carboxylated amino acids, histidines, lysines and N-terminal amines, which can potentially serve as catalysts of rearrangement and as nucleophiles in Michael additions. Thus the probability of a particuar fate of a peptidyl quinone will be exquisitely sensitive to its immediate environment, since in a solid state system, the interactions between neighboring groups will be constrained by the local three dimensional structure. The resolution of this conundrum is presently at the extreme limit of our experimental ingenuity and may require approaches in modeling the solid state quinone chemistry of purified DOPA proteins, rather than the traditional solution chemistries or compositional analyses of tanned structures.

ACKNOWLEDGMENTS

This work was supported in part by grants N00014-84K-0290 from the Office of Naval Research and 5R0IDE08058 from National Institutes of Health. We thank Stefan Samulewicz for work on byssal catecholoxidase, Steven Taylor for the NMR analysis, and Prof. George W. Luther for redox determinations.

REFERENCES

1. A. M. Kragh & J. Wootton in: "*Adhesion and Adhesives*," 2nd Edition, R. Houwink & G. Salomon, Eds., Elsevier, Amsterdam, Vol. 1, p. 141, 1965.
2. W. Nachtigall, "*Biological Mechanisms of Attachment*," Springer-Verlag, Heidelberg, (1974).
3. C. H. Brown, "*Structural Materials in Animals*," Halstead Press, London (1975)
4. J. Comyn, "*Developments in Adhesives*," A. J. Kinloch, Ed., Applied Science Publ., Barking, UK, Vol. 2, P 279 (1981).
5. H. Lipke, M. Sugumaran & W. Henzel in: "*Advances in Insect Physiology*," M. J. Berridge, J. E. Treherne & V. B. Wigglesworth, Eds., Academic Press, New York, Vol. 17, P 1 (1985).
6. M. Sugumaran in: "*Advances in Insect Physiology*," P. D. Evans & V. B. Wigglesworth, Eds,. Academic Press, New York, Vol. 21, p 179 (1988).
7. J. H. Waite, Comp. Biochem. Physiol. B, **97**, 19 (1990).

8. J. H. Waite, in: "*The Mollusca*," K. M. Wilbur, Ed., Academic Press, New York, Vol. 1, p 467 (1983).
9. J. D. Smyth & J. A. Clegg, Exp. Parasitol, 8, 286 (1959).
10. J. H. Waite, Int. J. Biol. Macromol., 12, 139 (1990).
11. J. H. Waite, Int. J. Adhesives, 7, 9 (1987).
12. A. W. Sternson, R. McCreery, B. Feinberg & R. N. Adams, J. Electroanal. Chem., 46, 313 (1973).
13. D. C. S. Tse, R. L. McCreery & R. N. Adams, J. Med. Chem., 19, 37 (1976).
14. K. Marumo & J. H. Waite, Biochem. Biophys. Acta, 872, 98 (1986).
15. A. Tsugita, T. Uchida, H. W. Mewes & T. Ataka, J. Biochem., 102, 1593 (1987).
16. J. H. Waite, Anal. Chem., 56, 1935 (1984).
17. J. Cabanes, A. Sanchez-Ferrer, R. Bru & F. Garcia-Carmona, Biochem. J., 256, 681 (1988).
18. L. E. Arnow, J. Biol. Chem., 118, 531 (1937).
19. D. W. Barnum, Anal. Chim. Acta, 89, 157 (1977).
20. J. H. Waite & M. L. Tanzer, Anal. Biochem., 111, 131 (1981).
21. E. J. Stamhuis, in: "*Enamines*," A. G. Cook, Ed., Dekker, New York, p 101 (1969).
22. K. Noda, Y. Shimohigashi & N. Izumiya, in: "*The Peptides*," E. Gross & J. Meienhofer, Eds., Academic Press, New York, Vol. 5, p 285 (1983).
23. G. Dryhurst, K. M. Kadish, F. Scheller & R. Renneberg, "*Biological Electrochemistry*," Academic Press, New York, Vol. 1, p 116 (1982).
24. R. F. Boyer, H. M. Clark & A. P. LaRoche, J. Inorg. Biochem., 32, 171 (1988).
25. M. G. Peter, Angew. Chem. Int. Ed. Engl., 28, 555 (1989).
26. J. S. Coleman, L. P. Varga & S. H. Mastin, Inorg. Chem., 9, 1015 (1970).
27. M. Sugumaran, Biochemistry, 25, 4489 (1986).
28. F. M. Ortiz, J. T. Serrano, J. N. R. Lopez, R. V. Castellanos, J. A. L. Teruel & F. Garcia-Canovas, Biochim. Biophys. Acta, 957, 158 (1988).
29. G. H. Alt, in: "*Enamines*," A. G. Cook, Ed., Dekker, New York, p 115 (1969).
30. E. M. Oltz, R. C. Bruening, M. J. Smith, K. Kustin & K. Nakanishi, J. Am. Chem. Soc., 110, 6162 (1988).
31. R. J. Anderson & R. J. Stonard, Can. J. Chem., 57, 2325 (1979).
32. R. J. Stonard & R. J. Anderson, J. Org. Chem., 45, 3687 (1980).
33. R. J. Stonard & R. J. Anderson, Can. J. Chem., 58, 2121 (1980).
34. K. Azumi, H. Yokosawa & S. Ishii, Biochemistry, 29, 159 (1990).
35. L. C. Dorsett, C. J. Hawkins, J. A. Grice, M. F. Lavin, P. M. Merefield, D. L. Parry & I. L. Ross, Biochemistry, 26, 8078 (1987).
36. B. Koning & J. H. Dustmann, Naturwissenchaften, 72, 659 (1985).
37. P. Yaish, A. Gazit, C. Gilon & A Levitzki, Science, 242, 933 (1988).
38. S. O. Anderson, in: "*Comprehensive Insect Physiology, Biochemistry & Pharmacology*," G. A. Kerkut & L. I. Gilbert, Eds., Pergamon Press, Oxford, Vol. 3, p 59 (1985).
39. J. H. Waite & A. C. Rice-Ficht, Biochemistry, 26, 7819 (1987).
40. J. H. Waite & A. C. Rice-Ficht, Biochemistry, 28, 6104 (1989).
41. J. H. Waite, D. C. Hansen & K. T. Little, J. Comp. Physiol. B, 159, 517 (1989).

SYNTHETIC MUSSEL ADHESIVE PROTEINS

Divakar Masilamani, Ina Goldberg, Anthony J. Salerno, Mary A. Oleksiuk, Peter D. Unger, Deborah A. Piascik, and Himangshu R. Bhattacharjee

Allied-Signal, Inc.
Biotechnology Department
101 Columbia Road
Morristown, NJ 07962-1057

The development of a technology for the biological and chemical synthesis of analogs to the adhesive protein of the mussel *Mytilus edulis* is described. This protein consists mainly of the repeating decapeptide sequence Ala-Lys-Pro-Ser-Tyr-Hyp-Hyp-Thr-Tyr-Lys.

Recombinant DNA techniques were used to create and clone a synthetic gene encoding a polydecapeptide analog protein. This protein was successfully expressed in *Escherichia coli* and has an apparent molecular weight of 25 kilodaltons. The chemical synthesis of the protein was achieved in two steps. The decapeptide sequence was first prepared in an automated synthesizer and then polymerized in a second step to form the protein with the molecular weight ranging from 35 to 65 kilodaltons.

The sequence of oxidation steps which converts the mussel adhesive protein to underwater glue was investigated using model peptides. The three-dimensional structure of the protein was derived from the decapeptide sequence with the help of computer modeling studies. The oxidation chemistries were then applied to the three-dimensional protein structure to develop a mechanistic picture, at the molecular level, showing how the mussel attaches itself to solid surfaces.

INTRODUCTION

Structural proteins such as collagen, elastin, and spider silk contain repeating sequences of amino acids. In 1983, Waite discovered that the protein secreted by the phenol gland of the blue sea mussel *Mytilus edulis* consists of closely related decapeptide and hexapeptide sequences with a combined molecular weight of approximately 130 kilodaltons (kDa).[1] He further elucidated that this protein is transformed into glue through enzymatic oxidation to form the adhesive plaques that anchor the mussel to a variety of solid surfaces in a rough aquatic environment.[2] The consensus sequence of the decapeptide reported by Waite is shown below:

Ala-Lys-Pro-Ser-Tyr-Hyp-Hyp-Thr-Tyr-Lys

In 1988, Laursen made the observation that mussels inhabiting the turbulent oceanic environment secrete proteins with little variation from the consensus sequence whereas those found in placid inland lagoons show large variations.[3] Presumably, sea mussels, through a process of natural selection have arrived at an ideal sequence of amino acids that produce the most effective glue. The relationship of the amino acid sequence of this protein to its efficient functioning as a glue is thus clearly demonstrated.[4]

In addition to its ability to perform in an aggressive aqueous environment, the mussel adhesive protein is also attractive because it is nontoxic, durable and biocompatible. For these reasons, it will find extensive use in medical applications like wound closures, bone and dental repair and tissue bonding, which will eliminate the use of sutures during surgery.

Our objective in undertaking the research presented here is to develop a generic technology for the large-scale production of structural proteins. We chose the mussel glue protein to initiate our study because it represents a more complex system compared to other structural proteins and is, therefore, a challenging target for biological and chemical synthesis. Secondly it has afforded us the opportunity to examine how the consensus decapeptide sequences will translate into a three-dimensional structure and how this structure will lend itself to a post-translational chemical conversion to glue. Thus our quest is not only to find ways to produce the mussel adhesive protein but also to understand how it is transformed into a complex cement composite capable of adhering to any solid surface in an aqueous environment.

In this paper, we report our progress in synthesizing the mussel adhesive protein and developing a mechanistic model for its conversion to a water-compatible glue. Following the EXPERIMENTAL section the results are presented in three sections under RESULTS AND DISCUSSION: (1) Production of a Synthetic Mussel Adhesive Protein in Bacterial Systems, (2) Chemical Synthesis and Polymerization of Decapeptides, and (3) Mechanism of Glue Formation.

EXPERIMENTAL

1. Biological Methods

1A. Bacterial Strains and Plasmids

E. coli strains DC1138 and IG110 and plasmid pJL6 have been previously described.[5] Briefly, pJL6 encodes ampicillin resistance and harbors a λ p_L promoter upstream from the cloning region. Plasmid pAV7 is a derivative of pJL6 which contains a synthetic StyI cloning site. Strain IG110 harbors a defective lambda prophage with a *c*I857 allele and an *rpoH*165 mutation which inhibits proteolysis of heterologous proteins. The T7 expression system including plasmid pET-3a and *E. coli* strain BL21/DE3-(plysS) were obtained from W. Studier.[6] The DE3 prophage contains the gene for the T7 polymerase, plysS prevents any fortuitous transcription of the cloned gene, and pET-3a contains the T7 promoter upstream of a cloning region. AS002 (BL21/DE3(plysS)Δ(srlR-recA)306::Tn10) was constructed by P1 transduction of the recA allele from DC1138.

1B. Assembly of a Synthetic Gene Encoding a Polyphenolic Analog Protein

The preparation of the decapeptide gene cassette is outlined in Figure 1. Two complementary pairs of oligodeoxynucleotides (oligos) were prepared by automated synthesis on an Applied Biosystems 380B DNA synthesizer. The first pair encodes one repeat of the decapeptide consensus amino acid sequence of the bioadhesive protein. The second pair also contains one decapeptide sequence which has incorporated a StyI restriction site into its DNA. The cassette oligos and the linker oligos were separately annealed to each other over a temperature range of 70°C → 45°C. The double-stranded cassette- and linker segments were then combined in a ratio of 20:1, respectively, and further annealed over a range of 45°C → 25°C. Nicks between adjacent oligos were covalently sealed with DNA ligase. The heteroduplexes were then restricted with StyI to generate gene cassettes of various sizes. The cassettes were size-fractionated over a Sepharose 4B column and the high-molecular-weight fractions were pooled. They were then cloned into the StyI restriction site of expression vector pAV7.

1C. Expression of the Polydecapeptide Analog Protein

E. coli strain IG110 was transformed with pAG3 and the vector control pAV7. IG110(pAV7) and IG110(pAG3) were grown separately to about 10^8 cells/mL in 10 mL LB-ampicillin broth at 30°C. The cells were then filtered and resuspended in 10 mL of M63 salts including glucose, vitamin B_1 and all amino acids except proline. Each culture was divided into two 5 mL aliquots and each was incubated at 30°C or 41°C for 10 min. Two μCi of [^{14}C] proline/mL of medium were then added to each culture and the incubation continued for 20 min. Cells were chilled in ice water, then centrifuged, and the cell pellets were resuspended in 0.15 mL of 50 mM Tris-HCl, 2% β-mercaptoethanol, 0.5% mixed alkyltrimethylammonium bromide (CTAB) and 1 mM phenylmethylsulfonyl fluoride (PMSF). Cells were broken open by sonication for 3 ten-second intervals. Protein concentration was determined by the Bradford assay and 100 μg of each sample were analyzed by electrophoresis in a 15% polyacrylamide gel containing 150 parts acrylamide to 1 part bis-acrylamide. The gel and buffer consisted of 0.9 M acetic acid, 2.5 M urea and 0.01% CTAB. Gels were treated with En³Hance (New England Nuclear) as described by the manufacturer and proteins were visualized by overnight exposure to x-ray film.[7]

Alternatively, IG110(pAV7) and IG110(pAG3) were grown to 10^8 cells/mL in LB broth at 30°C. Then, the cultures were shifted to 41°C and samples were taken at 0.5, 1, 2 and 3 h. Cells were recovered, prepared for analysis and processed as described above. This time, however, proteins were stained with Coomassie brilliant blue to detect the punitative polydecapeptide.

1D. Transmission Electron Microscopy of Intracellular Polydecapeptides

Cultures of AS002(pET-3a) and AS002(pAG9) were grown in LB-ampicillin-chloramphenicol broth at 37°C to A_{550} = 1.2. To induce the expression

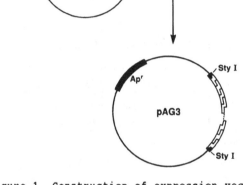

Figure 1. Construction of expression vector pAG3 containing a
synthetic polydecapeptide gene cassette.

system, isopropyl β-D-thiogalactopyranoside (IPTG) was added to 0.4 mM
and incubation was continued for 3 h. One mL samples were then pelleted
and fixed by adding 2.5% glutaralhyde in PBS (173 mM NaCl, 2.7 mM KCl,
4.3 mM NaHPO₄, 1.4 mM KHPO₄, pH 7.3) for 1 h. The cells were dehydrated
in a graded alcohol series and transferred to 100% acetone. The samples
were then embedded in Epon. After sectioning, they were stained in 4%
uranyl acetate for 10-15 min. Sections were examined with a Hitachi 800
electron microscope.

1E. Western Blotting of Polydecapeptide-Analog Protein

For the western analysis, cultures of AS002(pET-3a) and AS002(pAG9)
were prepared and induced with IPTG for 3 h. The cultures were then concentrated 20-fold into 0.5 sonication buffer and sonicated three times
for 10 s. The protein concentration in each sample was determined by the
Lowry Method. Aliquots containing 50 µg of protein were added to an equal
volume of 2x loading buffer and heated at 50°C for 4 min. The samples
were electrophoresed in a 15% polyacrylamide acetic-acid urea gel using

0.9 M acetic acid as running buffer. The gel was run with reversed polarity at 130 V for 3-4 h.

Cellular proteins were transferred to nitrocellulose in a Trans-Blot cell (Biorad). The gel was pre-equilibrated in transfer buffer (0.7% acetic acid) for 30 min. The transfer was then performed with reversed polarity at 30 v overnight. The nitrocellulose was incubated in blocking buffer (5% Carnation non-fat dry milk in PBS) at 37°C for 30 min. The nitrocellulose was then washed in PBS-Tween (0.05% Tween-20 in PBS) three times for 10 min. A 25 mL solution of primary antibody (12.5 µL polyglue antiserum, 250 µL heat-inactivated horse serum, 24.7 mL PBS-Tween) was added for 1 h. After washing in PBS-Tween three times for 5 min, the secondary antibody solution (goat anti-rabbit IgG coupled to horseradish peroxidase, Cappel) was added for 0.5 h. The nitrocellulose was then washed in PBS-Tween three times for 5 min and PBS once to remove residual Tween. The blot was developed using diaminobenzidine as enzyme substrate.

2. Analytical Procedures

2A. Enzyme Oxidation

Mushroom tyrosinase was obtained from Sigma Chemical Co., and used without further purification. The enzyme was dissolved in 0.1 M sodium buffer, pH 7, to a concentration of 1 µg/µL, and the amount used in any given experiment depended on the amount of substrate present. Most experiments required 100-200 units of the enzyme. The enzyme solutions were prepared fresh on the day of use. The specific activity of the lot used in these experiments was 4.9 units/µg freeze-dried powder. Since the enzyme requires oxygen for activity, most experimental solutions were sparged with room air using a pasteur pipette at regular intervals during the course of the exposure.

2B. NMR Studies

13C, 19F and 1H NMR spectra were obtained on a Varian XL-400 spectrometer. The spectra were obtained using a double precision data acquisition with a 45 degree flip angle and a pulse recycle time of 0.5 s. Typically, spectral data were collected in overnight runs. NMR spectra were not only used to determine sample purity (primarily 1H and 13C NMR) but also to monitor the course of removal of side chain protecting groups (19F NMR), and to follow enzymatic oxidation of substrate materials (1H).

2C. Spectrophotometric Analysis

Oxidation of oligopeptides was followed in real time using spectrophotometric analysis. Oxidation was initiated by adding about 150 units of enzyme solution to a "blanked" cuvette containing 1 mL of substrate peptide (0.075 mM) dissolved in 0.1 M phosphate buffer, pH 7. Reaction mixtures were then scanned over the 190-600 nm range, and spectra were recorded at timed intervals. Air was bubbled through the reaction mixtures between scans or at 5 min intervals.

2D. Circular Dichroism Analysis

Circular dichroism (CD) analysis of the Glue-13 20-mer was performed by Professor Ken Breslauer, Department of Chemistry, Rutgers University.

Samples were dissolved in water at a concentration of 2 mM, and CD spectra (190-270 nm) were taken at 5°C intervals between 5 and 25°C, and at 10°C intervals from 25°C to 65°C, The data were deconvoluted with the computer program PROSEC, and interpreted by S. Krimm.

2E. Computer Modeling

Several of the amino acid sequences were modeled using a VAX 8600 computer, with BIOGRAF (Biodesign, Inc.) software. Best conformations were determined by energy minimization calculations. Energy minimization calculations included the following contributions: bonds, angles, torsions and inversions. Van der Waals, electrostatic and hydrogen bonding interactions were also included, and calculations were performed both with and without water shells. The minimized peptide conformations were displayed and manipulated for clarity on Evans & Sutherland 3-D interactive graphics terminals.

RESULTS AND DISCUSSION

1. Production of a Synthetic Mussel Adhesive Protein in Bacterial Systems

1A. Background

Much effort has focused on the synthesis of polyphenolic protein using recombinant DNA technology. This approach has several advantages. It provides structurally perfect proteins with distinct molecular weights. Furthermore, it is possible to produce these proteins in large quantities by fermentation. Workers at Genex Corporation have successfully isolated a cDNA clone obtained from the phenol glands of live mussels.[8] This clone has been transferred into a yeast expression system and has produced polyphenolic protein in its precursor form at a level of about 5% of total cell protein.

Our research has centered on the production of a mussel glue analog protein in bacteria. The primary objective was the practical implementation of a gene cassette approach to the biological preparation of adhesive polymers. A synthetic DNA sequence has been constructed which codes for tandem repeats of the decapeptide precursor. This sequence has been cloned into *E. coli*. Our results show that a polydecapeptide analog protein of 25 kDa was successfully expressed from the synthetic gene.

1B. Design and Cloning of Synthetic Polydecapeptide-Analog Genes

To create a model system for the biosynthesis of polydecapeptide analogs in bacteria, we designed two complimentary pairs of oligos (Figure 1). Each pair contains one repeat of the ten-amino-acid core sequences. The codons were chosen for the coding strand with the expectation of maximizing gene expression.[9] The linker oligonucleotides contain a restriction site for the enzyme StyI. This enzyme was selected based on the following criteria: The restriction site maintains the reading frame, it causes only minimal disruption to the amino acid sequence (proline → alanine in the third position), and it is asymmetric, allowing for strictly head-to-tail ligations. The last property is particularly important for the insertion of a gene cassette into the expression vector as well as for the tandem arrangement of several gene cassettes. The simultaneous annealing of linker oligos and cassette oligos gave rise to a panel of gene cassettes of varying lengths. Colonies containing large gene cassettes were identified by hybridization to one of the cassette oligos. The size of the inserts was confirmed by restriction analysis with StyI and gene cassettes ranging in size from 120 base pairs (bp) to 600 bp were obtained. The 600 bp clone was designated pAG3 and chosen for further study.

1C. *In Vivo* Expression of the Synthetic Polydecapeptide Gene

The *in vivo* expression of the synthetic glue gene contained in plasmid pAG3 was demonstrated using a whole-cell labeling protocol which determined the incorporation of [^{14}C] proline into cellular proteins. Cultures of IG110(pAV7) and IG110(pAG3) were incubated at either 30°C (uninduced) or 41°C in order to induce λp_L promoter (Figure 2A). Autoradiography demonstrated that only the IG110(pAG3) culture at 41°C and not the other three cultures showed a highly labeled protein band. Other protein bands from all four cultures were only faintly visible at exposure times which allowed the unique band from the IG110(pAG3) culture to be readily detected.

In another experiment, whole-cell protein extracts were directly analyzed in a 15% polyacrylamide acetic acid-urea gel after induction at 41°C for various periods of time (Figure 2B). Proteins were visualized by staining with Coomassie brilliant blue. A faint but unique protein band was visible in IG110(pAG3) lanes that could not be detected in the IG110-(pAV7) lanes. This protein band co-migrates with the novel protein band observed by radiolabeling techniques. The presumed polydecapeptide band has a mobility relative to histone H1 that is consistent with its theoretical size of 25 kDa. The intense radiolabeling observed in the analysis above is an indication of the relatively high (30%) proline content of the polydecapeptide.

1D. Electron Microscopic Analysis of Intracellular Polydecapeptide

Since the yields of polydecapeptide using the λp_L promoter in IG110 were rather low, we decided to evaluate the T7 expression system.[6] The 600 bp glue cassette was inserted between the BamHI and NdeI restriction sites of vector pET-3a immediately downstream from the T7 promoter, giving rise to plasmid pAG9. The T7 RNA polymerase is encoded by phage DE3 under control of the *lacUV5* promoter present as a lysogen on the bacterial chromosome. Cultures of AS002(pET-3a) and AS002(pAG9) were induced with IPTG as described in experimental section 1D and processed for electron microscopy. The results are demonstrated in Figure 3. Sections of AS002(pET-3a) (Figure 3A) show round-to-bullet-shaped bacteria indicative of normal *E. coli* cells. The white centers are chromosomal material which strains poorly with uranyl acetate and the outside layer consists of ribosomes (black dots) and protein. In contrast, AS002(pAG9) cells (Figure 3B) are greatly elongated. The chromosomal material is displaced towards one side and the rest of the cell is filled with uniformly stained material. This material presumably represents inclusion bodies of polydecapeptide which is produced at high yields.

1E. Western Blotting of Polydecapeptide

We wished to confirm the identity of the 25-kDa protein as the product of the synthetic polydecapeptide gene. A Western blot was performed using rabbit antiserum raised against chemically synthesized polydecapeptide conjugated to bovine serum albumin. An indirect immunoassay method

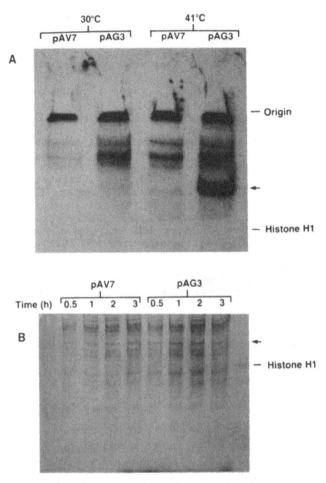

Figure 2. Expression of polydecapeptide in *E. coli*. The arrow
 indicates the position of the polydecapeptide
 visualized by the incorporation of [¹⁴C]proline
 (panel A) or by staining with Coomassie brilliant
 blue (panel B).

was used to detect polydecapeptide-specific proteins with the secondary
antibody being goat anti-rabbit IgG coupled to horseradish peroxidase.
The data are shown in Figure 4. A band was observed only in the lane (3)
containing protein extract of AS002(pAG9) and was absent in the adjacent
lane (2) containing AS002(pET-3a) extract. This band comigrated with a
band observed for purified polydecapeptide (lane 1).

2. Chemical Synthesis and Polymerization of Decapeptides

2A. Preparation of Decapeptide

Since structural proteins are made of repeating amino acid sequences,
they can also be produced chemically by first synthesizing the specific
peptide sequence and polymerizing it in a subsequent step. However, un-

A AS002(pET-3a)

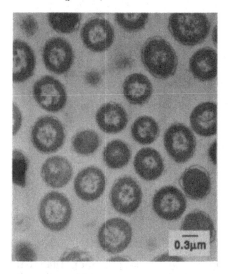

B AS002(pAG-9)

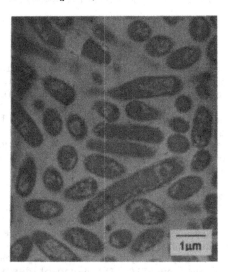

Figure 3. Electron microscopic analysis of polydecapeptide overproduction in *E. coli* AS002. Cells containing the vector alone (panel A) or the polydecapeptide gene cassette (panel B) were induced with IPTG and examined for the presence of intracellular protein aggregates.

like the biological synthesis which provides structurally perfect protein of distinct molecular weight, the chemical method results in mixtures of proteins with a wide distribution of molecular weights.

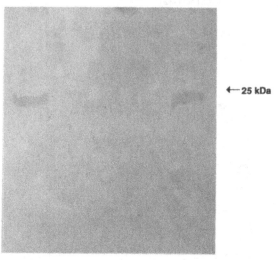

Figure 4. Immunoblotting of polydecapeptide. Lane 1, purified
polydecapeptide; 2, protein extract from AS002-
(pET-3a); 3, protein extract from AS002(pAG9).

We have developed synthetic capabilities to chemically prepare families of decapeptide analogs to repeating units of the polyphenolic protein of the sea mussel *M. edulis*. This was accomplished by preparing
model oligopeptides using an Applied Biosystems model 430A automated
peptide synthesizer employing modified Merrifield chemistries on phenyl-
acetamidomethyl (PAM) resins and t-BOC amino acids.[10] Some of the decapeptides were prepared with sidechain groups either blocked (e.g., ε-
amino groups of lysine residues as trifluoracetamides or in an unreactive
state (e.g., phenylalanine residues substituted for tyrosine residues).
Decapeptides with TFA-blocked lysine side chains were prepared by using
N-α-t-BOC-N-ε-TFA-lysine during solid phase syntheses.

All prepared oligopeptides are summarized in Table 1. Using the short
hand nomenclature of "Glue" peptides, we have already synthesized several
hundred milligrams of purified Glue-2, -3 and -5, Glue 7 to Glue 9, Glue-
12 and 13, as well as Glue-25 and Glue-26. Product yields for the crude
peptide cleaved from the PAM resin were routinely above 70% with this
procedure, and final decapeptide purities after preparative HPLC were
normally over 98% as assayed by HPLC. The decapeptides were characterized
by amino acid analysis and fast atom bombardment mass spectroscopy
(FAB/MS). Besides molecular weights, FAB/MS also gave sequence information, obviating in most cases the need to use conventional gas phase
sequencing techniques as a structural check for these peptides. Details
of the decapeptide synthesis were published earlier.[11]

2B. Polymerization

Since solid phase synthesis of peptides cannot achieve sufficiently
high molecular weights to confer desirable adhesive properties on the
peptides, we have polymerized the synthetic peptides using a coupling
reagent, diphenylphosphoryl azide [DPPA, $(C_6H_5O)_2P(O)N_3$]. DPPA is a
stable, non-explosive liquid (bp 157°C) and was originally reported in
1972 as a convenient reagent for inducing formation of amide bonds.[12] The

Table 1. Decapeptides synthesized by Merrifield's Automated Synthesizer.

H₂N-ALA-LYS-PRO-SER-TER-4-HYP-4-HYP-THR-TYR-LYS-OH	(Glue-2)
H₂N-ALA-LYS-PRO-SER-TYR-4-HYP-4-HYP-THR-TYR-LYS-OH	(Glue-5)

H₂N-ALA-LYS-PRO-SER-TYR-4-HYP-4-HYP-THR-TYR-LYS-OH (Glue-5)
    /   /
  TFA   TFA

H₂N-ALA-LYS-PRO-SER-PHE-4-HYP-4-HYP-THR-TYR-LYS-OH (Glue-7)

H₂N-ALA-LYS-PRO-SER-TYR-4-HYP-4-HYP-THR-PHE-LYS-OH (Glue-8)

H₂N-ALA-LYS-PRO-SER-PHE-4-HYP-4-HYP-THR-PHE-LYS-OH (Glue-9)

H₂N-LYS-PRO-SER-TYR-4-HYP-4-HYP-THR-TYR-LYS-ALA-OH (Glue-12)
    /   /
  TFA   TFA

H₂N-[ALA-LYS-PRO-SER-TYR-4-HYP-4-HYP-THR-TYR-LYS]₂-OH (Glue-13)

H₂N-THR-TYR-LYS-ALA-OH (Glue-25)
       /
     TFA

H₂N-THR-TYR-LYS-ALA-OH (Glue-26)

amide bonds are preferentially formed in the presence of a strong organic base, such as triethylamine, and in aprotic solvents, such as dimethyl-sulfoxide (DMSO), at ambient temperatures. The advantages of DPPA-promoted amidation reactions are racemic-free condensations of amino acids and minimal side reactions with -OH groups when they are present in unprotected peptide side chains. Successful peptide polymerization has been carried out with this agent using Glue-2, Glue-5 and Glue-12. Glue-2, (where the ε-amino groups of the two lysines are unprotected), formed a polymer with a low average molecular weight of the order of 7 kDa. Presumably, the polymerization has occurred not only between the amine and acid termini but also between the ε-amino groups and terminal acid groups resulting in a crosslinked product of low molecular weight. When Glue-5 and Glue-12 in which the ε-amino groups of the lysines are protected by trifluoroacetyl groups, were polymerized, the average molecular weights (determined by the ninhydrin assay) were in the range of 35-65 kDa. Because the ε-amino groups of lysine are blocked, these polymers will be linear. Deblocked polymers were used to measure the adhesive strength.[13]

3. Mechanism of Glue Formation

3A. Background

The polyphenolic protein extracted from the phenol gland does not possess adhesive properties. However, it becomes a glue when it is blended with collagen and subjected to enzymatic oxidation.[14] The participation of collagen in glue formation has been confirmed by Waite.[15] When the mussel finds a solid surface on which to anchor itself, it must first displace the water from the surface and spread the glue composite to establish a foothold. This is possible only if the glue composite contains a large number of hydrogen bonding domains such as hydroxyl groups and amines. Once contact is established with the surface, the glue must cure or "set" in order to form a plaque, which involves extensive cross-linking.[14] The surface interactions contribute to the adhesiveness of the glue while crosslinking contributes to its cohesiveness.

The transformation of the mussel adhesive protein to glue involves a sequence of chemical events in which timing is a critical element. The redox enzyme (catechol oxidase) plays an important role. Recently, Waite has observed that the catechol oxidase is present at very high concentrations in ribbed mussels.[15] At these concentration levels, it acts more like a reagent than a catalyst. Waite has also established that the enzyme shows an apparent molecular weight of 38 kDa and does not contain any metal atoms. Presumably it possesses a quinone cofactor at levels as high as one quinone/mg of enzyme.

To develop a mechanistic scheme for glue formation at the molecular level, it is important to know which amino acids in the mussel adhesive protein and collagen are oxidized and to follow the fate of these amino acids in subsequent secondary reactions such as crosslinking, complexation, degradation, etc. Once the sequence of chemical events is established, it can be applied to the three-dimensional structures of mussel glue protein and collagen to develop a rational model for glue formation.

Proline is the only amino acid common to both the mussel adhesive protein and collagen. Out of the three prolines present in the consensus sequence of the mussel adhesive protein, two are in the form of hydroxyproline. Since biological organisms do not have codons for hydroxyprolines, they are formed via a post-translational oxidation process. The hydroxylation of the prolines in mussel adhesive protein and collagen makes them both compatible for blending with each other to form a polymer composite. This oxidation must occur at the early stages of glue formation because it is essential for establishing the initial attachment to solid surfaces.[14]

In addition to proline, tyrosine is the only other amino acid in the mussel adhesive protein that undergoes oxidation. The enzymatic oxidation of tyrosine has been extensively studied and documented.[16] The conversion of tyrosine to L-Dopa and L-Dopaquinone is well known. Less well known is the fact that L-Dopaquinone can further rearrange to quinone methide and α,β-unsaturated catechol.[17] L-Dopa is an efficient complexing agent for transition metal ions.[18] L-Dopaquinone, on the other hand, is extremely reactive in forming Michael addition products with nucleophiles.[14,19] It has been speculated for more than thirty years that crosslinking between the ε-amino group of lysine and L-Dopaquinone in the mussel adhesive protein is responsible for the cohesiveness of the mussel glue. However, no one has isolated or identified the crosslinked species thus far.

3B. Enzymatic Oxidation of Tyrosine Residues

We studied the enzymatic oxidation of tyrosine using the decapeptide with the consensus sequence as the substrate. Since catechol oxidase from sea mussels is not available, we instead used mushroom tyrosinase for our studies. The oxidations were carried out in a phosphate buffer at ambient temperatures. The reactions were followed using UV spectroscopy and products were separated by HPLC and identified by FAB/MS. The oxidations were also followed in an NMR tube using D_2O as the solvent. When the decapeptide Glue-2 (see Table 1) with the consensus sequence was oxidized in the presence of ascorbic acid, mono-dopa was the predominant product. The enzyme selectively oxidized the tyrosine at position 9.[20] Glue-7, in which the tyrosine at position 5 is replaced by phenylalanine, was oxidized at the same rate as Glue-2 thus indicating that it is the tyrosine at position 9 that the enzyme prefers to oxidize. Glue-8, in which the tyrosine at position 9 is replaced by phenylalanine, did not undergo oxidation. Glue-9, where both tyrosines are replaced by phenylalanine, was also inert.

In the absence of ascorbic acid, Glue-2 underwent extensive polymerization during oxidation. Presumably, the mono-dopa is oxidized to the mono-dopaquinone which then undergoes polymerization via Michael addition. The methyl group of alanine showed a downfield shift in the H'NMR spectrum during polymerization implying that the terminal amino group is involved in polymer formation.

Glue-12, an analog of Glue-2, is obtained by relocating the alanine from the amine end of Glue-2 to the acid terminus. Further, the ε-amino groups of the two lysines are blocked in this decapeptide by trifluoroacetyl (TFA) groups. In an enzymatic oxidation, Glue-12 was first converted to the mono-dopa at position 8. However, after three hours, a new product was formed in more than 90% yield. The molecular weight of this product was twelve mass units higher than that of Glue-12. Since the ε-amino groups of the lysines are blocked, we concluded that an intramolecular Michael addition had occurred resulting in a cyclic quinone-peptide product. However, we could not establish whether the amine terminus or the acid end was involved in the cyclization.

In order to confirm that amino groups indeed participate in Michael additions, we studied the oxidation of a simpler model. The tetrapeptide (Glue-25) which has the same sequence (NH_2-Thr-Tyr-Lys(TFA)-Ala-OH) as the last four amino acids of Glue-12 was synthesized. Upon oxidation, the formation of the L-Dopa derivative was evident from HPLC/MS studies. After six hours, a broad peak indicating polymer formation was observed in the HPLC chromatogram. After 26 hours, the polymerization was complete. The oxidation of Glue-25 was also followed in the NMR tube. All the hydrogens in the tetrapeptide were clearly assigned in the NMR spectrum. Changes were first observed in the aromatic protons of tyrosine. Simultaneously, an upfield shift of the α-proton of the threonine adjacent to the terminal amine group was also observed. This is a clear indication that the polymer is formed by the Michael addition of the terminal amino group to the aromatic ring of dopaquinone. Since the methyl protons and the α-proton of alanine do not show any change in the NMR spectrum, we are certain that the acid terminus is not involved in the polymerization. Further, the β-protons of tyrosine were not sufficiently altered to indicate the participation of a quinone methide in the Michael addition. However, after 15 hours, there was a reduction in their size thus implying that the quinone methide may be formed subsequent to the Michael addition. Surprisingly, after 24 hours, the aromatic protons were greatly reduced. At the same time a white precipitate was formed in the NMR tube. (A similar precipitate was also observed during the oxidation of Glue-2.) This precipitate was dissolved in a sodium bicarbonate solution and was regenerated by acid before being freeze dried. The infrared spectrum of the solid showed an increase in the acid carbonyl functionality, and a dramatic decrease in its aromaticity. Presumably, the amine-substituted aromatic group opens up to form muconic acid derivatives which show lower solubility under our experimental conditions. Oxidations of substituted orthoquinones to muconic acid are well documented.[21]

It is clear from the above experiments that the enzymatic oxidation of tyrosine is rather complex. The observed sequence of reactions is summarized in Figure 5. Tyrosine [1] is first slowly oxidized to L-Dopa [2]. The oxidation of L-Dopa to dopaquinone [3] occurs somewhat more

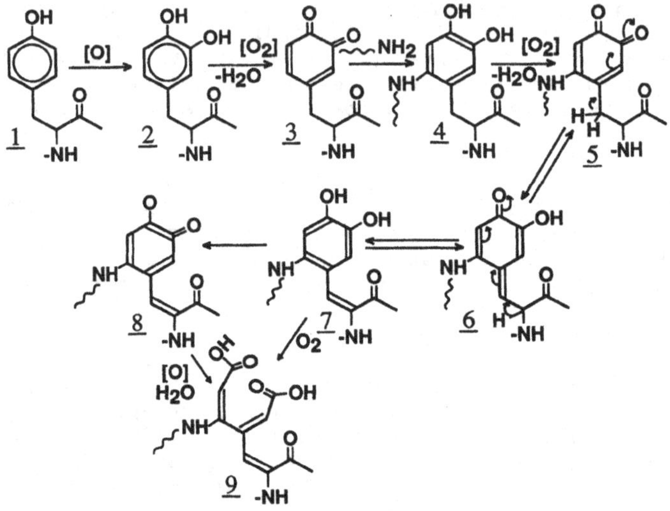

Figure 5. The fate of tyrosine during the enzymatic oxidation of mussel adhesive protein.

quickly, while the Michael addition of an amine to dopaquinone appears to be instantaneous. This result explains why dopaquinone is seldom observed as an intermediate. The Michael product [4] is converted to its corresponding orthoquinone [5]. This particular intermediate is reactive and can isomerize to the quinone methide [6] and an α-aminocaffeic acid derivative [7]. Compounds [5], [6] and [7] are susceptible to ring opening to form muconic acid derivatives [9].[21] These reactions may occur sequentially or separately. Compounds [5] to [9] are capable of undergoing further crosslinking through nucleophilic addition. The muconic acid derivative is particularly interesting as a surfactant compatibilizing the hydrophobic parts of the protein polymer with the hydrophilic environment.

3C. Computer Modeling Studies

The chemical reactions discussed above are not by themselves sufficient to develop a mechanistic scheme for underwater glue formation unless they are reviewed in the context of the three-dimensional structure of the mussel adhesive protein. The structure of this protein however is not available. We therefore decided to derive its three-dimensional structure from the consensus decapeptide sequence in an incremental fashion using computer modeling and chemical intuition. It is well-known that the primary amino acid sequence determines the structure of a protein.[4] However, predicting three-dimensional structures from an amino acid sequence is a risky venture unless there are experimental guide posts that support or reject the predicted structures.

We started with the consensus sequence represented by Glue-2. In the computer modeling studies, Glue-2 was deconvoluted by heat simulation and allowed to cool slowly. Energy minimization calculations were performed in order to predict the favored conformation based on bond stretching, bond bending, torsions, inversions and non-bonded interactions due to electrostatic, van der Waal's and hydrogen-bonding forces. As shown in Figure 6, the lowest energy conformation assumed by Glue-2 displays a large bend, which enfolds tyrosine-5, with Pro(3)-Ser(4) on the amino

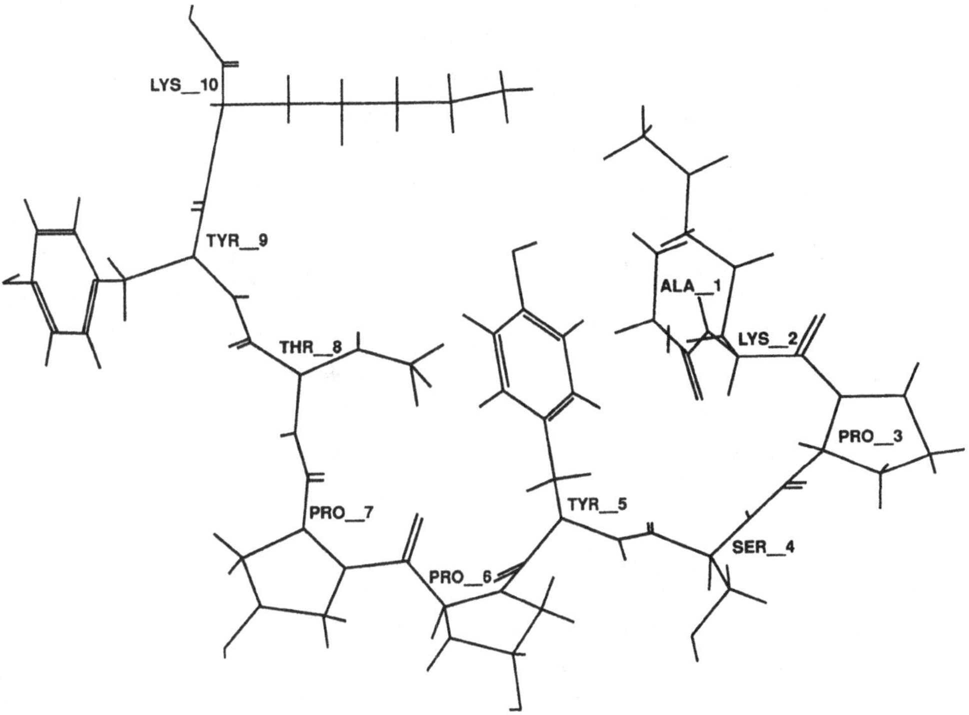

Figure 6. The preferred conformation of the consensus decapeptide (Glue-2).

side and Hyp(6)-Hyp(7)-Thr(8) on the acid side, resembling two arms
shielding the Tyr-5 residue. No such shield, however, is found around the
Tyr-9 residue which is rotated out away from the Tyr-5 bending domain.
This conformation is in excellent agreement with the experimental evidence discussed earlier, suggesting that the rate of oxidation of Tyr-9
is several times higher than that of Tyr-5. Based on these computer
modeling results, the steric shielding around the Tyr-5 should be sufficient to retard its oxidation by mushroom tyrosinase, a restraint not
imposed on the unhindered Tyr-9. However, the catechol oxidase produced
by the sea mussel may not discriminate between the two tyrosines.

The correlation of the predicted structure of Glue-2 with the experi-
mentally observed difference in the oxidation rates of its two tyrosines
gave us confidence in our models. We extended the computer modeling
studies to Glue-13, an eicosapeptide consisting of two consensus sequences in tandem. The energy-minimized structure is shown in Figure 7.
The basic conformation seen with Glue-2 is retained for each of the repeat segments, forming a double bend into which the Tyr-5 and Tyr-15 are
caged. A breakdown of the bonded vs. nonbonded interactions reveals that
nonbonded interactions, particularly electrostatic and hydrogen bonding,
are the major forces stabilizing the Glue-13 secondary structure. For
example, hydrogen bonding of the carbonyl between Hyp-6(16)-Hyp-7(17) and
the hydroxyl proton of Thr-8(18) appears to augment the left-handed
twists and also to reinforce and stabilize the β-turns of the Hyp-Hyp
ensembles.[22] An interesting conclusion from the computer modeling studies
of the Glue-13 is the left-handed helical structure between Thr-7 and
Lys-12. This prediction is supported by CD spectra of Glue-13. The CD
spectra were collected in two separate experiments over the temperature

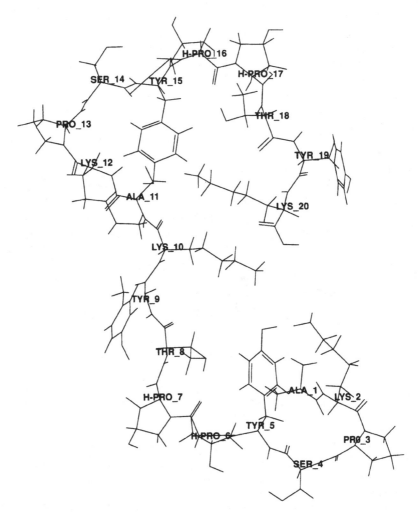

Figure 7. The preferred conformation of the eicosapeptide (Glue-13).

ranges of 5-55°C or 5-65°C. The original spectra (Figure 8) appear to support at least one regular conformation, a tight left-handed helix with 2.5 residues per turn.

Encouraged by these results, we extended our computer modeling studies to the 30-mer made of three consensus sequences in tandem. The basic conformation of the eicosapeptide (Glue-13) is retained. The 30-mer forms three curved pockets into which Tyr-5, Tyr-15 and Tyr-25 are enclosed. The 30-mer also displayed a complete loop into which Lys-10, Lys-20 and Lys-30 extend in a linear fashion. On the other hand, Lys-2, Lys-12 and Lys-22 as well as Tyr-9, Tyr-19 and Tyr-29 are rotated away from the interior.

From the computer modeling studies, we were able to recognize a regular repeating pattern in the mussel adhesive protein. It is easy to see this pattern if the consensus sequence of the polymer is rewritten as shown below. We can view the protein as consisting of two types of re-

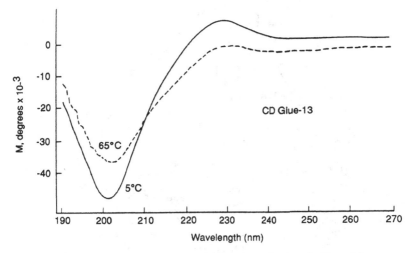

Figure 8. The circular dichroism (CD) of Glue-13.

peating pentapeptide segments giving rise to a loosely coiling macro-
structure (Figure 9). The first segment contains three prolines. These
amino acids disrupt regular protein structures. The first proline causes
this segment to bend while the two at the end of this segment cause it to
undergo a β-turn.[22] In this segment, all hydroxyl groups are arranged on
the outside while tyrosine at the center of the segment will project into
the loop. The second pentapeptide segment is a left-handed helix with two
complete turns. In this helical structure the hydrophobic methyl groups
of threonine and alanine are on the cavity side along with the central
lysine while the tyrosine and the second lysine occupy the opposite side
of the helical structure.

-[(Pro-Ser-Tyr-Hyp-Hyp)-(Thr-Tyr-Lys-Ala-Lys)]$_n$-

When viewed from the top, we can see three tyrosines and three
lysines per coil projecting into the cylindrical cavity while three other
tyrosines and lysines are placed outside the coil (see Figure 9). Two
kinds of crosslinking are feasible: one inside the coil rigidifying the
helical structure and the second outside the coil interlocking the
different spiral loops. In each loop there are three segments with
hydroxyl groups projecting outside the ring that play a critical role in
glue formation.

Based on the model shown in Figure 9, the following preliminary picture of the attachment process of the mussel emerges: when the mussel
makes contact with a solid surface with its foot and forces out the surface water, it deposits the adhesive precursor made of mussel adhesive
protein and collagen along with a large excess of the enzyme catechol
oxidase. The hydroxylation of prolines on the periphery of the mussel
adhesive protein and collagen will compatibilize these two proteins and
enhance their surface adhesion. These interactions will be followed by a
crosslinking between lysine and tyrosine, which contributes to the cohesiveness of adsorbed material. The tyrosines then undergo further oxidations to muconic acid derivatives which help to bind the glue material
even more tightly to the surface converting it to a plaque. As the mussel
pulls away from the plaque, it forms a byssal thread made of bundles of

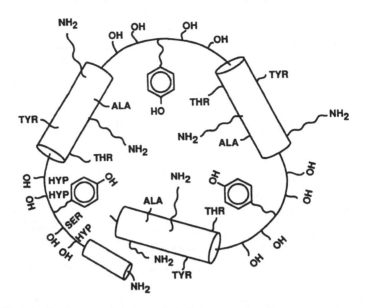

Figure 9. A structural model for mussel adhesive protein.

hydroxylated mussel glue protein reinforced by hydroxylated collagen. The divalent cations present in sea water presumably play a role in bringing the muconic acids from different protein strands together during byssus formation.[23]

CONCLUSIONS

We have designed and cloned into *E. coli* a synthetic DNA sequence coding for the polydecapeptide repeat of the bioadhesive protein of the mussel *M. edulis*. Codons were chosen in order to maximize expression of the synthetic protein. A 600 bp gene cassette was identified and successfully expressed in *E. coli* giving rise to a 25 kDa protein. In addition, we have synthesized the mussel adhesive protein chemically by first preparing the consensus decapeptide sequence and polymerizing it in the presence of DPPA. The molecular weight of the chemically synthesized protein ranged from 35-65 kDa. Antibodies raised against the chemically synthesized protein also recognized the one produced biologically from the gene cassette-containing cells. It is therefore clear that the proteins from these two sources must share antigenic domains indicating that they have similar structural conformations.

A loosely coiling macro-structure was assigned to the mussel adhesive protein. This structure was derived from the consensus decapeptide sequence using computer modeling studies as well as supporting data from oxidation rates, CD, NMR and IR spectroscopy. The transformation of the mussel adhesive protein to glue involves a series of post-translational processes. First, the protein is blended with collagen and the prolines in both proteins are hydroxylated on the outer surface. The hydroxylated polymer composite would adhere strongly to solid surfaces. The tyrosines in the mussel adhesive protein are then oxidized triggering crosslinking reactions. Further oxidation converts the aromatic groups to muconic acid derivatives which, in the presence of sea water, will be in the form of ionic salts. Such salt formation will help to strengthen the surface adhesion of plaques and promote aggregation of the various protein strands in the assembly of byssal threads.

ACKNOWLEDGMENTS

Carol Gross, Donald Court, and William Studier kindly provided *E. coli* strains and plasmids. We wish to thank Mina Gabriel, Ray Brambilla, David Hindenlang, and Kennedy O'Brien for performing the HPLC, NMR, MS and IR analyses, respectively. We are also grateful to Deirdre Furlong and Vivian Kramer for help with the electron microscopy. Marylou Grumka is acknowledged for typing the manuscript.

This work was supported in part by Office of Naval research contract N00014-86-C-0484.

REFERENCES

1. J. H. Waite, J. Biol. Chem., 2911 (1983).
2. J. H. Waite, Biol. Rev. Cambr. Philos. Soc., 58, 209 (1983).
3. R. A. Laursen, M. J. Conners, X. T. Shen & J. J. Ou, SIM News Suppl., 38(4), 18, 5 (1988).
4. J. U. Bowie, J. F. Reidhaar-Olson, W. A. Lim & R. T. Sauer, Science, 247, 1306 (1990).
5. I. Goldberg & A. J. Salerno, Gene, 80, 305 (1989).
6. F. W. Studier & B. A. Moffatt, J. Mol. Biol., 189, 113 (1986).
7. R. A. Lasky, Methods Enzymol., 65, 363 (1980).
8. R. L. Strausberg, D. M. Anderson, D. Filpula, M. Finkelman, R. Link, R. McCandliss, S. A. Orndorff, S. L. Strausberg & T. Wei in: "Adhesion from Renewable Resources," R. W. Hemmingway, A. H. Conners, S. J. Branham, Eds., American Chemical Society, Washington, DC, 1989, Chapter 32, p. 453.
9. H. A. de Boer & R. A. Kastelein in: "Maximizing Gene Expression," W. S. Reznikoff & L. Gold, Eds., Butterworth, Boston, 1986, Chapter 8, p. 225.
10. Applied Biosystems Peptide Synthesizer User Bulletin, No. 3, August, 1985; also B. J. Bergot, R. L. Noble and T. Gerser, Op. Cit., No. 16, September, 1986.
11. M. D. Swerdloff, S. B. Anderson, R. D. Sedgwick, M. K. Gabriel, R. J. Brambilla, D. M. Hindenlang & J. I. Williams, Int. J. Peptide Protein Res., 33, 318 (1989).
12. T. Shioiri, K. Ninomiya & S. I. Yamada, J. Amer. Chem. Soc. 94, 6203 (1972); N. Nishi, B.-I. Nakajima, N. Hasebe and J. Naguchi, Int. J. Biol. Macromol., 2, 53 (1980); B.-I. Nakajima & N. Nishi, Polymer J., 13, 183 (1981).
13. H. R. Bhattacharjee, Paper presented at the ACS National Meeting, Los Angeles, CA, September, 1988.
14. J. H. Waite, Chemtech, 692 November, 1987.
15. J. H. Waite, Annual Report, Office of Naval Research Contract No. N00014-84-K-0290, October 1, 1989.
16. E. Lindner & C. A. Dooley, in: "Proc. 3rd Int. Biodegradation Symp.," J. M. Sharpley & A. M. Kaplan, Eds., Applied Science Press, Barkley, England, 1976, p. 465-494.
17. M. Sugumaran, H. Dali, H. Kundzicz & V. Semensi, Bioorg. Chem., 17, 443 (1989).
18. C. G. Pierpont & R. M. Buchanan, Coord. Chem. Rev., 38, 45 (1981).
19. C. R. Tindale, Aust. J. Chem., 37, 611 (1984).
20. K. Marumo & J. H. Waite, Biochem. et. Biophys. Acta., 872, 98 (1986).

21. M. M. Rogic' & T. R. Demmin, in: "*Aspect of Mechanism and Organo-metallic Chemistry*," J. M. Brewster, Ed., Plenum Press, New York, 1978, p. 141-168.
22. P. Y. Chou & G. D. Fasman, J. Mol. Biol., **115**, 135 (1977).
23. V. J. Morris, Proc. Polymeric Materials: Science and Engineering, **62**, 482-487 (1990).

POLY(VAL[1]-PRO[2]-ALA[3]-VAL[4]-GLY[5]): A REVERSIBLE, INVERSE THERMOPLASTIC

D. W. Urry, J. Jaggard, K. U. Prasad, T. Parker and
R. D. Harris

Laboratory of Molecular Biophysics, School of Medicine
The University of Alabama at Birmingham
P. O. Box 300/University Station
Birmingham, Alabama 35294

Initial characterization of the sequential polypentapeptide, poly(VPAVG) an analog of the polypentapeptide of elastin poly(VPGVG), is reported. It undergoes a concentration
dependent inverse temperature transition, in which it is
soluble in water below 25°C and aggregates on raising the
temperature. When crosslinked by 20 Mrads of γ-irradiation to
give X[20]-poly(VPAVG), an elastomeric matrix is formed. At
25°C, this material exhibits an elastic modulus of 1.5 x 10[5]
dynes/cm², similar to that found at that temperature for X[20]-
poly(VPGVG). On raising the temperature to 37°C, however, the
elastic modulus increases three orders of magnitude to 3 x
10[8] dynes/cm² which is two orders of magnitude greater than
obtained for X[20]-poly(VPGVG). This reversible hardening on
raising the temperature defines X[20]-poly(VPAVG) as a reversible, inverse thermoplastic. This property makes it an interesting new material to add to the already described set of
bioelastic materials. The potential applications and exten-
sions of applications made possible by this additional bioelastic material are briefly considered.

INTRODUCTION

The most striking repeating sequence in elastin is poly(Val-Pro-Gly-
Val-Gly), poly(VPGVG), which repeats eleven times in bovine elastin without a single substitution,[1] and in porcine elastin with but one substitution.[2] This material is soluble in water below 25°C but aggregates and
separates into a viscoelastic phase on raising the temperature to 40°C.
Crosslinking of the viscoelastic phase by means of γ-irradiation results
in an elastomeric matrix which exhibits thermomechanical[3] and chemomechanical transduction.[4,5] The contractile process of thermomechanical
transduction, like the temperature elicited aggregation (coacervation) in
the absence of crosslinking, is the result of an inverse temperature
transition in which the polypeptide folds optimizing intramolecular and
intermolecular hydrophobic contacts. Making the polypentapeptide more
hydrophobic, e.g., by synthesizing poly(Ile[1]-Pro[2]-Gly[3]-Val[4]-Gly[5]) in

which there is the addition of one CH₂ moiety per pentamer, causes the transition to occur at a lower temperature,[3] whereas making the polypeptide less hydrophobic, as is the case for poly(VPGG) or for poly(Val[1]-Pro[2]-Gly[3]-Ala[4]-Gly[5]), causes the transition to occur at a higher temperature.[3] Furthermore, introducing a glutamic acid (E) residue, as in poly-[4(VPGVG),(VPGEG)], allows the temperature of the transition to be changed by changing the pH where COOH is more hydrophobic (less polar) than COO⁻.[6] Thus, when crosslinked, the Glu-containing polypentapeptide exhibits mechanochemical coupling;[4] it contracts on lowering the pH and relaxes on raising the pH to near 7, and it is capable of picking up weights that are a thousand times its dry weight.

We here report another interesting analog that has the unique property of being an inverse thermoplastic. The analog is poly(Val[1]-Pro[2]-Ala[3]-Val[4]-Gly[5]), abbreviated as poly(VPAVG) or the L-Ala[3]-polypentapeptide. Previous replacement of the Gly[5] residue by L-Ala[5] destroyed elasticity resulting in a granular precipitation on raising the temperature of an aqueous solution.[7] The D-Ala[5] analog slowed the rate of coacervate formation but did not result in a functional γ-irradiation crosslinked matrix.[8] The D-Ala[3] analog led to a stiffer, more crystalline, matrix.[9] As shown below, the L-Ala[3] analog can be formed into a matrix of low elastic modulus, but on raising the temperature through the inverse temperature transition, it hardens into a plastic state which again softens on lowering the temperature. The elastic modulus was observed to change by three orders of magnitude from 10^5 to 10^8 dynes/cm² on raising the temperature 10°C from 25°C to 35°C.

EXPERIMENTAL

1. Peptide Synthesis

Boc-Val-Pro-OBzl and Boc-Val-Gly-OBzl were prepared as previously described.[10]

1A. Boc-Val-Pro-OH (I)

Boc-Val-Pro-OBzl (30.0g, 0.074 mol) was hydrogenated to obtain 21.40g (yield 91.6%) of (I).

1B. Boc-Ala-Val-Gly-OBzl (II)

Boc-Val-Gly-OBzl (10.93g, 0.03 mol) was deblocked with HCl/Dioxane and coupled to Boc-Ala-OH (5.66g, 0.03 mol) using EDCI with HOBt. The reaction was worked up by acid and base extractions to obtain 11.4g (yield 87.3%) of (II).

1C. Boc-Val-Ala-Val-Gly-OBzl (III)

Compound (II) (9.0g, 0.02 mol) was deblocked using TFA and coupled to Boc-Val-Pro-OH (I) (6.28g, 0.02 mol) using EDCI with HOBt to obtain 12.20 g (yield 96.7%) of (III).

1D. Boc-Val-Pro-Ala-Val-Gly-ONp (IV)

Compound (III) (5.0g, 0.008 mol) was hydrogenated and reacted with bis-PNPC (3.46g, 1.5 equiv.) in pyride (40 mL). When the reaction was complete, the product was worked up by acid and base extractions to obtain 4.0g (yield 79.8%) of (IV).

1E. Poly(Val-Pro-Ala-Val-Gly) (V)

Compound (IV) (4.0g, 0.006 mol) was deblocked using TFA, and one molar solution of the TFA salt in DMSO was polymerized for 18 days using 1.6 equiv. of NMM as base. The polymer was dissolved in water, dialyzed using 3,500 mol wt. cut-off dialysis tubing and lyophilized. The product was base treated with 1N NaOH (2 equiv. per pentamer), dialyzed using 50 kD cut-off dialysis tubing for one week and lyophilized to obtain 1.7g (yield 46.8%) of the title compound, poly(VPAVG). This gives a polypeptide chain of 600 residues or more. The carbon-13 nuclear magnetic resonance spectrum of poly(VPAVG), compound V, is shown in Figure 1 where the assignments of resonances are given. This, with the absence of extraneous peaks, verifies the synthesis.

Abbreviations: Boc, tert-butyloxycarbonyl; OBzl, benzyl ester; EDCI, 1-ethyl-3-dimethylaminopropyl carbodiimide; HOBt, 1-hydroxybenzotriazole; bis-PNPC, bis(4-nitrophenyl) carbonate; TFA, trifluoroacetic acid; DMSO,

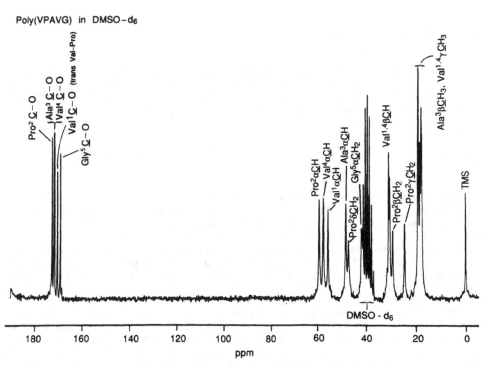

Figure 1. Carbon-13 nuclear magnetic resonance spectrum for poly(VPAVG). All assignments are given demonstrating the correct synthesis, and the absence of extraneous peaks verifies the purity of the synthesis.

dimethylsulfoxide, NMM, N-methylmorpholine; Ala, alanine; Gly, glycine; Pro, proline; Val, valine.

2. Concentration Dependence of Inverse Temperature Transition

The temperature of the inverse temperature transition was determined for poly(VPAVG) as a function of concentration of polypentapeptide by means of the development of turbidity on raising the temperature. The temperature dependent aggregation was followed as turbidity at 300 nm using a CARY 14 spectrophotometer equipped with a 300 Hz vibrator in order to minimize settling during the temperature scan. Concentrations were varied from 100 mg/mL down to 0.15 mg/mL and the temperature was scanned from below 20°C to 60°C or more.

3. Cross-linking of Poly(VPAVG)

A quantity of 149.27 mg of poly(VPAVG) was dissolved in 375 μL of Type 1 H_2O at 4°C in a cryotube. A Plexiglas pestle with a circular channel 7mm wide and 0.7 deep was inserted into the cryotube such that the material flowed into the channel. The sample was then centrifuged at 3000 rpm for 2 hours at room temperature. Thereafter, the sample was cross-linked by 20 Mrad of γ-irradiation from a cobalt-60 source, producing an elastomeric strip, which is referred to as X^{20}-poly(VPAVG).

4. Stress-strain Apparatus

Stress-strain data were obtained using an instrument built in this laboratory and previously described.[11] To accommodate high force ranges required with the present material, a UL4-2 load cell accessory was attached to the Statham UC-2 transducer extending the transducer force range to 1 kg and beyond.

RESULTS

The temperature profiles for turbidity formation (TPτ) curves for poly(VPAVG) are given in Figure 2 for different concentrations. These curves demonstrate the concentration dependence for the onset of polypentapeptide chain association, that is, for the onset of the inverse temperature transition. This aggregation is reversible. On lowering the temperature below 25°C, the sample clears and the temperature profile for turbidity formation can be repeated. This polypentapeptide differs from the naturally occurring repeating pentamer sequence of elastin poly-(VPGVG), by the addition of a single CH_2 moiety per pentamer. Curiously, in this case the temperature of the transition is not decreased on this addition of hydrophobicity whereas it is markedly decreased on adding the CH_2 moiety to the Val[1] residue to result in poly(IPGVG) or poly(LPGVG) where I = Ile and L = Leu. Perhaps this is due to the observation that the difference in hydrophobicity, as measured for example in the Nozaki and Tanford scale[12], is much less for Gly and Ala than it is for Val and Ile or Leu.

When a concentration of approximately 400mg/mL of poly(VPAVG) in water is shaped into a band and γ-irradiation crosslinked, an elastomeric

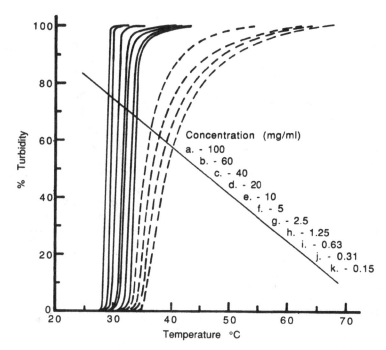

Figure 2. Temperature profiles for aggregation for poly-
(VPAVG) as a function of concentration ranging from
100 mg/mL to 0.15 mg/mL.

band is prepared. As shown in Figure 3A, at 25°C the elastic modulus is
1.5 x 10^5 dynes/cm². When the same strip of material is heated to 35°C,
the initial elastic modulus has increased dramatically to 2.1 x 10^8
dynes/cm² (see Figure 3B). This is a remarkable change in elastic modulus. The material has been thermally transformed from a fragile material
that readily tears to a very tough material in which a band can support a
weight that is 50,000 times its dry weight. Significantly, this process,
which is that of an inverse thermoplastic, is entirely reversible. Instead of a thermoplastic which becomes softer on raising the temperature
and hardens on lowering the temperature, this material does exactly the
inverse: it hardens on raising the temperature to body temperature and
softens on lowering the temperature. As may be seen by the series of
curves in Figure 3C, however, this material does not exhibit an ideal
elasticity in its hardened state. There is both a significant hysteresis
and each stress/strain cycle is displaced without readily seating into an
exactly reproducible cycle. The abrupt increase in elastic modulus as the
temperature is raised from 25°C to 35°C is reproducible. The rate of
change of elastic modulus with temperature is truly striking.

The temperature dependence of length at fixed force (2g), $(\partial l/\partial T)_f$,
for X^{20}-poly(VPAVG) is given in Figure 4. The most rapid change of length
with change in temperature occurs near 25°C when the temperature scan is
from low to high. At temperatures below the transition, there occurs a
more gradual, nearly linear, decrease in length with increasing temperature from about 10°C to nearly 20°C. In Figure 5 are photographs of a
piece of X^{20}-poly(VPAVG), which had been previously used in mechanical
studies, contracted at elevated temperature near 37°C and swollen at low
temperature. Overall, the change in length on raising the temperature
approaches a factor of 2.2. Interestingly, on lowering the temperature

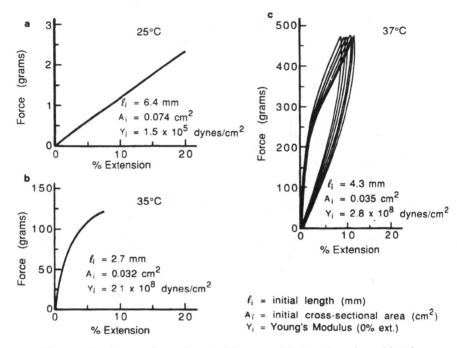

Figure 3. Stress/strain studies on 20 Mrad γ-irradiation crosslinked poly(VPAVG), designated as X^{20}-poly-(VPAVG), in H_2O. (A) At 25°C, the elastic modulus is 1.5×10^5 dynes/cm². (B) At 35°C, the elastic modulus obtained from the initial slope is 2.1×10^8 dynes/ cm². (C) At 37°C, series of stress/strain curves are obtained showing a rather uniform displacement with each subsequent cycle. Very large forces are reached, about 470 grams, for a 0.035 cm² cross-sectional area of sample. The sample between clamps with a dry weight of the order of 10 mg was able to develop forces sufficient to lift weights some 50,000 times its dry weight. The elastic modulus obtained from the initial slope was 2.8×10^8 dynes/cm².

from 35°C, the increase in length is delayed by 4°C as seen in Figure 4. This delay in onset of the transition on lowering the temperature is reproducible and occurs even when the scan rate is 2 hrs./°C.

A strip of X^{20}-poly(VPAVG) which had been used in the mechanical studies is shown in Figure 5 near 37°C when it is contracted and near 20°C where it is largely swollen.

DISCUSSION

In the examination of previous analogs of poly(VPGVG), addition of one CH_2 per pentamer as in poly(IPGVG) gave the same conformation and elastic modulus but resulted in lowering the temperature of the transition by some 20°C.[3] On the other hand, addition of one CH_2 moiety as in D-Ala³-polypentapeptide, abbreviated as poly(VPA'VG), did lower the tran-

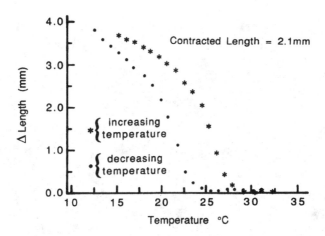

Figure 4. Temperature dependence of length for X^{20}-poly-(VPAVG) in water. On raising the temperature from near 10°C to 30°C, the sample contracts to about one third of the low temperature length with the major change centered near 25°C. On lowering the temperature, the increase in length is again near a factor of three, but the major change in length occurs at a temperature some 3°C to 4°C lower.

sition temperature by a few degrees, but was reported also to increase the elastic modulus by a factor of two.[9] In the present analog, the added CH_2 moiety caused the transition temperature to increase a few degrees but remarkably caused, at 37°C, an increase in the elastic modulus of two orders of magnitude to 10^8 dynes/cm². This greater elastic modulus is a result of the transition, as is apparent in Figure 3 where the force required for a given % extension and for the same cross-section of material is more than 100 times greater at 35°C than at 25°C. Both X^{20}-poly-(VPGVG) and X^{20}-poly(VPAVG) have elastic moduli in the range of 10^5 dynes/cm² at temperatures below the transition.

Extrapolating from the poly(VPGVG) studies, it is expected that the temperature of the inverse temperature transition for this interesting material can also be modified by changes in hydrophobicity and, for example, that the marked contraction and development of force of X^{20}-poly-(IPAVG) as for X^{20}-poly(IPGVG) might occur near 10°C.[3] This would make it possible for this material to have its contraction centered at any temperature between 0°C and 70°C where in the latter case the composition could be, for example, X^{20}-poly(VPAAG).

It has already been demonstrated that changes in sodium chloride concentration can change the transition temperature for poly(VPAVG). In fact, the same change in NaCl concentration leads to the same change in temperature for poly(VPGVG) as for poly(VPAVG) but the endothermic heat of transition of the latter is twice that of the former (C. H. Luan and D. W. Urry, in preparation). That being the case, it can also be anticipated that addition of solvents such as dimethylsulfoxide and ethylene glycol[13] would lower the temperature of the transition and increase the elastic modulus further. Thus, it is expected that X^{20}-poly(VPAVG) would exhibit solvent-based chemomechanical transduction. Clearly, the greater loads that X^{20}-poly(VPAVG) can lift would constitute an advantage, and one might wonder whether greater efficiency might be achievable. The latter possibility, however, must be tempered by two factors: one is the

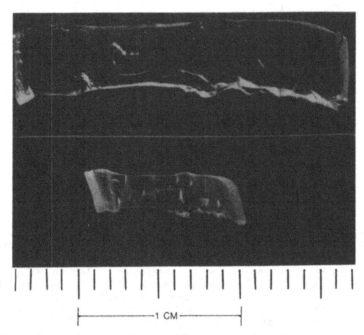

Figure 5. Strip of X^{20}-poly(VPAVG) in a largely concentrated
state (below) and a mostly swollen state (above).
This sample had been used in a mechanical study in
which there some 4 mm between clamps in the concentrated state.

increased heat required to drive the endothermic transition of contraction and the second is due to the hysteresis seen in Figure 4. If chemomechanical transduction follows the same pattern as thermomechanical
transduction, larger changes in concentration of a chemical will be required to reverse the process just as greater changes in temperature are
required to reverse changes in length. Thus, stepping the temperature
from 22°C to 27°C can bring a shortening and a lifting of a weight but,
based on Figure 4, reversing the process would require decreasing the
temperature from 27°C to 17°C.

Another significant issue that has yet to be addressed is whether
poly(VPAVG) can be designed to exhibit polymer-based chemomechanical
transduction. For example, can X^{20}-poly[4(VPAVG),(VPAEG)] function as a
reversible mechanochemical engine. The resolution of these issues will be
required before a more complete understanding of the potential commercial
applications of the L-Ala[3] analog of the polypentapeptide of elastin is
at hand. Nonetheless, based on what is known for poly(VPGVG) and its
other analogs, it is of interest to consider the particular advantages of
poly(VPAVG).

The potential applications of poly(VPAVG) run the spectrum of those
already under consideration for poly(VPGVG) and its other analogs, but,
with the added features of greater strength and elastic modulus at temperatures above those of the inverse temperature transition, interesting
variations become possible. The medical applications will depend on properties of biocompatibility and biodegradability which have yet to be
determined, but the following discussion will assume that these are basi-
cally similar for poly(VPGVG) and poly(VPAVG). Qualitative expectations

might be noted, however, before proceeding. Biodegradability is to be expected though it could still be slowed due to the more tightly packed poly(VPAVG) at 37°C. Biocompatibility still would be expected to be promising with the yet simple amino acid composition though a relatively less dynamic, more rigid structure could lend itself more readily to function as an epitope. While PG is the most common naturally occurring dipeptide in the elastic fiber of mammals, the sequence PA does occur.

With regard to prevention of adhesions, equivalently crosslinked poly(VPAVG) is a tougher material with which to work and could be made into thinner sheets which would yet be strong and adequately compliant. In the potential application of synthetic arteries, the greater strength and elastic modulus would allow for a porosity which would provide for cellular migration through the matrix and yet retain the required compliance. A similar situation would obtain for synthetic ligament. In both cases, the perspective would be for a temporary scaffolding with the correct compliance, which would allow the invading natural cells to rebuild the natural tissue. As for artificial sphincters, the material could be sealed in a ring with irrigation capability from the outside of the body. Here the strength of the material could prove advantageous.

The properties of poly(VPAVG), when a drop of aqueous solution is gotten on the skin, are worth noting because it immediately coats the skin so tightly that even scraping with a fingernail leaves a coating. Rinsing with water, however, is sufficient to remove the coating. This property of tightly coating the skin raises the possibility of considering this polypentapeptide as a base for cosmetic applications.

Whether used as a prosthetic artery, ligament, etc., a desirable property is biodegradation, but for different applications and different sites, the desired rate of degradation may differ. Varied rates of degradation may be achieved for poly(VPAVG) and for any of the already described bioelastic materials by a number of ways, both chemical and physical. A more porous, less dense, matrix would be a physical means to achieve a greater rate of degradation due to accessibility of increased surface area, whereas a chemical means would be to introduce more readily enzymatically cleaved sequences or to introduce, as required and where appropriate, ester linkages which spontaneously hydrolyzed in tissue fluids as can occur for poly(lactic-glycolic acid) materials.[14,15]

With regard to industrial applications, its capacity to function as a free energy transducer is of interest. Its greater strength, e.g., the capacity to lift weights that are tens of thousands of times greater than the dry weight of the material, would extend the possible applications involving thermomechanical and chemomechanical transduction. Two issues that require addressing in this class of applications are the efficiency and the rates of contraction and relaxation.

CONCLUSIONS

Because of the large elastic moduli that crosslinked poly(VPAVG) can exhibit being two orders of magnitude greater than for similarly treated poly(VPGVG), this new analog provides for widening the range of applications of elastomeric polypeptide biomaterials. As the elastic matrix becomes harder with increasing elastic modulus as the temperature is raised through that of the inverse temperature transition and as the process is reversible, this new material may be referred to as a reversible, inverse thermoplastic.

ACKNOWLEDGMENTS

This work was supported in part by NIH Grant No. HL-41198 and the Office of Naval Research Contract No. N00014-89-J-1970. The authors are pleased to acknowledge Richard Knight of the Auburn University Nuclear Science Center for carrying out the γ-irradiation crosslinking.

REFERENCES

1. H. Yeh, N. Ornstein-Goldstein, Z. Indik, P. Sheppard, N. Anderson, J. C. Rosenbloom, G. Cicila, K. Yoon, & J. Rosenbloom, Collagen and Related Research, 7, 235-247 (1987).
2. L. B. Sandberg, J. G. Leslie, C. T. Leach, V. L. Torres, A. R. Smith, & D. W. Smith, Pathol. Biol., 33, 266-274 (1985).
3. D. W. Urry, J. Protein Chem., 7, 1-34, 81-114 (1988).
4. D. W. Urry, B. Haynes, H. Zhang, R. D. Harris, & K. U. Prasad, Proc. Natl. Acad. Sci. USA, 85, 3407-3411 (1988).
5. D. W. Urry, R. D. Harris, & K. U. Prasad, J. Am. Chem. Soc., 110, 3303-3305 (1988).
6. D. W. Urry, Intl. J. Quantum Chem.: Quantum Biol. Symp., 15, 235-245 (1988).
7. D. W. Urry, T. L. Trapane, M. M. Long, & K. U. Prasad, J. Chem. Soc., Faraday Trans. I, 79, 853-868 (1983).
8. D. W. Urry, T. L. Trapane, S. A. Wood, J. A. Walker, R. D. Harris, & K. U. Prasad, Int. J. Pept. Protein Res., 22, 164-175 (1983).
9. D. W. Urry, T. L. Trapane, S. A. Wood, R. D. Harris, J. T. Walker, & K. U. Prasad, Int. J. Pept. Protein Res., 23, 425-434 (1984).
10. K. U. Prasad, M. A. Iqbal, & D. W. Urry, Int. J. Pept. and Protein Res., 25, 408-413 (1985).
11. D. W. Urry, R. Henze, R. D. Harris, K. U. Prasad, Biochem. Biophys. Res. Commun., 125, 1082-1088 (1984).
12. Y. Nozaki & C. Tanford, J. Biol. Chem., 246, 2211-2217 (1971).
13. C-H Luan, J. Jaggard, R. D. Harris, & D. W. Urry, International Journal of Quantum Chemistry: Quantum Biology Symposium, 16, 235-244 (1989).
14. Y. V. Moiseev, T. T. Daurova, O. S. Voronkova, K. Z. Gumargalieva, & L. G. Privalova, J. Polym. Sci. Polym. Symp., 66, 269 (1979).
15. G. E. Zaikov, Rev. Macromol. Chem. Phys., C25, 551-597 (1985).

REQUIREMENT FOR A 1-μm PORE CHANNEL OPENING DURING PERIPHERAL NERVE

REGENERATION THROUGH A BIODEGRADABLE CHEMICAL ANALOG OF ECM

I. V. Yannas,[1] A. S. Chang,[1] S. Perutz,[1] C. Krarup,[2]
T. V. Norregaard,[3] and N. T. Zervas[4]

(1) Fibers and Polymers Laboratory, Massachusetts Institute
    of Technology, Cambridge, MA 02139
(2) Division of Neurology, Brigham and Women's Hospital
    Boston, MA 02115
(3) The Deaconess Hospital, Boston, MA 02214
(4) Department of Neurosurgery, Massachusetts General
    Hospital, Boston, MA 02114

A continuing study of preferences of elongating axons and
Schwann cells for specific matrix features has revealed an
apparently critical structural requirement. Well-defined,
chemical analogs of ECM based on a collagen-glycosaminoglycan
(CG) copolymer were used to bridge a 10-mm gap between cut
ends of the rat sciatic nerve. The nerve stumps and the CG
matrix bridging them were ensheathed in a silicone tube.
Electrophysiological properties of regenerating motor nerve
fibers innervating the plantar flexor muscles were serially
monitored over 40 weeks following surgery. The results suggest that functional recovery of motor function requires the
presence of an average pore diameter of order 1 μm. These
results pose novel questions about the nature of cell-matrix
interactions during nerve regeneration.

INTRODUCTION

Chemical analogs of the extracellular matrix (ECM) provide a new
probe into the complex processes of development. The analogs synthesized
so far are graft copolymers of collagen and one each of several glycosaminoglycans (GAGs) and are, admittedly, very simple models of the chemically complex ECMs. Nevertheless, careful control of the crosslink density has yielded macromolecular networks which degrade under the action
of tissue enzymes over a period that ranges from days to weeks. The essential transience of the ECM during development and wound healing can
thereby be deliberately adjusted and its effect on cell-matrix and cell-
cell interactions can be studied. Furthermore, control of the volume
fraction of pores and of the average pore diameter has yielded gels with
levels of hydration and specific surface which extend over wide ranges.

By contrast with casually reconstituted collagens, the collagen analogs of the ECM described in this report interact in an active and highly

specific manner with cells which migrate from adjacent tissues into the implanted analogs. The activity of certain analogs is recognized by their ability to induce tissue regeneration in a well-defined animal lesion (wound) in which it has been previously demonstrated that regeneration does not occur spontaneously. At least five physicochemical features distinguish templates from biologically inactive collagens, namely, the chemical composition of the macromolecular network (collagen/GAG ratio), the density of crosslinks between the macromolecules, the average diameter of pores in the range 1-1000 μm, the fraction of collagen which is present in a highly crystalline (banded) form and the volume fraction of water. In addition, templates may be seeded with cells prior to implantation, a procedure which has been found to affect their biological activity equally as strongly as do physicochemical manipulations.[1]

A pilot histological study was previously performed using CG matrices to regenerate severed rat sciatic nerves.[2] The graft consisted of a silicone tube ensheathing a 15-mm section of CG copolymer. Because of the large gap length, the graft could not reside entirely in one leg. Cross-anastomoses were accordingly performed by transecting both sciatic nerves and grafting the right proximal end with the left distal end across the back of the animal. This morphological study showed that regeneration of a richly myelinated nerve occurred across the entire 15-mm distance when the gap was filled with the highly porous copolymer. No axons formed when the copolymer was omitted in control studies. The histological studies also showed well-developed vascularization within, as well as distal to, the gap.[2]

In the present study, we have studied regeneration by use of an electrophysiological procedure. Some advantages of the latter over a morphological study are the possibilities of monitoring regeneration in the same animal over the duration of the regeneration process and of a functional, rather than a structural, procedure for measurement of the extent of regeneration. A disadvantage of the use of electrophysiology is lack of reproducibility of neurological signals from a pathway which has the complex anatomy of the cross-anastomosed sciatic nerve. This difficulty has been circumvented by decreasing the length of the gap down to 10 mm, an implant length which can be accommodated entirely within the rat femur.[3,4]

EXPERIMENTAL

Collagen-glycosaminoglycan copolymers were prepared from bovine hide collagen which had been precipitated from acid dispersion by chondroitin-6-sulfate.[1] The suspended coprecipitate was injected into a silicone tube which was immersed in a cold bath. The resulting ice crystals were subsequently sublimed by exposure to vacuum and a porous structure resulted with a pore volume fraction of about 0.99. Cylindrical grafts with average pore diameters from 5 μm to 300 μm were prepared by controlling the bath temperature and tube entrance velocity. Grafts with average orientation of pore channel axes either predominantly along the axis of the cylinder or along the radial were prepared by adjusting the dominant direction of heat flux during freezing. Four groups of matrices which had average pore diameters of 5, 10, 60, and 300 μm were prepared with pore channels oriented along the axis of the tubes. Calculation gave total surface areas of 2700, 1400, 230, and 46 mm^2/mm graft length, respectively.

The degradation rate was decreased by crosslinking CG matrices in 0.25% glutaraldehyde. Degradation rates in collagenase were assayed *in vitro* using a procedure which has been described elsewhere.[1] Results of the assay are reported in enzyme units which increase as the degradation rate increases. A slowly degrading matrix was prepared by immersion in glutaraldehyde for over 24 hours, yielding a polymer with an average molecular weight between crosslinks, M_c, of 12 kD. The untreated preparation, which degraded rapidly, exhibited an M_c of 60 kD. All results reported here were obtained with the untreated, rapidly degrading, matrix.

Adult female Sprague-Dawley rats (Charles River Laboratories, Inc.) weighing 230-250 g were anesthetized with intraperitoneal injections of sodium pentobarbital (35 mg/kg body weight). Under a surgical microscope the sciatic nerve was transected with microscissors at mid-thigh and grafted with a CG copolymer graft that separated the nerve stumps by 10 mm. The grafts were silicone tubes 20 mm in length with a 1.5 mm inner diameter. The central 10 mm consisted of a CG matrix, allowing 5 mm on each end to insert the nerve stumps. Each stump was abutted against the CG matrix and was secured in the silicone tube by three stitches of 10-0 nylon (Ethilon) through the epineurium.

The electrophysiological procedure provided an evaluation of the reinnervation of plantar muscles after transection and grafting of the sciatic nerve. Electrical stimuli were applied to the nerve via percutaneous needle electrodes, both proximal and distal to the transection site, and the resulting compound muscle action potentials (CMAPs) elicited in the plantar muscles of the foot were recorded using a concentric needle. The time between the stimulation of the nerve and the onset of the elicited CMAPs, termed the distal motor latency, was measured.[5]

RESULTS

When first detected, the distal motor latencies were about 2-3 times higher than normal (Figure 1).[5] The latter shows the decay of the distal motor latency following hip stimulation of nerves grafted with a CG implant which had pores with an average diameter of 60 μm, strongly axial orientation of pore channels and rapid degradation rate (R = 180 ± 20 enzyme units). A tabulation of steady state electrophysiological data appears in Table 1.

Table 1. Long Term Electrophysiological Measurements (by week 40).

Matrix Characterization			Distal Motor Latency (msec)	
Average Pore Size μm	Average Pore Axis Orientation	Degradation Rate, e.u.	From Ankle	From Hip
5	Axial	180 ± 20	1.67 ± 0.20	3.04 ± 0.45
10	Axial	180 ± 20	1.84 ± 0.20	3.43 ± 0.49
60	Axial	180 ± 20	2.01 ± 0.28	3.74 ± 0.44
300	Axial	180 ± 20	2.16 ± 0.35	4.04 ± 0.58
Intact nerve	–	–	1.58 ± 0.17	2.47 ± 0.17

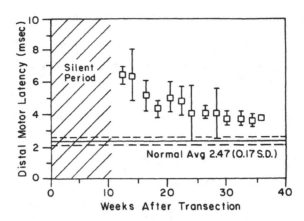

Figure 1. Decrease with time of distal motor latency following hip stimulation of nerves grafted with a CG
implant which had pores with an average diameter of
60 μm, strongly axial orientation of pore channels
and rapid degradation rate (R = 180 ± 20 enzyme
units).

DISCUSSION

The question posed by the data is intriguing. The functional recovery
of sciatic nerve following bridging of a gap by collagen/GAG copolymers
was found to <u>increase</u> continuously as the average pore diameter of the
substrate <u>decreased</u> from 300 μm to 5 μm (Table 1). Intuitively, it seems
inevitable that there should be a lower limit of the pore diameter, a
sort of "barrier effect", below which no elongation of growth axons
should occur and the recovery should consequently suffer. This inevitable
lower limit must be lower than 5 μm, since our data strongly suggest
improved recovery even below that level. Let us suppose that the lower
limit is about 0.1 μm, several times smaller than the diameter of a growing axon.[6,7] But, if so, there must be a critical pore size, between 5 μm
and 0.1 μm, at which the recovery is maximized. What is that critical
pore size for a degradable matrix at which the functional recovery of the
sciatic nerve is maximized? And, what is its developmental significance?

The characteristic diameter of an immature, nonmyelinated axon tip is
of order 1 μm.[6,7] Other factors remaining constant, a decrease in pore
size of a CG matrix is accompanied by an increase in specific surface of
pore channels (cm^2/g). For example, we calculate that a decrease in pore
size from 300 μm to 5 μm of the CG copolymers synthesized in our laboratory is accompanied by an approximately 50-fold increase in specific
surface of the substrate. Therefore, we hypothesize that (a) the optimal
pore diameter for nerve regeneration is of order 1 μm, (b) a pore diameter of about 1 μm is critical because it provides opportunity for maximum
contact between the surface of the immature, non-myelinated axon and the
inner surface (lumen) of the pore channel of the matrix and (c) when the
pore diameter becomes much smaller than about 1 μm, axonal elongation is
physically hindered.

Maximum contact between axon tip and matrix surface may be required
in order to achieve maturation of the axon, with accompanying myelination
and increase in axon diameter. It is conceivable that nerve growth factor
and laminin are adsorbed on the matrix surface, and that it is contact of

278

the axon tip with these adsorbed substances which promotes axonal elongation through contact with the matrix. Work is in progress to test this
"sleeve" hypothesis.

ACKNOWLEDGMENTS

This research was supported by NSF Grant EET-8520548.

REFERENCES

1. I. V. Yannas, E. Lee, D. P. Orgill, E. M. Skrabut and G. F. Murphy,
   Proc. Natl. Acad. Sci. USA, **86**, 933 (1989).
2. I. V. Yannas, D. P. Orgill, J. Silver, T. V. Norregaard, N. T. Zervas
   and W. C. Schoene in: "*Advances in Biomedical Polymers,*" C. G.
   Gebelein, Ed., Plenum Publishing Co., NY, pp. 1-9 (1987).
3. I. V. Yannas, C. Krarup, A. Chang, T. V. Norregaard, N. T. Zervas and
   R. Sethi, Soc. Neurosci. Abstr., **13**, 1043 (1987)
4 I. V. Yannas, A. S. Chang, C. Krarup, R. Sethi, T. V. Norregaard and
   N. T. Zervas, Soc. Neurosci. Abstr., **14**, 165 (1988).
5. A. S. Chang, I. V. Yannas, C. Krarup, R. Sethi, T. V. Norregaard and
   N. T. Zervas, Proceedings Polymeric Materials: Science and
   Engineering, **59**, 906-910 (1988).
6. C.-B. Jeng and R. E. Coggeshall, Brain Research, **326**, 27-40 (1985).
7. C.-B. Jeng and R. E. Coggeshall, Brain Research, **345**, 34-44 (1985).

THE DEVELOPMENT OF COLLAGEN NERVE CONDUITS THAT PROMOTE PERIPHERAL NERVE

REGENERATION

Shu-Tung Li, Simon J. Archibald, Christian Krarup* and
Roger D. Madison

Colla-Tec, Inc.
Plainsboro, NJ 08536

Department of Neurosurgery
Duke University Medical Center
Durham, NC 27710 and

* Neurophysiology Laboratory, Harvard Medical School
Brigham and Women's Hospital
Boston, MA 02115

We have investigated the repair of peripheral nerves in
animal models using tubular guiding conduits. The materials
used to fabricate the nerve conduits and their physicochemical and mechanical characteristics can influence the extent,
rate and morphology of regeneration. Permeability of the
conduit membranes is one parameter which seems to play an
important role in nerve regeneration. In the present study,
two types of nerve conduits were developed from bovine tendon
collagen with distinctly different permeabilities. The permeability of the conduit membranes was determined by diffusion
of various sized molecules across these membranes. One type
of conduit had pores which only allowed small molecules such
as glucose to pass (small pore collagen conduits). The other
type had pores which were readily permeable to macromolecules
such as bovine serum albumin (large pore collagen conduits).
The large pore collagen conduits supported nerve regeneration
to a greater degree than the small pore collagen conduits
when tested in mice to bridge 4 mm gaps of the sciatic nerve.
Studies in rats and primates suggested that large pore collagen conduits worked as effectively as nerve autografts in
terms of physiological recovery of motor and sensory responses. The results of *in vitro* and *in vivo* studies of these
conduits represent a significant step towards our specific
aim of developing suitable off-the-shelf prostheses for clinical repair of damaged peripheral nerves.

---

* Current address: Laboratory of Clinical Neurophysiology
Rigshospitalet, Blegdamsvej 9, 2100
Copenhagen, Denmark

INTRODUCTION

Clinical practice for repair of human peripheral nerve injuries favors either direct anastomosis with microsutures of the severed nerve endings or autografting if the damage is severe enough that part of the nerve is lost.[1] Unfortunately, the clinical results of peripheral nerve repair are often disappointing.[2-4] Therefore, a search for an alternative repair method has been continuously pursued by many investigators.

The concept of repair with a tubular guiding conduit has been the one receiving the most extensive investigations.[5] A number of tubular devices including natural and synthetic polymeric materials have been tried in recent years.[6-13] Each of these materials offer some advantages in terms of physical and mechanical characteristics. However, none of these materials have performed ideally in terms of peripheral nerve regeneration *in vivo*. An indepth evaluation is needed of key design parameters that will enhance nerve regeneration within tubular guiding conduits. These include: (1) the materials used for manufacturing the guiding conduit; (2) the susceptibility of the materials to biodegradation and the rate of degradation under various repair conditions; (3) the effect of nerve growth promoting factors; (4) the mechanical properties and (5) the permeability of the conduit. A logical way to develop a guiding conduit would be to systematically evaluate each of the parameters listed above in order to optimize the design requirements and the efficacy of the device *in vivo*.

In an effort to develop a nerve guiding conduit, we have specifically focused on two of the key parameters listed above, namely the material and the permeability of the conduit. We selected type I collagen from bovine Achilles tendon for nerve conduit fabrication because it causes minimal inflammation responses in humans.[15-17] Type I collagen is biodegradable and can be chemically modified to alter its physicochemical, mechanical and biological properties. Thus, type I collagen is a logical material to meet the requirements for peripheral nerve repair.

Studies from several laboratories, including our own, have shown that conduits which are permeable to cells and/or macromolecules support axonal regeneration to a greater degree than conduits that are impermeable or only permeable to ions and small molecules.[14,18-21] Two types of nerve conduits from bovine tendon collagen were developed which have distinctly different permeability properties to further study the effect of macromolecular communication on nerve regeneration. Results of *in vitro* and *in vivo* studies of these two types of collagen nerve conduits are presented with the aim of developing suitable off-the-shelf prostheses for clinical repair of damaged peripheral nerves. Preliminary reports of these experiments have been presented elsewhere.[14,19,22-25]

MATERIALS AND METHODS

Collagen derived from bovine deep flexor (Achilles) tendon was used for nerve conduit fabrication. Tendons were first cut into thin slices and then purified by a combination of enzymatic and chemical treatments to reduce the non-collagenous moieties while retaining the integrity of the fibrillar structure for mechanical strength. Chemical analysis of this material showed a minimal level of hexosamine content indicating most of the glycoproteins were removed by the purification steps.[17]

The steps in nerve conduit fabrication procedure are as follows: (1)

collagen fibrils were uniformly dispersed in an acidic media, pH 2.5-3.0 at a final concentration of 0.70% (w/v) of collagen; (2) the dispersed fibrils were coacervated by adjusting the pH to the isoionic point of the collagen; (3) the coacervated fibrils were uniformly coated onto a mandrel; (4) the collagen coated mandrels were allowed to dry under controlled conditions to obtain conduits with different morphologies and permeability properties; (5) the dried conduits were then subjected to a chemical crosslinking procedure to optimize the *in vivo* stability and (6) the crosslinked conduits were sterilized.

Permeability was determined for the collagen conduits by diffusion of various sizes of molecules across the conduit membranes. Test molecules were studied ranging in molecular weight from glucose (MW = 180) to β-Galactosidase (MW = 5.4 x 10^5). Known amounts of various concentrations of test molecules were sealed into the lumen of the conduits which were then incubated in a known volume of 0.05M Tris buffered solution, pH 7.4 at 4°C for 24 hours. Spectrophotometric methods were used to quantify the amount of material that had diffused across the membranes and entered the extra-tubular space. The other *in vitro* parameters studied included shrinkage temperature and swelling of the collagen conduit.

The method for measuring the extent of crosslinking by shrinkage temperature is as follows: One end of a 5 cm long collagen conduit was clamped at a fixed point. The other end of the conduit was then clamped and a small weight was attached through a pulley. The sample conduit was immersed in a phosphate buffer solution bath, pH 7.4 and the sample length recorded. The temperature of the solution was increased at a rate of 1°C/min. The length of the sample was continuously recorded until a constant length was reached, representing full denaturation of the collagen. The shrinkage temperature of the collagen conduit was defined as the temperature corresponding to 50% of the maximum change in length of the sample.

Swelling of the collagen conduit in a phosphate buffer solution was determined by weighing a 4 cm sample and immersing the sample in a 0.05M Tris buffered solution, pH 7.4 for one hour. The swelling of the collagen conduit was defined as the weight of solution uptake per weight of the conduit (g/g).

Scanning electron micrographs of the lumenal surfaces and cross-sectional areas of the collagen conduits were obtained from Structure Probe, Inc., Metuchen, NJ.

Previously published animal models[19,23,26] were utilized to study: (1) the effect of permeability of the conduit membrane on peripheral nerve axonal regeneration and (2) the effectiveness of collagen conduits to promote physiological recovering of motor and sensory nerve potentials.

To study the effect of nerve conduit permeability on axonal regeneration, we quantified the number of myelinated axons which regenerated through both types of nerve conduits. The transected sciatic nerve of thirteen adult male mice (C57 BL/6J) was used as a model, with four mice in each of two experimental groups and five mice serving as non-transection controls. All operations were performed under deep anesthesia. The left sciatic nerve was exposed and transected at the mid-thigh level and the proximal and distal stumps were inserted into either large pore or small pore collagen conduits 6 mm long, 1 mm inner diameter, and secured with a single 10/0 microsuture. The final nerve gap distance was 4 mm. At the end of four weeks, animals were sacrificed for histological evaluations using standard techniques.[26] One micron transverse sections were cut and stained with alkaline toluidine blue. The number of myelinated axons in these sections was determined with a computer-controlled system. Control animals had their sciatic nerves dissected out and processed exactly as experimental animals for the quantification of the normal number of myelinated axons.

For electrophysiological studies, we used rats and primates. Briefly, in the rat model, forty adult male Long Evans rats received transection of the left sciatic nerve under deep anesthesia. The transected nerve was then repaired by either (1) direct anastomosis (n = 10) in which the proximal and distal stumps were sutured with two 10/0 microsutures; (2) nerve autograft (n = 10), in which a 4 mm nerve segment was removed, reversed, rotated 180° and sutured to the proximal and distal nerve stumps with two 10/0 microsutures at each junction and (3) collagen nerve conduit repair (n = 10), in which a 4 mm segment of nerve was removed and the proximal and distal nerve stumps were sutured into a large pore collagen conduit with a single 10/0 microsuture connecting the tube wall and perineurium of the nerve stump 1 mm from the transected face of the nerve. The negative control group (n = 5) had 4 mm segments of nerve removed but no repair was performed. Five rats were entered as normal group without surgery. After four and twelve weeks, animals from each group were prepared for electrophysiological studies. Platinum wire stimulating electrodes were placed directly on the sciatic nerve 5 mm proximal to the transection site and evoked EMG responses were recorded and averaged within the gastrocnemius muscle sites.

In the primate model, median nerves of six young adult male monkeys (*Macaca Fasicularis*) were transected, under deep anesthesia, and repaired by either 4 mm nerve autografts or large pore collagen conduits with a final gap of 4 mm. Four monkeys received bilateral transections and two monkeys received unilateral transections. Five autografts and five large pore collagen conduits were implanted in ten nerve sites over a period of six months. Using standard clinical procedures, an analysis of motor and sensory nerve function was performed for the median nerve and its respective targets, including the abductor pollicis brevis muscle and the sensory innervation of the second digit.

RESULTS

The two prototypes of collagen conduits developed for this study are shown in Figure 1. The surfaces and cross-sectional morphologies of these conduits are shown in Figure 2. The small pore conduit membrane was comprised of a densely packed, collagenous material uniformly distributed throughout the cross-section of the membrane. The lumenal surface appeared similar to a film. The morphology of the large pore conduit was quite different. The outer layer consisted of large interconnecting open pores with a laminated appearance. The inner layer was composed of densely packed collagen fibers that are intertwined in a multi-layer mesh structure.

Table 1 summarizes the permeability of the conduits to a series of molecules of varying dimensions. The small pore conduit membrane allowed the diffusion of glucose molecules (7A), but not the diffusion of myoglobin molecules (38A). The large pore conduit membrane allowed the diffusion of myoglobin and bovine serum albumin (BSA) (68A), but the rates were slower than that of glucose. The diffusion of β-Galactosidase (215A) was restricted, indicating the approach of the upper limit of the pore size of the conduit membrane (see Table 2).

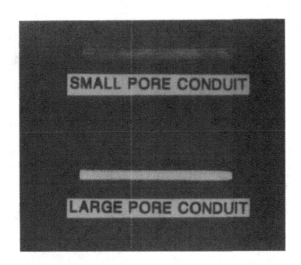

Figure 1. Appearance of the two types of collagen nerve conduits; (top) small pore conduit; (bottom) large
pore conduit.

Table 2 summarizes the results of various *in vitro* studies on the
collagen conduits. The swelling of the large pore conduit is two to three
times that of the small pore conduit. This is consistent with the open
structure of the large pore conduit. The large pore conduits had a
shrinkage temperature in the range of 53-69°C; indicating that conduits
were adequately crosslinked to bridge a 4 mm nerve gap.[19]

Table 3 summarizes the mouse studies. There was a tremendous increase
in the number of regenerated myelinated axons in the large pore conduits
as compared to the small pore conduits, even though both were fabricated
from the same collagen material (p < 0.05, Student-Newman-Keuls test).

Table 1. Results of permeability studies.*				
Molecules	M.W.	Size (A)	% Permeated in 24 hours	
			Small Pore	Large Pore**
				82A	85A
Glucose	180	7	88±8	95±13	84±7
Myoglobin	16,900	38	5±6	63±33	78±12
BSA	68,000	68	0	69±38	62±24
β-Galactosidase	5.4x10^5	215	–	28±14	29±17

*Results are expressed as average ± standard deviation.
**82A and 85A represent different collagen preparations.

The sizes of the molecules were calculated
based on these assumptions: (1) these molecules
exist in spherical form in aqueous condition; (2)
the intrinsic viscosity of the proteins in aqueous
salt solutions is 3.4 cm/g.[27]

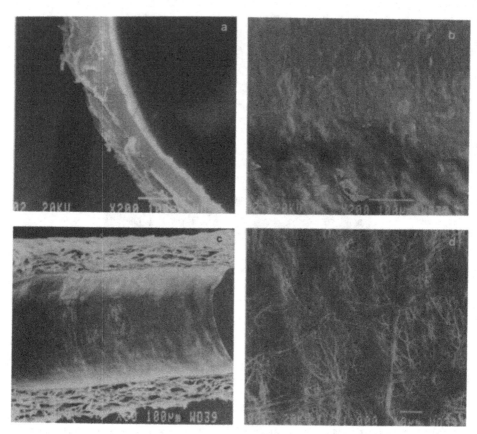

Figure 2. Scanning electron micrographs of the two types of
collagen conduits; (A) cross-sectional view, small
pore conduit; (B) lumenal surface view, small pore
conduit; (C) cross-sectional view, large pore conduit; (D) lumenal surface view, large pore conduit.

The mean number of regenerated axons found in the large pore group is
similar to normal control numbers of myelinated axons in mouse sciatic
nerve. Further studies are needed to determine whether this effect actually represents an increase in the number of regenerating axons or simply
an increase in the extent of axonal branching within the conduit.

| | Small Pore Conduit | Large Pore Conduit** | |
		82A	85A
Pore Size (A)	38	215	215
Swelling (g/g)	1.7±0.1	3.4±1.2	5.8±1.5
Shrinkage Temperature (°C)	75±4	69±3	53±2

Table 2. Summary of *in vitro* results.*

*Results are expressed as average ± standard deviation
**82A and 85A represent different collagen preparations

Table 3. Summary of mice study.*	
	Number of Myelinated Axons
Control (n = 5)	3851 ± 196
Small Pore Conduit (n = 4)	1036 ± 850
Large Pore Conduit (n = 4)	3963 ± 1990**

*Results are expressed as average ± standard deviation.
**Significantly higher than small pore conduits, p is <
0.05 Student-Newman-Keuls Test.

These encouraging results from the large pore conduits prompted us to carry out further investigations using rat and primate models of nerve repair. We compared the effectiveness of the large pore conduits to nerve autografts in terms of physiological recovery of sensory and motor responses following peripheral nerve repair.

Figure 3 summarizes the results of the rat studies. The maximum peak-to-peak amplitude was measured from the average EMG response of each muscle examined at four and twelve weeks. The four week results showed a statistically significant difference between the direct anastomosis and the other surgical groups ($p < 0.01$, Student-Newman-Keuls). This was expected since there was no gap distance in the direct anastomosis. By twelve weeks, there was no statistically significant difference between any of the surgical groups. However, all surgical groups produced a significantly higher response than the negative control ($p < 0.05$, Student-Newman-Keuls).

The pooled results of the EMG amplitudes and the latencies of the evoked EMG for median nerves in six monkeys repaired by either a collagen conduit or a nerve autograft are shown in Figure 4. A replicated measure of analysis of variance showed no statistically significant difference at the $p = 0.05$ level. There was a rapid return of the nerve conduction to the baseline level indicating that the recovery of the conduction time for the fastest conducting motor fibers was similar for both repair groups. The pooled results of the sensory amplitudes and the latencies of the sensory potentials are shown in Figure 5. The recovery of the sensory amplitudes was slower than the motor EMG amplitudes for both the repaired groups. There was no significant difference in the recovery of the sensory amplitudes between the nerve conduit and autograft repairs with both recovering to approximately 20% of baseline levels. There was also no significant difference in the conduction time of the sensory fibers between the two repair groups.

DISCUSSION

The conventional repair of severed peripheral nerves with direct suture or autografting can result in neuroma formation and poor functional recovery.[2-4] Attempts in the past several decades to provide a device to facilitate nerve regeneration have resulted in minimal success. The recent advances in material science and technology have provided an opportunity to investigate this important clinical problem with a more systematic approach.

One of the major requirements for a particular medical device is its

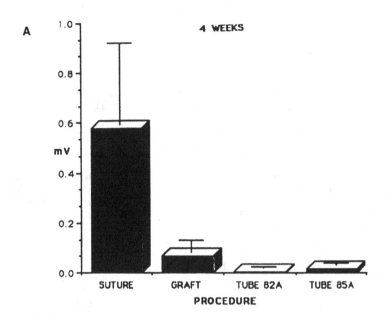

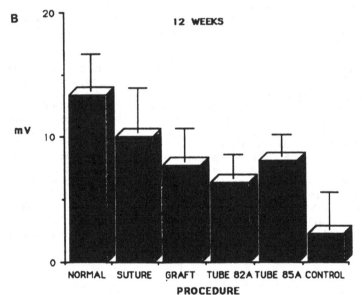

Figure 3. Average peak-to-peak maximum amplitude of the evoked EMG response from rat gastrocnemius muscle at four and twelve weeks following sciatic nerve transection and repair; (A) 4 week data; (B) 12 week data. Tube 82A and Tube 85A represent different collagen preparations.

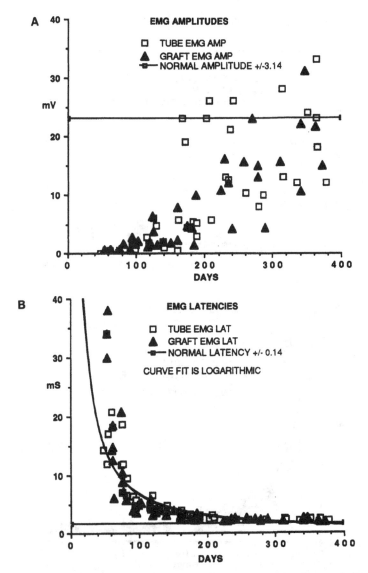

Figure 4. Scatterplot of the evoked EMG amplitudes and latencies of the abductor pollicis brevis (APB) muscle
following monkey median nerve transection and repair by either large pore conduit (Tube 85A) or
nerve autograft to bridge a 4 mm deficit; (A)
evoked EMG amplitudes; (B) EMG latencies.

physical form. The physical form of a device limits the growth or regeneration of a particular tissue or organ within its boundaries. In the case
of nerve regeneration, the boundary should be tubular so that growth of
axonal fibers is directed in one dimension. Some advantages of repair
with a conduit versus other nerve repair methods are: (1) the conduit
will limit the growth of fibrogenic cells into the repair site and thus
avoid excessive fibrosis and scar formation; (2) the conduit can provide
directional guidance to the regenerating axons and can prevent axonal

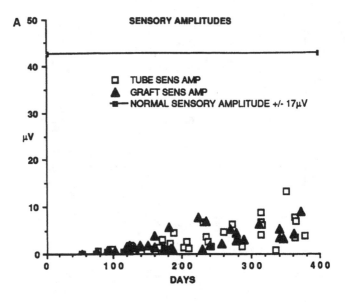

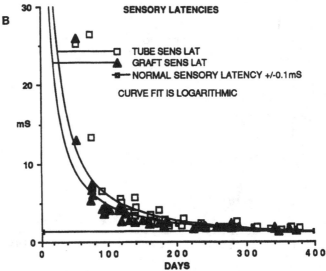

Figure 5. Scatterplot of the sensory amplitudes and latencies of the monkey median nerve repaired by either large pore conduit (Tube 85A) or nerve autograft obtained by stimulating the index finger and recording the evoked potential at the level of the wrist; (A) evoked sensory amplitudes; (B) latencies of the sensory potentials.

escape into the area surrounding the repair site and (3) nerve growth promoting factors from the injury site may become concentrated within the lumen of the conduit and serve to facilitate nerve growth. In addition, the conduit can be used as a test chamber *in vivo* to investigate various factors that either promote or inhibit nerve growth.

The physical form of a device only plays a partial role in facilitating nerve regeneration. The material selection for the device fabrication is also critical. This has been shown in the *in vivo* results obtained from tubular devices comprised of synthetic materials and natural materials.[26] Although both have tubular configurations, the extent and morphology of the regeneration were different. In addition, different results were obtained from resorbable and permanent devices.

It is generally recognized that resorbable materials are superior to non-resorbable materials, since the presence of permanent foreign materials in the body provokes host defense systems and leads to chronic inflammation. Among resorbable materials, we prefer the use of type I collagen over synthetic polymers due to its unique physicochemical and biological properties. The properties of collagen can be judiciously manipulated to fulfill the specific needs for a particular medical application. For example, depending on the length of the gap to be bridged, the *in vivo* stability of the material can be controlled such that the newly regenerated axons are given proper support by the collagen scaffold while the conduit is continuously being resorbed by the body. This can be done either by controlling the extent of intermolecular crosslinking or by controlling the density and thickness of the membrane.

Additionally, it has been recognized that the permeability of the conduit membrane plays an important role in facilitating nerve regeneration. The results of the *in vivo* studies of the two types of collagen conduits clearly showed that when the membrane of the conduit was made permeable to macromolecules the size of BSA, the extent of axonal regeneration after four weeks of implantation was three to four times larger than the regeneration found in those collagen conduits that were only permeable to small molecules such as glucose. Various neurotrophic factors comparable to BSA in size, should be able to readily penetrate the conduit membrane and reach the regenerating axons to perform their supportive biological functions. The results of this study suggest that cellular communication at the macromolecular level facilitates axonal growth and regeneration, thus constituting another major requirement in designing a device for nerve repair.

The most important finding of the rat and primate studies is that the large pore collagen conduits appear as effective as an autogenous nerve graft for peripheral nerve repair over small gap distances. We are presently investigating the use of large pore conduits to bridge longer nerve gaps in primate models.

The results of this study show that the collagen-based, permeability-controlled conduits satisfy several key design requirements for effective peripheral nerve repair applications. The collagen conduits should find widespread use for experimental studies of nerve regeneration. We are hopeful that this technique will be useful for clinical nerve repair in the future.

ACKNOWLEDGEMENTS

We wish to thank Ms. Debbie Yuen for technical assistance and Ms. Kimberly Serluco for typing this manuscript. We also wish to thank Dr. Jeremy Shefner, Bruce Bonsack and Nancy Dobratz who assisted with the rat and primate studies. Portions of this work were supported by NIHNS22404 (RM).

REFERENCES

1. Sir Sydney Sunderland, *"Nerves and Nerve Injuries,"* Churchill-Livingstone, Inc., New York, 1978, pp. 483-652.
2. A. L. Dellon in: *"Evaluation of Sensibility of Re-education of Sensation in the Hand,"* Williams and Wilkins, Baltimore, 1981, pp. 193-202. "Results of Nerve Repair in the Hand."
3. V. I. Young, C. R. Wray and P. M. Weeks, Ann. Plast. Surg., 5, 212-215 (1980). "The Results of Nerve Grafting in the Wrist and Hand."
4. W. C. Beazley, M. A. Milek and B. H. Reiss, Clin. Ortho. Rel. Res., 188, 208-216 (1984). "Results of Nerve Grafting in Severe Soft Tissue Injuries."
5. Noriaki Suematsu, Microsurgery, 10, 71-74 (1989). "Tubulation for Peripheral Nerve Gap: Its History and Possibility."
6. G. Lundborg, R. H. Gelberman, F. M. Longo, H. C. Powell and S. Varon, Neuropathology Exp. Neurology, 41, 412-422 (1982). *"In Vitro* Regeneration of Cut Nerves Encased in Silicone Tubes."
7. E. Nyilas, T. H. Chiu, R. L. Sidman, E. W. Henry, T. H. Brushart, P. Dikkes and R. Madison, Trans. Amer. Soc. Artif. Internal Organs, 29, 307-313 (1983). "Peripheral Nerve Repair with Bioresorbable Prosthesis."
8. B. G. Uzman and G. M. Villegas, in: *"Nerve, Organ and Tissue Regeneration: Research Perspectives,"* F. J. Seil, Ed., Academic Press, New York, 1983, pp. 109-124. "Peripheral Nerve Regeneration Through Semipermeable Tubes."
9. Y. Restrepo, M. Merle, J. Michon, B. Folliguet and D. Petry, Microsurgery, 4, 105-112 (1983). "Fascicular Nerve Graft Using an Empty Perineurial Tube: An Experimental Study in the Rabbit."
10. H. Molander, O. Engkvist, J. Hagglund, Y. Olsson and E. Torebjork, Biomaterials, 4, 276-280 (1983). "Nerve Repair Using a Polyglactin Tube and Nerve Graft: An Experimental Study in the Rabbit."
11. W. Collin and R. B. Donoff, J. Dental Res., 63, 987-993 (1984). "Nerve Regeneration Through Collagen Tubes."
12. I. V. Yannas, D. P. Orgill, J. Silver, T. Norregaard, N. T. Ervas and W. C. Schoene, Polymeric Materials Science and Eng., 53, 216-218 (1985). "Polymeric Template Facilitates Regeneration of Sciatic Nerve Across a 15 mm. Gap."
13. P. Aebischer, R. F. Valentini, P. Dario, C. Domenici and P. M. Galletti, Brain Res., 436, 165-168 (1987). "Piezoelectric Guidance Channels Enhance Regeneration in the Mouse Sciatic Nerve After Axotony."
14. S. T. Li, Soc. Neurosci. Abst., 13, 1042 (1987). "Porous Collagen Nerve Conduits for Nerve Regeneration: In Vitro Characterization Studies."
15. L. S. Cooperman, V. Mackinnon, G. Bechler, B. B. Pharriss, Aesth. Plast. Surg., 9, 145-151 (1985). "Injectable Collagen: A Six Year Clinical Investigation."
16. M. D. Stein, L. M. Salkin, A. L. Freedman and V. Glushko, J. Periodontal, 56, 35-38 (1985). "Collagen Sponge as a Topical Hemostatic Agent in Mucogingival Surgery."
17. F. Ceravolo and S. T. Li, Int. J. of Oral Implant., 4, 15-18 (1988). "Alveolar Ridge Augmentation Utilizing Collagen Wound Dressing."
18. C. B. Jeng and R. E. Coggeshall, Brain Res., 408, 239-242 (1987). "Permeable Tubes Increase the Length of the Gap that Regenerating Axons Can Span."
19. R. Madison, Soc. Neuroscience Abst., 2, 1042 (1987). "A Tubular Prosthesis for Nerve Repair: Effects of Basement Membrane Materials, Laminin and Porosity."
20. R. F. Valentini, P. Aebischer, S. R. Win and P. M. Galletti, Trans. Soc. Biomaterials, p. 8 (1987). "Role of Transmural Permeability

Characteristics in Peripheral Nerve Regeneration Through Synthetic Tubular Biomaterials."

21. P. Aebischer, V. Guenard and S. Brace, J. Neurosci., **9**, 3590-3595 (1989). "Peripheral Nerve Regeneration Through Blind-ended Semipermeable Guidance Channels: Effect of the Molecular Weight Cutoff."

22. C. Krarup, S. J. Archibald, R. L. Sidman, A. Sabra and R. Madison, Soc. Neurosci. Absts., **14**, 204 (1988)."Primate Peripheral Nerve Repair by Entubulation: An Electrophysiological Study."

23. S. J. Archibald and R. Madison, Soc. Neurosci. Absts., **14**, 204 (1988). "Functional Recovery Following Rat Sciatic Nerve Regeneration Through Collagen Nerve Guide Tubes: Comparison of Direct Anastomosis, Nerve Graft and Entubulation Repair."

24. S. J. Archibald, C. Krarup, J. Shefner, B. Bonsack and R. Madison, Soc. Neurosci. Absts., **15**, 125 (1989). "Semipermeable Collagen-based Nerve Guide Tubes Are as Effective as Standard Nerve Grafts to Repair Transected Peripheral Nerves: An Electrophysiological Study in the Non-human Primate."

25. S. T. Li, S. J. Archibald, C. Krarup and R. Madison, Polymeric Materials Science and Engineering, 62, 575-582 (1990). "Semipermeable Collagen Nerve Conduits for Peripheral Nerve Regeneration."

26. R. Madison, C. DaSilva, P. Dikkes, R. Sidman and T. H. Chiu, Exper. Neurology, **95**, 378-390 (1987). "Peripheral Nerve Repair: Comparison of Biodegradable Nerve Guides Versus Polyethylene Tubes and the Effect of a Laminin-Containing Gel."

27. C. Tanford, *"Physical Chemistry of Macromolecules,"* Wiley and Sons, Inc., New York, 1961, pp. 317-456.

POLYMERIC REAGENTS FOR PROTEIN MODIFICATION

Danute E. Nitecki and Lois Aldwin

Cetus Corporation
1400 53rd Street
Emeryville, CA 94608

Protein modification with polymeric reagents is carried out to extend *in vivo* circulation, to reduce antigenicity or to improve physical properties, such as solubility. We have developed polymeric reagents for such modifications. One group is based on polyproline. Polyproline exists in aqueous solution as a relatively rigid structure, polyproline II. Its single amino group allows derivatization to yield reagents for either amino groups or sulfhydryl groups on the protein. Development of various polyethylene glycol active esters allowed us to compare the reativities of those esters in aqueous buffers, since polyethylene glycol imparted water solubility to esters which are not normally water soluble. We investigated the influence on relative reactivity of polyethylene glycol esters by comparing leaving groups (i.e., the hydroxyl component) and by comparing relative electrophilicity of carbonyl moiety.

INTRODUCTION

Protein modification with polymeric reagents is undertaken for a variety of reasons. Some of these are increased solubility,[1] increased half-life *in vivo*,[2] modified bioactivity,[3] and reduced antigenicity.[4] We will describe the development and use of protein modifying reagents based on polyproline [H(PRO)$_x$-OH], and monomethoxy-polyethylene glycol (monomethylpolyoxyethylene, PEG), a molecule with basic formula CH_3-O-(CH_2-CH_2-O-)$_n$-H.[5] Both of these reagents were chosen for their water solubility and well-defined chemical functionality.

EXPERIMENTAL

Polyproline is a polyamide with one N-terminal secondary amine function and one carboxylic acid terminus. It is known to exist in polar medium, such as aqueous buffers, in the form of polyproline-II. The proline residue is arranged in a trans conformation into a left handed helix of 3 residues per turn and an axial translation of 3.12 A per residue.[6]

Biotechnology and Polymers, Edited by C.G. Gebelein
Plenum Press, New York, 1991

There is considerable physical and biological data showing that the molecule is a very rigid rod and it is not folded into a globular form, like the proteins. For this reason, the effective size of polyproline is very large, about 10 times its nominal molecular weight compared with globular proteins. This is an important consideration in permeability of such molecules *in vivo*, such as kidney filtration. We have attached polyproline to one of our recombinant proteins, interleukin-2, (IL-2), 15,000 D. This immuno-stimulant is small enough to be rapidly cleared by the kidney, as healthy kidneys clear globular proteins with molecular weights below 30,000 D. We have worked with commercially available poly-L-proline of average 7000 D.

Proteins are, of course, highly polyfunctional molecules. In order to avoid damage to the protein, the reagents selected should react through a well defined functional locus on the protein.

First, we sought to introduce a sulfhydryl-reactive group onto the protein, such as maleimido, to react with a sulfyhydryl introduced on the N-terminus of polyproline. Polyproline was successfully reacted with 2'-pyridyl-dithio propionyl active ester and the resulting 2'-pyridyl-dithio-3-propionyl-polyproline was reduced to HS-3-propionyl-polyproline. This reagent, assayed to contain free sulfhydryl, was reacted with IL-2 protein containing an average of 1.75 maleimido groups per IL-2 molecule. The reaction between maleimido and sulfhydryl groups is usually rapid, independent of pH between 5-7 and nearly quantitative, yet no coupled product was obtained in this case (see Scheme 1).

We suspected that the relatively large and rigid HS-polyproline has difficulty in reacting with the few maleimido groups on the protein because of the low frequency of effective collisions between the two molecules. Therefore, we have introduced a flexible spacer between the reactive group and polyproline, at the same time converting maleimido-reactive sulfhydryl group into an active ester[7] group (reactive with amino function on the protein; IL-2 molecule has 11 NH2 groups). The two changes, introduction of the flexible spacer and increase of reactive groups on the protein yielded a satisfactory polyproline-IL-2 product. This product showed full *in vitro* bioactivity (see Scheme 2).

The second protein modifying group of reagents we sought are based on monomethoxy-polyethylene glycol (M-PEG). Polyethylene glycol has been attached to proteins by using *sym*-trichlorocyanuric acid and by conver-

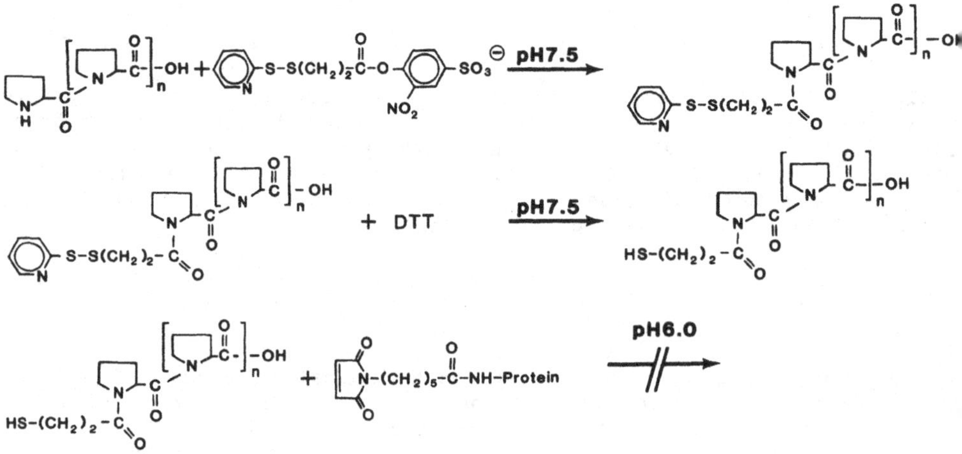

Scheme 1. Initial Reaction Sequence

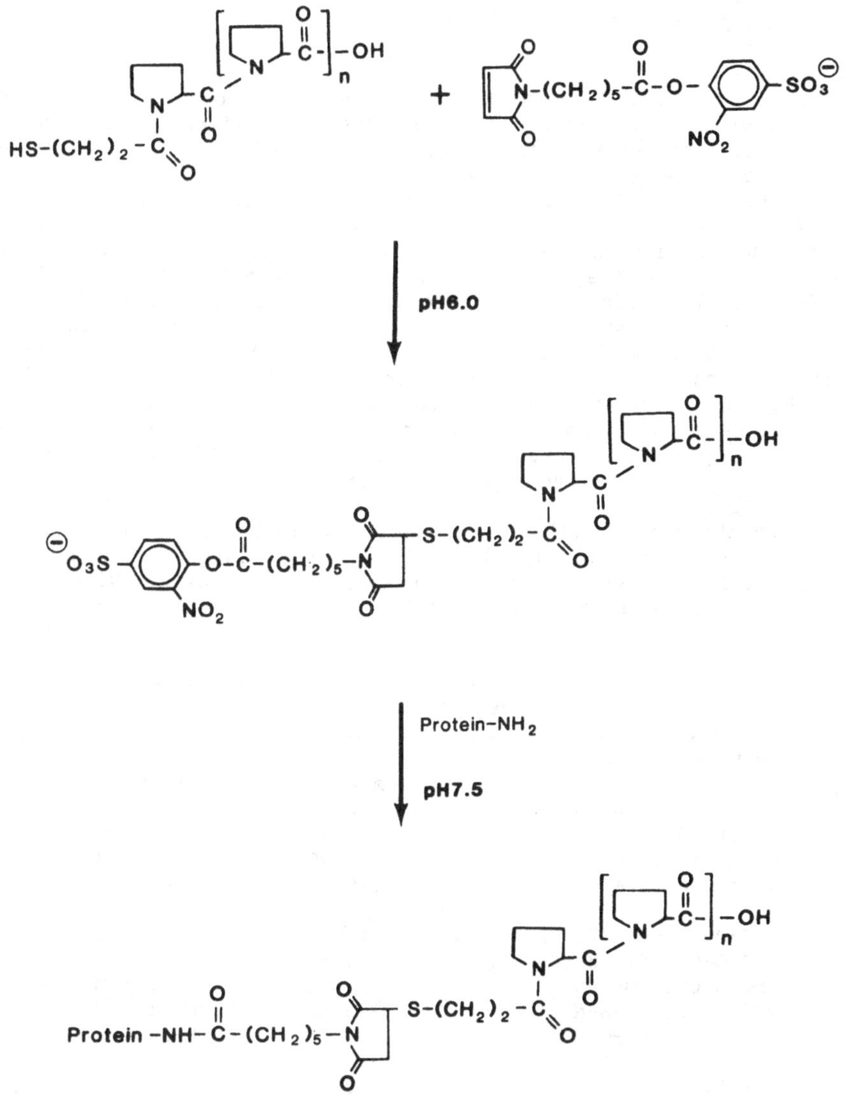

Scheme 2. Revised Reaction Sequence

sion of the terminal hydroxy group to a carboxylate followed by preparation of an active ester reagent.[4,5]. There is much data in the literature on activities of the esters in organic solvents but the data for aqueous solutions is rather sparse. The ultimate purpose, of course, is to find a reagent which is most selective to reaction with amino group of a protein (aminolysis) and does not react appreciably with hydroxide ions or water (hydrolysis). Generally, one would expect that the higher the reactivity of a reagent, the less likely it is to be selective. This is an important point, because purification of PEG-substituted protein from the two unreacted species (protein and PEG) is not trivial. Polyethylene glycol is water soluble, and this property allowed for us to look at reactivities in aqueous solutions of esters that normally are not water soluble (such as p-nitro-or o-nitrophenyl).[8]

RESULTS AND DISCUSSION

The structures and hydrolysis and aminolysis rates of some investigated M-PEG and other esters used in protein modification are shown in Table 1. Rates of reaction are influenced both by the nature of the leaving group and the pK_a of the acid moiety. In the rates of biotin active esters, the k_{OH^-} in M^{-1} min^{-1} (i.e. rate constants for hydroxide caused hydrolyses) are 1.46 x 10^4 for sulfo-N-hydroxysuccinimide, 2.28 x 10^4 for N-hydroxysuccinimide, and 8.00 x 10^2 for hydroxy-2-nitrobenzene-4-sulfonic acid. The respective $t_{1/2}$'s for hydrolysis are 26.9 min., 43.4 min., and 320 min. This order of magnitude difference is also reflected in the aminolysis rate constants (using as model amine 6-aminocaproic acid) which are: 5.01 x 10^4, 7.71 x 10^4 and 3.16 x 10^3 M^{-1} min^{-1}, respectively. The rate constant ratios for amine over hydroxide become 3.38, 3.43, and 3.95, which confirms that an improved ratio, and thereby better selectivity for protein amino groups can be obtained with an ester, such as HNSA, that reacts more slowly, and is less sensitive to buffer-catalyzed hydrolysis, as compared to N-hydroxysuccinimide ester.

The relative reactivities of esters are influenced not only by the nature of the leaving group (i.e. phenol or N-hydroxysuccinimide), but also by the substituent on the carbonyl side of the ester. Within the α-halo-acetic acids the reactivity of the esters is parallel to relative

Compound[a]	$t_{1/2}$ hydrolysis (min)	k aminolysis[d] ($M^{-1}min^{-1}$)
Biotin-O-NHS	43.4[b]	7.71 x 10^4
Biotin-O-SNHS	26.9[b]	5.01 x 10^4
Biotin-O-HNSA	320[b]	3.16 x 10^3
PEG-O-CO-CH$_2$-CH$_2$-CH$_2$-CO-NHS	40[b]	4.38 x 10^4
PEG-O-CO-CH$_2$-CH$_2$-CO-NHS	25.5[b]	1.02 x 10^5
PEG-O-CH$_2$-CO-NHS	0.98[b]	1.94 x 10^6
PEG-O-CH$_2$-CO-O-No	125.0[b]	1.82 x 10^4
PEG-O-CH$_2$-CO-O-Np	21.5[b]	1.19 x 10^5
Br-CH$_2$-CO-NHS	0.08[b]	
Br-CH$_2$-CO-HNSA	0.57[b]	
Cl-CH$_2$-CO-HNSA	0.42[b]	
I-CH$_2$-CO-HNSA	2.1[b]	
Br-CH$_2$-CO-NH-(CH$_2$)$_5$-CO-HNSA	404[b]	
PEG-O-CO-CH$_2$-CH$_2$-CH$_2$-CO-Np	165[c]	
PEG-O-CO-O-Np	270[c]	

Table 1. Hydrolysis and Aminolysis of Esters.

(a) Abbreviations: NHS - N-hydroxysuccinimide; No, ortho-nitrophenyl; Np, para-nitrophenyl; HNSA, 1-hydroxy-2-nitrobenzene-4-sulfonic acid sodium salt; SNHS, sulfo-N-hydroxysuccinimide.
(b) 0.08 M Pi, pH 7.5, 27°C.
(c) 0.1 M Pi, pH 8.5, ambient temperature.
(d) Reaction with 6-aminocaproic acid, pH 7.5, 0.08 M Pi, 27°C.

acidities of the parent acids, and of course, determined by electronegativity of the halogen; if halogen is further removed as in $Br-CH_2-CO-NH-(CH_2)_5-CO-HNSA$, the hydrolysis time is similar to the corresponding biotin ester.

In the polyethylene glycol series, similar effects are apparent. The ester carbonyl, removed by three methylene groups from ether oxygen is some 40 times less reactive than ester carbonyl containing only one methylene group separating it from ether oxygen. Interestingly enough, p-nitrophenyl ester carbonate is even slower to hydrolyze than the corresponding succinate ester.

REFERENCES

1. F. M. Veronese, R. Largajolli, E. Boccu, C. A. Benassi, and O. Schiavon, Appl. Biochem. Biotech., 11, 141 (1985).
2. M. J. Knauf, D. P. Bell, Z.-P. Luo, J. D. Young, and N. V. Katre, J. Biolog. Chem., 263, 15064 (1988).
3. N. V. Katre, M. J. Knauf, and W. J. Laird, Proc. Natl. Acad. Sci. USA, 84, 1487 (1987).
4. F. F. Davis, A. Abuchowski, T. van Es, N. C. Palczuk, K. Savoca, R. H.-L. Chen, and P. Pyatak, in: "Biomedical Polymers," Academic, New York, 1980, pp. 441-451.
5. J. M. Harris, J. Macromol. Sci. Rev. Macromol. Chem. Phys., C25, 325-373 (1985).
6. Reviewed by E. Katchalski, A. Berger, and J. Kurz, in: "Aspects of Protein Structure," G. N. Ramachanaran, Ed., Academic Press, Inc., New York, NY, 1963, p. 205.
7. L. Aldwin and D. E. Nitecki, Anal. Biochem., 164, 494 (1987).
8. D. E. Nitecki, L. Aldwin, and M. Moreland, in: "Peptide Chemistry 1987," T. Shiba and S. Sakakibara, Eds., Protein Research Foundation, Osaka, 1988, p. 243.

PREPARATION OF SEMISYNTHETIC ENZYMES BY CHEMICAL MEANS

David E. Albert, Mary B. Douglas, Marcia A. Hintz,
Christopher S. Youngen, and Melvin H. Keyes*

Anatrace, Inc.
1280 Dussel Drive
Maumee, OH 43537
and
Department of Biochemistry, Medical College of Ohio
P.O. Box 10008, Toledo, OH 43699-0008

A method to generate catalytic activity by conformational
modification of proteins is discussed and several examples
elaborated. Semisynthetic amino acid esterases and glucose
isomerases are discussed briefly. The most recent studies,
which are described in more detail, concern semisynthetic
fluorohydrolases prepared by conformational modification of
bovine pancreatic ribonuclease (RNase) as well as other proteins. RNase, modified with hexamethylphosphoramide (HMPA),
was derivatized with diimidates of chain lengths from one to
eight carbon atoms to determine which chain length produced
the maximum fluorohydrolase activity. The highest activity is
observed when RNase is crosslinked with dimethyl pimelimidate. This derivative operates over a pH range of 6.5 to 8.0
with an optimum pH of approximately 7.5 and hydrolyzes
phenylmethylsulfonylfluoride (PMSF) as well as the potent
acetylcholinesterase inhibitor, diisopropylfluorophosphate
(DFP). The mean fluorohydrolase activity, after chromatography on G-15 Sephadex to remove reactants, was 0.8 ± 0.2 U/mg.

GENERATION OF ENZYMATIC ACTIVITY

The two characteristics of enzymes which blend together to give catalytic activity are molecular recognition and rate acceleration. More than
forty years ago, Pauling suggested that enzymes are complementary to an
activated complex or transition state of the substrate to be reacted.[1] In
the decades that followed, chemists attempted to design and synthesize
molecules possessing a catalytic site which mimics the transition state.
The interest in synthetic and semisynthetic enzymes stems from the desire
to understand how natural enzymes work and to create catalysts with properties that are superior to natural enzymes.

---

* Address correspondence to Dr. Melvin H. Keyes at Anatrace, Inc.

Biotechnology and Polymers, Edited by C.G. Gebelein
Plenum Press, New York, 1991

1. Limitations of Natural Enzymes

The techniques to extract, purify, utilize and derivatize enzymes have been developed and refined over many years. Inadequacies of these techniques do not limit the use of enzymes as much as the scope of reactions catalyzed by naturally occurring, readily available enzymes. Of the three thousand plus enzymes known to exist, only a handful are available commercially in large quantities.

Even if an enzyme catalyzes a commercially useful reaction, it may possess characteristics which are not advantageous in the industrial process. For example, isomerization of glucose to fructose is the largest industrial application of immobilized enzymes in the United States, yet the properties of the glucose isomerase used in the industrial process are far from ideal.[2] It requires the use of transition metal ions, such as Mg^{+2}, which necessitates the use of ion exchange columns to remove metal ions from the product. The initial hydrolysis of starch is accomplished by the aid of glucoamylase in an acid environment. Unfortunately, glucose isomerase is most effective in the alkaline region requiring the pH to be increased.[2]

2. Enzyme Engineering

All the tools necessary for the synthesis and production of novel enzymes not found in nature are available today. Many enzymes are being modified by replacing one or two residues within the primary structure and observing the changes in activity and characteristics. Even with these small alterations in the primary structure, the results are often quite unexpected. Researchers are altering whole regions of the protein structure or building an entirely new protein. Unfortunately the sequences that would generate new enzyme activities remain a mystery. Until this mystery is solved, the successes will be in the modification of existing enzymes to increase their activities or to make their characteristics more amiable to commercialization.

The extraction and purification of the engineered protein is often difficult and time consuming. Since the protein is not always folded properly in the host microorganism, the conditions required for folding to the native state must be devised in addition to the extraction and purification scheme.

3. Catalytic Antibodies

More than forty years ago, Pauling implied a method to generate new enzyme activity:[1] "...an enzyme has a structure closely similar to that found for antibodies, but with one important difference, namely that the surface configuration of the enzyme is not so closely (complementary) to its specific substrate as is that of an antibody to its homologous antigen, but is instead complementary to (the activated complex)."

In 1969 Jencks stated the case for development of catalytic activity in antibodies more directly:[3] "One way to (synthesize an enzyme) is to prepare an antibody to a haptenic group which resembles the transition state of a given reaction. The combining sites of such antibodies should be complementary to the transition state and should cause an acceleration by forcing bound substrates to resemble the transition state."

Attempts to isolate catalytic antibodies in the seventies and early eighties proved unsuccessful.[4] When any reaction was observed with antibodies selected to possess a catalytic site, it was stoichiometric rather than catalytic. These early failures could have been due to the antibody methodology, or, quite possibly, to the transition state analog not closely resembling the true transition state.

More recent work, presumably with more careful attention to the selection of the transition state analog, have isolated catalytic antibodies. A patent application filed in 1983 by Schochetman and Massey of IGEN Inc. states that immunization of mice with synthetic immunogens together with hybridoma technology would lead to monoclonal antibodies with catalytic activity.[5] Two papers were published in 1986 by research groups who independently produced very similar results using phosphonate transition state analogs. These analogs produced monoclonal antibodies possessing activity to the corresponding esters.[6,7] Examples of catalytic antibodies which catalyze several different reactions have been reviewed in the literature.[4,8]

The success of catalytic antibodies appears to be based on the very careful selection of the transition state analog. The design of the analog, which can be a formidable task in organic chemistry, is made all the more difficult since the exact structure of the transition state is not known. The time required to develop and process the monoclonal antibody adds to the problem. Another approach to bypass the synthesis of the transition state analog is to induce an antibody to the substrate molecule and modify the active site by protein engineering.[4]

4. Protein Derivatives

One method to prepare semisynthetic enzymes involves substituting one segment of a natural enzyme with a synthetic peptide; this semisynthetic enzyme will probably possess catalytic activity similar to the starting enzyme with altered kinetics.[9-13] Of even greater interest in terms of commercial applications would be a modified enzyme that possesses new catalytic activities not found in the native enzyme. An approach to generate such semisynthetic enzymes reported by Kaiser, et. al, involves the covalent linkage of flavin derivatives to the active site of papain to induce oxidoreductase activities.[14-17] In this paper, we describe a novel approach to induce new catalytic activities in proteins, by means of conformational modifications.

5. Generation of Catalytic Activity by Conformational Modification of Proteins

A systematic approach to induce new catalytic activities in proteins has been developed in our laboratory.[18-23] Our method consists of perturbing the native conformation of the protein and subsequent binding of an analog to the active site of an enzyme whose activity is to be generated. The product of this method is known as a catalytic conformationally modified protein (CCMP). Although the mechanism of the conformational modification method remains unknown, the flexibility of protein structure, and ligand-induced conformational modifications, probably play a role in the process.[24-38]

5A. Description

The steps in the process are:

(a). A readily available protein is chosen as the starting material.
(b). The protein conformation is modified by changes in pH, temperature, ionic strength or some other condition which perturbs the secondary and/or tertiary structure.
(c). The perturbed protein is then exposed to a modifier which is a template of the desired active site.
(d). The newly formed conformation is preserved by crosslinking through the side chains of the protein. (e). The modifier and excess crosslinking reagent are then removed by dialysis or gel filtration.

Surprisingly this simple procedure has been shown to yield catalytic enzyme-like materials for several reactions. These materials can be purified in the same way as natural proteins are extracted from a crude biological fluid. In addition the properties of these catalysts are in many cases different than their natural enzyme counterpart. The pH optimum is often changed, the cofactor requirement of the native enzyme may not be present, and the substrate specificity is usually quite different. Of course, these materials when prepared from readily available protein can easily be produced in large quantities. Several examples of catalytic conformationally modified proteins (CCMP) are described below.

5B. Examples of CCMP

The first two examples of CCMP described herein, have been studied and reported elsewhere in the literature. Only the highlights of these studies are given. The last example, fluorohydrolases, includes our latest results and are given in more detail.

Amino acid esterases were generated from RNase by using a variety of indole derivatives as modifiers and, perturbing the conformation by titration to an acid pH.[23,29] Crosslinking with glutaraldehyde was used to stabilize this new conformation. When assayed with L-tryptophan ethyl ester, the modified RNase was found to possess two pH optima: one at 6 and the other at 7.5. Purification of the crude reaction mixture demonstrated the presence of two types of amino acid esterases which account for the two pH optima. After the conformational modification process, the native activity of RNase is lower. After purification of the crude mixture no native RNase activity can be measured in the fractions containing amino acid esterase activity.

Initially the modifier consisted of indole propionic acid dissolved in water at a concentration of 1 mM. Increasing the concentration of indole propionic acid up to the limit of its solubility increases the initial activity of the modified RNase. The use of modifiers such as 5-hydroxyindole-3-acetic acid and 5-hydroxyindole-3-acetamide, which are more water soluble than indole propionic acid, increased the activity even further.

Several literature studies indicate that the mechanism of crosslinking proteins with glutaraldehyde changes dramatically with the pH of the reaction.[40,42] Using 5-hydroxyindole-3-acetamide as the modifier (10 mM) and perturbation at pH 3.0, several samples of RNase were crosslinked in the pH range of 3 to 8. The highest activity is obtained when RNase is crosslinked at pH 3 to 4, the same pH used for perturbation.

Even though the production of high fructose corn syrup is made possible by the use of glucose isomerase, some properties of the native enzyme are not desirable for commercial application.[2] The enzyme requires a metal ion cofactor and has a pH optimum in the neutral or slightly alkaline region. We prepared semisynthetic glucose isomerases from concanavalin A, hexokinase and apoglucose oxidase, none of which require a cofactor.[43] In addition the catalyst prepared from concanavalin A was most active at pH 5. Such a glucose isomerase, active in the acid region, is advantageous, since the initial hydrolysis of starch catalyzed by glucoamylase is also performed in the acid region.

In the balance of this paper, we describe the application of our method of chemical modification to the generation of fluorohydrolases.[3] The most common substrate to measure fluorohydrolase activity is diisopropylphosphorofluoridate (DFP). Although DFP is not found in nature, Mazur reported the enzymatic hydrolysis of this highly toxic organofluorophosphate by an enzyme isolated from hog kidney.[44] Since then, diisopropyl phosphorofluoridate fluorohydrolase (DFPase, E.C. 3.8.2.1) has been extensively studied and well characterized. In addition, enzymes with organofluorophosphate hydrolyzing activity have been shown to be widely distributed phylogenetically and sources are known from bacteria, protozoa, invertebrates, and vertebrates.

Until now, two major types of DFPases have been described. The "Mazur-type" designates a dimeric Mn^{2+} stimulated enzyme having a molecular weight of 62,000 daltons isolated from hog kidney. Fairly high concentrations of this type of DFPase is also found in *Escherichia coli* and mammalian tissues.[45] The second type (squid-type) has a molecular weight of 26,000 daltons and is found in the optic ganglia, giant axon, hepatopancreas, and salivary gland of cephalopods. Unlike the "Mazur-type" enzyme, this DFPase is unaffected or slightly inhibited by Mn^{2+} at about $10^{-3}M$. Despite intensive investigations, the natural substrate(s) for DFPase has not been identified, leaving the physiological role of this enzyme unclear.

CCMPs have been prepared, which have the ability to catalyze the hydrolysis of organofluorophosphates such as diisopropylphosphorofluoridate and phenylmethylsulfonylfluoride, using several different starting proteins.[46] In addition, semisynthetic fluorohydrolases prepared by the conformational modification of RNase with HMPA were crosslinked with diimidates of chain lengths from C_1 to C_8 to determine the optimum crosslink for the maximum fluorohydrolase activity.

MATERIALS AND METHODS

1. Materials

RNase from bovine pancreas was purchased from Biozyme Laboratories International, San Diego, CA. Aldolase, bovine serum albumin (BSA), DFP, egg albumin, PMSF and soybean trypsin inhibitor were purchased from Sigma Chemical Company, St. Louis, MO. Carbonic anhydrase from bovine erythrocytes was obtained from United States Biochemical Corp., Cleveland, OH. The organophosphates were purchased from either Aldrich Chemical Company, Inc., Milwaukee, WI. or Alfa Products, Danvers, MA. Diimidoester crosslinking reagents were obtained from Anatrace, Inc., Maumee, OH while Sephadex™ G-15 was purchased from Pharmacia Inc., Piscataway, NJ. All other reagents were of analytical grade.

2. Conformational Modification

The starting protein (bovine RNase) is chromatographed on a G-15 Sephadex gel column with 1 mM HCl or 5 mM phosphoric acid as the eluent. The concentration of the eluted RNase, approximately 2 mg/ml, is determined using the absorbance coefficient of 7.3 at 280 nm for a 1% solution. The perturbed protein is reacted with a modifier, such as HMPA, for 15 minutes and the pH titrated to 8.5 with 0.05 M NaOH. The initial activity of the modified protein solution is measured at pH 8.5 prior to crosslinking.

3. Crosslinking and Purification of Fluorohydrolase

Following conformational modification, a diimidate crosslinking reagent is added at a 20:1 molar ratio (diimidate/protein) and the solution incubated for 3 hours at room temperature followed by incubation at 5°C for 17 hours. To remove excess bifunctional reagent and modifier, a 5 ml aliquot of this solution is applied to a column (1.5 x 100 cm) of G-15 Sephadex which has been equilibrated with 5 mM phosphoric acid. Elution was accomplished with the same buffer at a flow rate of 1.0 ml per minute and the protein peak was detected by monitoring the absorbance at 280 nm.

4. Assay Methods

Activity measurements were accomplished by incubating control and assay solutions at 30°C (unless specified) in a circulating water bath. The hydrolysis of DFP was monitored by measuring the formation of F^-. An Orion 901™ microprocessor ionanalyzer and an Orion fluoride sensitive electrode calibrated to 1.0 and 5.0 mM NaF standards was used to record F^- concentration at 1 minute intervals. The hydrolysis of PMSF was measured by the fluoride electrode as well as by high performance liquid chromatography (HPLC). For the HPLC assay, a stainless steel column (3 mm x 25 cm) packed with an amine support (Baker #7034-0) and equilibrated to pH 4.0 with 0.04M sodium acetate buffer containing 0.1M NaCl was used to measure the hydrolysis product phenylmethylsulfonic acid (PMSA). The PMSA is detected at 220 nm with a UV monitor.

In a separate assay, the rate of hydrolysis of each substrate was analyzed using an automated Sargent-Welch pH Stat system (Sargent-Welch Scientific Company, Skokie, IL). By stepwise addition of a previously standardized 0.01N NaOH solution, the acid generated upon hydrolysis was measured. Titrations were carried out at pH 7.6 and 30°C.

A typical control solution contains 100 mM MOPS (3-(4-morpholino)-propanesulfonic acid), pH 7.6, 5 mM potassium phosphate buffer, pH 7.6, and 5 mM DFP or PMSF in a final volume of 8 ml. The assay solution is prepared by adding modified protein to an identical solution.

RESULTS

We investigated eighteen (18) organophosphate compounds as potential modifiers. The initial activity toward DFP or PMSF was measured at pH 7.6, fifteen minutes after the addition of modifier, but before reaction with crosslinking reagents. Although the initial activity generated decays to zero in a few days, its measurement allows us to easily and quickly survey the potential modifiers before crosslinking the modified protein. Table 1 lists the initial activities and the modifiers used. Some of these organophosphates failed to generate any measurable activity for either substrate, some generated activity towards only one, while others generated good activity towards both substrates. The greatest activity was obtained with diisopropylphosphoric acid (DPA). With DFP as the substrate the activity was 0.25 U/mg and the catalyst had a molecular activity of 8.6 min.$^{-1}$; however, with PMSF the activity was 0.94 U/mg and the molecular activity was 17.4 min.$^{-1}$ measured over 13.5 minutes. DPA, which is not commercially available, was prepared by hydrolyzing diisopropyl phosphorofluoridate, 0.2 mole/liter, in a dilute sodium hydroxide solution. The second best activity was generated with diethylisopropyl-phosphonate (DEIP). Unfortunately this compound is very unstable and therefore difficult to utilize.

One modifier that was commercially available, inexpensive, and reasonably stable, hexamethylphosphoramide, worked very well with other proteins as well as RNase. Table 2 shows the results of modifying BSA, casein, egg albumin, and RNase with HMPA. Without modifier, no fluoro-hydrolase activity was detected for any of these proteins.

These CCMPs were not stable without crosslinking to maintain the active conformation. RNase was modified with HMPA as described above and derivatized with each of the eight commercially available aliphatic di-imidoesters. The activities measured using both DFP and PMSF are shown in Figure 1.

Conformationally modified RNase, derivatized with dimethyl pimelimi-date, had the highest fluorohydrolase activity toward DFP; however, the highest activity toward PMSF was obtained for dimethyl suberimidate crosslinked protein. The best overall activity for both substrates was obtained with the pimelimidate (C$_5$) crosslinked modified RNase. A plot of activity towards PMSF versus pH, after chromatography on G-15 Sephadex, shows that the optimum pH is approximately 7.5 (Figure 2). The values of

Table 1. Fluorohydrolase activity of modified RNASE.

Modifier	Activity (U/mg)	
	DFP	PMSF
Diisopropylphosphoric Acid	0.25	0.94
Diethylisopropylphosphonate	0.11	0.31
Ethyldiethylphosphonylformate	0.06	0.16
Diisopropylphosphite	0.05	0.74
Diethylcyanomethylphosphonate	0.05	0.12
Hexamethylphosphoramide	0.04	0.34
Diethylphenylphosphate	0.03	-
Diethylphosphite	0.02	0.14
Trimethylphosphonoacetate	0.02	0.11
Dimethylphosphite	-	0.33
Diisopropylmethylphosphonate	-	0.30

Diethylchloromethyl phosphonate, phenylmethyl sulfon-ate, pyridoxal-5-phosphate, tributyl phosphate, tri-ethyl phosphate and trimethyl phosphate gave no activity.

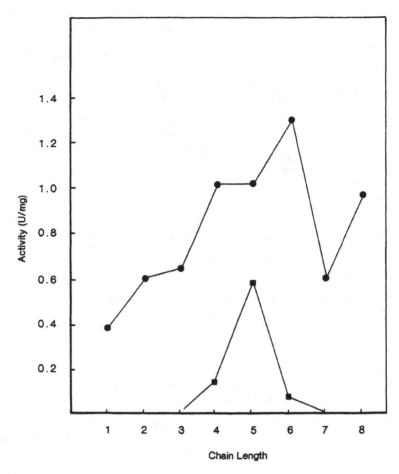

Figure 1. Fluorohydrolase activity of CCMPs prepared from RNase as a function of the dimension of the cross-linking span of the diimidate reagent. Activity was measured at pH 7.4 and 30°C using both DFP (■) and PMSF (o) as substrates. Each activity is the average of two independent assay methods.

Table 2. Fluorohydrolase activity of various proteins modified with HMPA.

Protein	Chromatographed in 1 mM HCl	
	Substrate	Mean Activity (U/mg) + S.E.
BSA	DFP	1.03 ± 0.30
Casein	PMSF	12.67 ± 1.33
Egg Albumin	DFP	-
	PMSF	0.49 ± 0.18
RNase	DFP	0.08 ± 0.03
	PMSF	0.44 ± 0.10

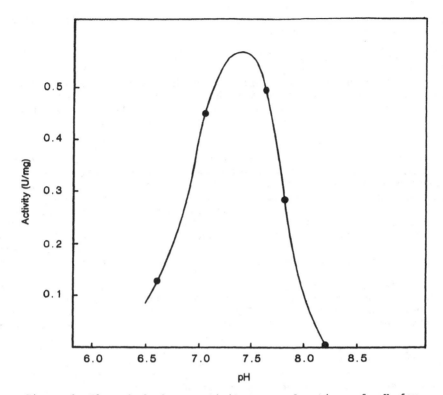

Figure 2. Fluorohydrolase activity as a function of pH for dimethyl pimelimidate crosslinked CCMP. Activity was measured using 5 mM PMSF as the substrate at 30°C. Each activity is the average of two separate measurements obtained by HPLC and fluoride electrode methods described in the text.

activity plotted in Figure 2 were calculated by averaging the results of the HPLC assay with those obtained using the fluoride electrode.

DISCUSSION

In previous articles, we described a procedure for the induction of fluorohydrolase activity in bovine albumin, casein, egg albumin, hexokinase and ribonuclease.[21,46] By perturbing these proteins at pH 3 and adding a modifier, significant activity towards DFP and PMSF was obtained. The initial activity was measured before protein crosslinking indicating that the active conformation was retained for several minutes in the presence of modifier at neutral pH. If the perturbed protein was chromatographed on G-15 to remove modifier, no activity could be detected.

Of the eighteen organophosphate compounds investigated as potential modifiers of RNase, eleven generated activity toward either DFP or PMSF. That eleven modifiers generated activity is surprising since they were not carefully selected to mimic the transition state. These organophosphate compounds are simply the commercially available ones. Apparently, the choice of modifier for this method is more lenient than in the case of catalytic antibodies. In addition the eighteen potential modifiers were evaluated in several days rather than the several weeks required to induce and purify a monoclonal antibody.

Diisopropylphosphoric acid produced the greatest activity (Table 1) while diethylisopropylphosphoric acid gave approximately half the specific activity of DPA and is unstable. Since DEIP is unstable and DPA is not commercially available, we used HMPA as the modifier for preparing and characterizing semisynthetic fluorohydrolases. Furthermore, HMPA has been shown to generate activity with other proteins as well as RNase (Table 2).

To maintain the active conformation and optimize the activity, RNase conformationally modified with HMPA was crosslinked with a series of aliphatic diimidoesters (C_1 to C_8). These diimidoesters were chosen for crosslinking since they react under mild conditions to yield derivatives with unaltered charge. With the surface charge unchanged, the protein derivatives are usually as soluble as the native protein.[47] In addition bifunctional aliphatic diimidoesters are commercially available in chain lengths of 1 to 8 carbon atoms, allowing the distance of the linkage after reaction to vary from approximately 3 to 13 angstroms.

The plot in Figure 1 of the activity versus the chain length, indicates that the length of the crosslink had a dramatic effect on the fluorohydrolase activity. Only CCMPs crosslinked with diimidoesters possessing four through six carbon atoms had activity toward DFP. Although, all CCMPs regardless of the chain length of the diimidoester had some activity toward PMSF, the optimum span was between 6 and 11 angstroms. The best overall activity was obtained with a diimidoester having a chain length of 5 carbon atoms (dimethyl pimelimidate). The resulting semisynthetic enzyme is active over a pH range of 6.5 to 8.0 with an optimum pH of approximately 7.5 (Figure 2).

To compare the catalytic efficiency of catalysts, it is helpful to compare the enhancement ratios (E.R.). E.R. is calculated by dividing the k_{cat} by the k_{uncat} (the rate constant for the uncatalyzed reaction). Enormous rate enhancements are achieved by enzymes. For example, hydrolytic enzymes often exhibit rate enhancements of 10^8-10^{12} compared with the spontaneous water-catalyzed or the acid/base-catalyzed reaction at about neutral pH.[48] For purposes of comparison, the k_{cat}, k_{uncat}, and E.R. values for two hydrolytic abzymes,[8] as well as CCMP fluorohydrolase chromatographed on G-15 Sephadex gel is presented in Table 3. The fluorohydrolase, chromatographed on G-15 Sephadex, has an E.R. four times that of the two abzymes. In addition, the enhancement ratios of natural DFPases from various sources as well as CCMP fluorohydrolase are presented in Table 4. In all cases, the initial E.R. (before any purification) is higher for semisynthetic fluorohydrolases than for any of the natural unpurified DFPases.

Table 3. Comparison of enhancement ratios of abzymes and semisynthetic fluorohydrolases.

Source of catalyst	k_{cat} (Min^{-1})	k_{uncat} (Min^{-1})	E.R.
MOPC 167 Antibodies	6.7×10^{-3}	8.6×10^{-6}	770
mAb 6D4 Antibodies	2.7×10^{-2}	2.8×10^{-5}	960
CCMP G-15 Sephadex	20.7	5.0×10^{-3}	4,140

Table 4. Comparison of enhancement ratios of natural DFPases and CCMP fluorohydrolases.

Source of Fluorohydrolase	Activity (U/g)	K_{cat} (Min^{-1})	K_{uncat} (Min^{-1})	E.R.
T. thermophilia[49] (brie)	11.1	0.8	5×10^{-3}	770
T. thermophilia[49] S-200 Sephacryl Fraction	104.0	7.8	5×10^{-3}	1,568
CCMP	72.8	20.7	5×10^{-3}	4,140
Hog Kidney[50]				
Homogenate	14.0	0.9	5×10^{-3}	174
Partially Purified	250.0	15.5	5×10^{-3}	3,100
Squid Nerve[50]				
Homogenized Ganglia	45.0	1.2	5×10^{-3}	234
Purified Ganglia	22,500	585.0	5×10^{-3}	1.17×10^5

The production of semisynthetic enzymes by conformational modification has been well documented in the literature.[18-23,39] The utility and versatility of this process is exemplified by the numerous starting proteins that have been successfully modified to produce new catalysts. Examples include the preparation of amino acid esterases, fluorohydrolases, and glucose isomerases. In many cases these semisynthetic catalysts have activities and turnover numbers comparable to or greater than

Table 5. Comparison of enhancement ratios of various semisynthetic enzymes.

Protein Source	Type of Activity	k_{cat} (Min^{-1})	k_{uncat} (Min^{-1})	E.R.
Apoglucose Oxidase	Glucose Isomerase			
1) Crude		60.4	7.5×10^{-5}	2.1×10^4
2) G-200 Sephadex		835.0	7.5×10^{-5}	1.1×10^7
Hexokinase	Glucose Isomerase			
1) Crude		0.4	7.5×10^{-5}	5.9×10^3
2) Cellobiose Gel		1.4	7.5×10^{-5}	1.9×10^4
RNase	Esterase			
1) Crude		0.3	4.2×10^{-4}	7.1×10^2
2) P-30 Gel		66.0	4.2×10^{-4}	1.6×10^5
RNase	Fluorohydrolase			
1) Crude		21.0	5.0×10^{-3}	4.1×10^3
2) G-75 Sephadex		110.0	5.0×10^{-3}	2.2×10^5

some natural enzymes. Table 5 lists the enhancement ratios calculated for some CCMPs. As expected, partial purification by gel filtration, affinity chromatography and /or ammonium sulfate fractionation results in significantly higher enhancement ratios.

Semisynthetic fluorohydrolase has activity comparable to the naturally occurring DFPases and does not require a divalent cation as does the "Mazur-type" enzyme. It operates over a pH range from 6.5 to 8.0 with the optimum at approximately 7.5. A comparison of natural and semisynthetic fluorohydrolase activity is shown in Table 4. An examination of the enhancement ratio (2.2×10^4) shows that the CCMP fluorohydrolase is a remarkably good catalyst. If the crosslinking reaction is further optimized to increase the yield of the dimer along with further purification steps, a stable highly efficient semisynthetic fluorohydrolase should be achieved.

ACKNOWLEDGMENTS

The authors wish to thank the Army Research Office (Contract No. DAAL03-86-C-0021) for their financial support of the fluorohydrolase project. We also express our thanks to the Department of Energy - Office of Basic Energy Science (Contract No. DE-AC02-81er12003) for support of the earlier amino acid esterase studies. Additional support was given by Owens-Illinios, Inc. and Anatrace, Inc.

We are grateful to Dr. Don Gray for his helpful comments during the preparation of this manuscript and to Mrs. Lois Staunton for her word processing efforts.

REFERENCES

1. L. Pauling, Am. Scient., **36**, 51 (1948).
2. M. H. Keyes & D. Albert, in: "*Mark-Bikales-Overberger-Manges: Encyclopedia of Polymer Science & Engineering,*" 6, John Wiley & Sons, Inc., New York, 1986, p 189-209.
3. W. P. Jencks, in: "*Catalysis in Chemistry and Enzymology,*" McGraw-Hill, New York, 1969, p 288.
4. M. J. Powell & D. E. Hansen, Protein Engineering, 3, 69 (1989).
5. G. Schochetman & R. J. Massey PCTWO, 85/02414 (1985).
6. A. Tramontano, K. D. Janda & R. A. Lerner, Science, **234**, 1566 (1986).
7. S. J. Pollack, J. W. Jacobs & P. G. Schultz, Science, **234**, 1570 (1986).
8. G. M. Blackburn, A. S. Kang, G. A. Kingsbury & D. R. Burton, Biochem. J., **262**, 381 (1989).
9. A. Komoriya & I. M. Chaiken, J. Biol. Chem., **257**, 2599 (1982).
10. C. DiBello, A. Lucchiari, O. Buso & M. Tonellato, Int. J. Pept. Protein Res., 23, 61 (1984).
11. M. Juillerat & G. A. Homdanberg, Int. J. Pept. Protein Res., **18**, 335 (1981).
12. G. A. Homdanberg, G A. Komoriya, M. Juillerar, I. M. Chaiken, Proc. VIth Am. Pept. Symp., 597, (1979).
13. R. Geiger, V. Teetz, V. Konig, & R. Obermeir, in: "*Semisynthetic Peptides & Proteins,*" R. E. Offord & C. DiBello, Eds., Academic Press, New York, 1978, p 141.
14. H. L. Levine & E. T. Kaiser, J. Am. Chem. Soc., **100**, 7670 (1978).
15. F. T. Slama, S. R. Oryganti & E T. Kaiser, J. Am. Chem. Soc., **103**, 6211 (1981).

16. E. T. Kaiser, H. L. Levine, T. Outski, H. E. Fried & R. M. Dupyere, in: "*Biomimetric Chemistry,*" D. Dolphin, et al., Eds, Am. Chem. Soc., Washington, DC, 1980, p 35.
17. H. L. Levine & E. T. Kaiser, J. Am. Chem. Soc., **102**, 342 **(1980)**.
18. M. H. Keyes, in: "*Biotec 1,*" C. P. Hollenberg, & H. Sahm, Eds., VCH Publishers, Inc., New York, 1987, p 137.
19. M. H. Keyes, in: "*Protein Engineering: Current Status, Proceedings of Bioexpo 86,*" Butterworth Publishers, Stoneham, MA, 1986, p 273.
20. M. H. Keyes, Protein Modification to Provide Enzyme Activity. U. S. Patents Nos. 4,716,116; 4,714,677; 4,714,676.
21. M. H. Keyes & D. E. Albert, "Proc. Polymeric Materials: Science & Engineering", **58**, 111 (1988).
22. M. H. Keyes, D. E. Albert & S. Saraswathi, in: "*Enzyme Engineering,*" A. I. Laskin, K. Mosbach, D. Thomas & L. B. Wincard, Jr. Eds., The New York Academy of Sciences, New York, 1986, p 201.
23. M. H. Keyes & S. Saraswathi, in: "*Polymeric Materials in Medication,*" C. G. Gebelein & C. E. Carraher, Jr. Eds., Plenum Publ. Corp., New York, 1985, p 249.
24. S. W. Englander & N. R. Kallenbach, Quarterly Rev. Biophys., **16**, 521, (1984).
25. M. R. Eftnik & C. A. Ghiron, Anal. Biochem., **114**, 199 (1981).
26. D. A. Torchia, Ann. Rev. Biophys. Bioeng., **13**, 125 (1984).
27. G. A. Petsko & D. Ringe, Ann. Rev. Biophys. Bioeng., **13**, 331 (1984).
28. J. A. Richardson, Adv. Protein Chem., **34**, 167 (1981).
29. T. Creighton, Prog. Biophys. Molec. Biol., **33** 231 (1978).
30. J. Janin & S. J. Wodak, Prog. Biophys. Molec. Biol., **42** 21 (1983).
31. W. S. Bennet & R. Huber, CRC Critical Rev. Biochem., **15** 291 (1984).
32. N. Citri, in: "*Adv. Enzymology,*" Vol. 37, A. Meister, Ed., John Wiley & Sons, 1973, p 397.
33. G. Weber, Adv. Prot. Chem., **29**, 1 (1978).
34. R. Wolfenden, Acc. Chem. Res., 5, 10 (1972).
35. G. E. Leinhard, Science, **180**, 149 (1973).
36. Z. Wasylewski & P. M. Horowitz, Biochem. Biophys, Acta., **701**, 12 (1982).
37. O. W. Howarth & I. Y. Lian, Biochemistry, 23, 3522 (1984).
38. J. B. Prenberg, J. M. Schaffert & H. H. Sussman, J. Biol. Chem., **211**, 327 (1981).
39. S. Saraswathi & M. H. Keyes, Enzyme Microbial Technol. **6**, 98 (1984).
40. P. Monsan, G. Puzo & H. Mazarguil, Biochemie, **57**, 1281 (1975).
41. R. Lubig, P. Kush, K. Roper & H. Zahn, Monatshifte Fur Chemie, **112**, 1313 (1981).
42. R. Koelsch, M. Fusek, Z. Hostomaska, J. Larch & J. Turkova, Biotechnology Letters, **8**, 283 (1986).
43. M. H. Keyes & D. E. Albert, in: "*Biomimetic Polymers,*" C. G. Gebelein, Ed., Plenum Press, New York, 1990, p. 115.
44. A. Mazur, Jr. Biol. Chem., **164**, 271 (1946).
45. F. C. G. Hoskin, Biochem. Pharm. 34, 2069 (1985).
46. M. H. Keyes & D. E. Albert, Proc. Polymeric Mater. Sci. Eng., **62**, (1990).
47. M. J. Hunter & M. L. Ludwig, Methods Enzymol., 25, 585 (1972).
48. G. P. Royer in: "*Fundamentals of Enzymology,*" John Wiley & Sons, New York, NY, 1982, p 39.
49. W. G. Landis, R. E. Savage & F. C. G. Hoskin, J. Protozool, 32, 517 (1985).
50. F. C. G. Hoskin, M. A. Kirkish & K. E. Steinmann, Fundam. Appl. Toxicol., **4**, S165 (1984).

REDESIGN OF PROTEIN FUNCTION: A SEMISYNTHETIC SELENOENZYME

Zhen-Ping Wu and Donald Hilvert

Departments of Chemistry and Molecular Biology
Research Institute of Scripps Clinic
10666 North Torrey Pines Road
La Jolla, California 92037

An artificial selenoenzyme, selenolsubtilisin, was prepared by chemical conversion of the active site nucleophile
(Ser 221) in the protease subtilisin into a selenocysteine.
The properties of the selenol group make it a useful and
general molecular probe of steric and electronic effects in
catalysis involving serine (or cysteine) side chains in proteins. In the case of subtilisin, the effects of the selenium
for oxygen substitution on the acylation and deacylation
steps that occur during substrate hydrolysis and on partitioning of the acyl enzyme intermediate between amines and
water were studied. Increases in aminolysis selectivity of up
to 14,000 fold were achieved with the selenoenzyme compared
to native subtilisin, suggesting that selenolsubtilisin could
be useful as a peptide ligase. Like the naturally occurring
selenoenzyme glutathione peroxidase, selenolsubtilisin also
has interesting redox properties. It catalyzes the reduction
of hydroperoxides by thiols. Comparison of the initial rates
for the oxidation of 3-carboxy-4-nitro-benzenethiol by t-
butyl hydroperoxide in the presence of the enzyme and a model
selenocompound indicates a 17,000 fold rate enhancement due
to protein binding.

INTRODUCTION

Enzymes are being used increasingly as practical catalysts in organic
synthesis due to their ability to promote highly selective transformations of complex molecules.[1] However, for specific reactions of interest,
enzymes are not always available. Rational design of proteins that mimic
the high rates and selectivities of enzymes is therefore of great significance. Moreover, characterization of these artificial enzymes may enhance our understanding of the mechanisms of molecular recognition and
catalysis.

The construction of enzyme-like catalysts requires the generation of
highly selective binding sites containing an appropriate constellation of
chemical groups. Since the de novo synthesis of enzymes with predictable

Biotechnology and Polymers, Edited by C.G. Gebelein
Plenum Press, New York, 1991

tertiary structures and high binding affinities for specific substrates remains a distant goal, naturally occurring proteins represent valuable starting points for the creation of new biocatalysts. Many proteins have been well characterized, and they can be redesigned either at the level of DNA using recombinant technology or post-translationally through chemical modification.

We are utilizing chemical methods to develop artificial selenoenzymes.[2] The molecular properties of selenium make it useful as a probe of steric and electronic factors that are important for the function of enzymes containing oxygen and sulfur nucleophiles. In addition, organoselenium chemistry is rich and of synthetic importance.[3] Artificial selenoenzymes with template-imposed selectivities might therefore find practical application as selective catalysts in organic synthesis. Finally, the redox properties of selenols make artificial selenium-containing enzymes potentially interesting models of naturally-occurring selenoenzymes like glutathione peroxidase.

The bacterial serine protease subtilisin (EC 3.4.21.14) is an excellent template for constructing artificial enzymes, because it is structurally and chemically well characterized.[4,5] We recently prepared the first artificial selenoenzyme, selenolsubtilisin, by chemically converting the active site alcohol in this enzyme into a selenol.[6] Ser 221 of subtilisin Carlsberg was activated by reaction with phenylmethanesulfonyl fluoride. Displacement of the resulting sulfonate by hydrogen selenide gave the desired selenol-containing enzyme. Some of the hydrolytic and redox properties of this semisynthetic enzyme are reviewed in this article.

SUBSTRATE HYDROLYSIS

Systematic variation of structure is a useful technique for determining the factors that contribute to the reactivity of chemical substances. The related properties of selenols, thiols and alcohols suggested that comparison of analogous serine-, cysteine- and selenocysteine-containing enzymes might shed light on the relative importance of steric and electronic factors in enzymatic catalysis, for example, by serine and cysteine proteases.

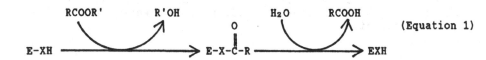

(Equation 1)

Serine and cysteine protease-catalyzed hydrolysis of esters and amides involves two steps: (1) attack of the active site nucleophile on the carbonyl at the cleavage site to form an acyl-enzyme intermediate, and (2) hydrolysis of the acylated enzyme to regenerate the catalyst (Equation 1). We wondered whether a selenol group (-SeH) could assume the role of the active site nucleophile. We found that selenolsubtilisin, like thiolsubtilisin, is a poor amidase.[7] Thus, N-succinyl-Ala-Ala-Pro-Phe-p-nitro-anilide is not hydrolyzed at all by the selenol-containing enzyme even though it is an excellent substrate for subtilisin itself. Activated esters, on the other hand, are substrates.

We studied the hydrolysis of p-nitrophenyl acetate (PNPA) in the presence of the reduced selenoenzyme. Initial rates were determined by following the absorbance change at 400 nm at pH 7.0 and 25°C, and were corrected to reflect the activity due to the selenol moiety alone. The kinetic data follow the Michaelis-Menten equation, and the steady state parameters are given in Table 1 together with data for the wild type enzyme and thiolsubtilisin.[7] As can be seen, the turnover number for thiol- and selenolsubtilisin are roughly the same magnitude, but are 40 and 60 times smaller than that of the native enzyme. Moreover, burst kinetics were observed for subtilisin and thiolsubtilisin, but not for the selenoenzyme. These results suggest that deacylation is the rate-limiting step for the unmodified enzyme and thiolsubtilisin with this substrate, while acylation limits the rate for selenolsubtilisin. Previous work showed that the acylation rate of thiolsubtilisin with other substrates is much reduced compared to native subtilisin.[8] Slower acylation of selenol- and thiolsubtilisin is consistent with the lower nucleophilicity toward carbonyl groups of selenols and thiols than alcohol.[9]

We studied deacylation of the three enzymes to examine the consequences of replacing serine with selenocysteine in greater detail. We prepared an authentic acyl-enzyme intermediate by treating reduced selenolsubtilisin with excess cinnamoylimidazole at pH 5. The resulting adduct, Se-cinnamoyl-selenolsubtilisin, (λ_{max} 308 nm) was purified by gel filtration at 4°C. Carlsberg O-cinnamoyl-subtilisin (λ_{max} 289 nm) and S-cinnamoyl-thiolsubtilisin (λ_{max} 310 nm) were prepared as previously described for the BPN'-enzymes.[7] Deacylation of the acyl-enzymes was followed spectroscopically at 25°C as a function of pH. In each case the observed rate increased in a sigmoidal fashion as the pH was raised. The data were fitted to a kinetic scheme requiring ionization of a catalytic group with a pK_a of approximately 7. This group is presumably His 64, which is known to be essential in the wild type enzyme. While the limiting rate constants for thiol- and selenolsubtilisin (0.08 and 0.04 min^{-1}) are within a factor of two, they are more than two orders of magnitude smaller than that of subtilisin (23 min^{-1}). Since the non-enzymatic hydrolysis of structurally analogous esters, thiolesters and selenolesters is usually comparable, these data presumably reflect steric constraints within the active sites of the thiol- and selenolenzymes rather than electronic effects.[10]

ACYL TRANSFER TO AMINES

Although selenolsubtilisin is essentially inactive in cleaving amides, it could still be useful as a ligase for preparing amides from esters and amines. In such a scheme the modified enzyme would be acylated with an activated ester and subsequently transfer its acyl group to a suitable nucleophile. However, efficient amide bond formation will require high selectivity for aminolysis with respect to hydrolysis of the

Table 1. Rate parameters for the hydrolysis of PNPA by subtilisin (O), thiolsubtilisin (S), and selenolsubtilisin (Se). Reactions were carried out in 100 mM phosphate buffer (pH 7.0) containing 5-10% acetonitrile at 25°C.

	O	S	Se
k_{cat}/K_m (min^{-1})	18	0.40	0.32
k_{cat}/K_m (M^{-1} min^{-1})	42,000	12,000	41

acyl-enzyme intermediate (Equation 2). Nonenzymatic aminolysis of selenolesters is considerably faster than that of structurally analogous esters and thiolesters, and it seemed likely that similar selectivity would exist for selenolsubtilisin.[10]

We consequently examined the partitioning of each of the cinnamoyl-enzymes between various amines and water. We found that the ratio of the rate constants for aminolysis and hydrolysis of the acyl-enzymes increases considerably in the order O < S < Se. Acyl transfer to butyl amine versus water for selenolsubtilisin, for example, is 14,000 times more efficient than for subtilisin, and 20 times more efficient than for thiolsubtilisin. These results demonstrate that we have successfully turned a protease into an acyl transferase with high selectivity for amines, simply by converting the active site hydroxyl group into a selenol. Kaiser and coworkers have already exploited the increased selectivity of thiolsubtilisin to catalyze peptide bond formation. Selenolsubtilisin, with its substantially higher selectivity for aminolysis, may indeed be useful as a specific peptide ligase. Practical application of this catalyst to the condensation of peptide segments is currently being pursued.

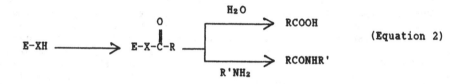

$$\text{(Equation 2)}$$

REDOX CHEMISTRY

The significance of artificial selenoenzymes extends beyond their potential to clarify structure-function relationships in hydrolytic enzymes. The redox and metal binding properties of selenols can be exploited to accomplish interesting chemistry as well. Artificial selenoenzymes may, for example, prove to be interesting analogs of naturally occurring selenoenzymes like glutathione peroxidase. The latter enzyme catalyzes the reduction of hydroperoxides by glutathione *in vivo*. Thus it removes the source of oxygen radicals which have been implicated in the aging process and various diseases, including heart disease and cancer.[12]

$$ROOH + 2R'SH \longrightarrow ROH + R'SSR' + H_2O \qquad \text{(Equation 3)}$$

Selenolsubtilisin, like glutathione peroxidase, is an efficient catalyst for reducing hydroperoxides in the presence of thiols (Equation 3). The rates for the reactions of hydroperoxides with thiols were determined spectroscopically or by HPLC in the presence of selenolsubtilisin at 25°C in aqueous buffer. Rate enhancements over background were observed for the reduction of several hydroperoxides like hydrogen peroxide, *t*-butyl hydroperoxide, and cumene hydroperoxide. The substrate specificity of the enzyme has not yet been optimized, but a number of thiols, such as glutathione, 4-nitro-benzenethiol, 3-carboxy-4-nitro-benzenethiol [1] and dihydrolipoamide, can be used as reductants. While glutathione itself was a relatively poor substrate (presumably because of poor binding to the catalyst), reaction of dihydrolipoamide (0.3 mM) with cumene hydroperoxide (1 mM) in the presence of selenolsubtilisin (6 μM) was more than two orders of magnitude faster than the background rate at 25°C and pH

5.5. The enzymatic reaction went to completion indicating multiturnover catalysis. The catalytic reaction is inhibited by alkylating the reduced enzyme with iodacetamide, demonstrating that the selenocysteine residue is catalytically essential. In preliminary experiments, we found that the redox activity of the selenolenzyme increases with decreasing pH, suggesting the possible participation of a protonated active site residue (perhaps His 64) in catalysis.

$$\frac{(E)_T}{V} = \phi_0 + \frac{\phi_1}{(R'SH)} + \frac{\phi_2}{(R'SH)^2} \qquad \text{(Equation 4)}$$

We studied the reduction of t-butyl hydroperoxide with [1] in some detail. This reaction was chosen because it is convenient to follow spectroscopically. The initial rates of the reaction (V_0) were determined at 25°C and pH 5.5 by monitoring the disappearance of thiol at 410-510 nm. To determine the steady state parameters, kinetic experiments were carried out by varying the concentration of one substrate at fixed concentrations of the other. Saturation kinetics were observed with [1] at several peroxide concentrations, and the kinetics data were fitted to equation 4.[13] The values of ϕ_0, ϕ_1 and ϕ_2 were determined to be 0.84 min, 4.3 x 10⁻⁵ M min and 2.4 x 10⁻¹⁰ M² min, respectively, at 0.25 mM t-butyl hydroperoxide.

The reaction rates as a function of hydroperoxide concentration in the presence of reduced selenosubtilisin can be described by the Michaelis-Menten equation. The apparent k_{cat} anf K_m values at 60 μM t-butyl hydroperoxide are 430 min⁻¹ and 160 mM, respectively.[14] Lineweaver-Burk plots of the kinetic data at several thiol concentrations gave parallel lines, indicating the involvement of covalent intermediates in the reaction. A possible kinetic mechanism consistent with all the data is shown in Equation 5.

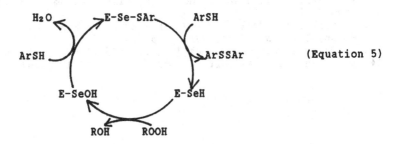

(Equation 5)

To assess the effect of protein binding on catalysis, we compared the initial rates of [1] oxidation by t-butylhydroperoxide in the presence of selenolsubtilisin and diphenyldiselenide, a nonenzymatic model compound. At low substrate concentrations the enzymatic reaction is roughly 70,000 times faster than the selenocystein catalyzed process.[14] Clearly, the redox activity of the selenium group is considerably increased in the active site of subtilisin. Further characterization of the redox chemistry of selenolsubtilisin is in progress. We believe that the information gained in this study will provide a better understanding of how natural glutathione peroxidase works.

CONCLUSION

Chemical modification of existing enzyme active sites provides a powerful strategy for creating entirely new enzymatic activities for use in research, industry and medicine. As described in this article we have successfully used this approach to prepare a semi-synthetic selenoenzyme with diverse chemical activities. Ongoing structural and chemical studies of selenolsubtilisin are likely to provide additional insight into the mechanism of molecular recognition and catalysis. We are currently applying this methodology to other active sites and to the development of catalysts for other transformations of interest.

REFERENCES

1. G. M. Whitesides and C. H. Wong, Angew. Chem. Int. Ed. Eng., 2, 617 (1985).
2. D. Hilvert and E. T. Kaiser, Biotech. Gen. Eng. Rev. Biochem., 5, 297 (1987); E. T. Kaiser, D. S. Lawrence and S. E. Rokita, Ann. Rev. Biochem. 54, 565 (1985).
3. K. C. Nicolaou and N. A. Petasis, "*Selenium in Natural Products Synthesis*," CIS, Inc., Philadelphia, PA, 1984.
4. J. Kraut, in: "*The Enzymes*," 3rd. ed., P. D. Boyer, Ed., Academic Press, New York, Vol. III, p 547, 1971.
5. F. S. Markland, Jr. and E. Smith, in: "*The Enzymes*," 3 rd ed., P. D. Boyer Ed., Academic Press, New York, Vol III, p 561, 1971.
6. Z. P. Wu and D. Hilvert, J. Am. Chem. Soc., 111, 4513 (1989).
7. K. E. Neet, A. Nanci, and D. E. Koshland, J. Biol. Chem., 243, 6392 (1968) L. Polgar and M. L. Bender, Biochemistry, 6, 610 (1967).
8. M. Philipp and M. L. Bender, Mol. Cell. Biochem., 51, 5 (1983).
9. J. O. Edward and R. G. Pearsond, J. Am. Chem. Soc., 84, 16 (1962); R. G. Pearson and J. Songstad, J. Am. Chem. Soc., 89, 1827 (1967).
10. S.-H. Chu and H. G. Mautner, J. Org. Chem., 31, 308 (1966).
11. T Nakatsuka, T. Sasaki, and E. T. Kaiser, J. Am. Chem. Soc., 109, 3808 (1987).
12. L. Floke, in: "*Free Radicals in Biology*," W. A. Pryor, Ed., Academic Press, Inc., New York, Vol. V, p 223, 1982.
13. K. Dalziel, Acta Chem. Scand., 11, 1706 (1957).
14. Z.-P. Wu and D. Hilvert, J. Am. Chem. Soc., 112, 5647 (1990).

SPACER EFFECTS ON ENZYMATIC ACTIVITY IMMOBILIZED ONTO POLYMERIC SUBSTRATES

Toshio Hayashi and Yoshito Ikada

Research Center for Biomedical Engineering
Kyoto University
Sakyo-ku, Kyoto 606, Japan

Papain, chymotrypsin, and lipoprotein lipase were covalently immobilized onto the surface of polymeric beads with spacers of different lengths. The enzymes immobilized with spacer gave an almost constant activity in marked contrast with the free enzymes whose activity monotonously decreased with the decreasing surface concentration. The activity of the immobilized enzymes for hydrolysis of a high molecular weight substrate greatly depended on the length of the spacer. The pH, thermal, and storage stabilities of the immobilized enzymes were higher than those of the free ones and the enzymes immobilized directly to the bead surfaces without any spacer gave the higher stability than those immobilized with spacer. The spacer effect on the activity could be explained in terms of the mobility of the immobilized enzymes.

INTRODUCTION

The new biotechnology consists of two techniques, breeding and processing. The former technique involves gene recombination and cell fusion, while large-scale cell culture and the bioreactor are the most important in the latter processing technique. The fundamentals of these techniques have mostly been well established during the past decade. Although poly(ethylene glycol) is essential in the cell fusion and polymeric substrates are often used for the large-scale cell culture, it will be probably in the bioreactor system that polymers become the most important among the biotechniques in the near future.

The bioreactor is generally composed of biologically active proteins and their bioinert carriers. A large number of works have been devoted to the polymeric carriers, especially to immobilization of the proteins to the carriers,[1] but it seems that most of the works have been carried out by biologists, biochemists, processing engineers, and biomedical engineers, but not by polymer chemists. Therefore, we have initiated a series of investigations on protein immobilization to polymer carriers, focusing on the change of biological activities of proteins by immobilization using different chemical methods. The accumulated knowledge will contribute to development of not only bioreactors, but also immunoaffinity

Biotechnology and Polymers, Edited by C.G. Gebelein
Plenum Press, New York, 1991

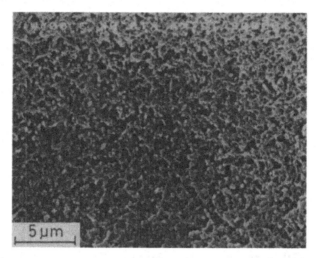

Figure 1. SEM of the surface of porous chitosan bead.

adsorbents and biosensors with higher efficacy.

Here we report the effects of spacers on the activity of enzymes covalently immobilized to polymeric beads. It will be described that the insertion of spacers between a protein and a carrier maintains the immobilized protein in less denatured state with less restricted mobility, leading to higher relative activity, but less stability than immobilization without spacers.

EXPERIMENTAL

1. Carriers and proteins

Polyacrolein microspheres (PAM) (Matsumoto Yushi Inc.)[2] of 0.4 μm diameter and chitosan beads (ChB) (Fuji Spinning Co. Ltd.)[3] of 300 μm diameters were employed as carriers. Aldehyde and amino groups are present on the surface of PAM and ChB, respectively. Figure 1 shows a scanning electron microphotograph (SEM) of the ChB bead. As the proteins to be immobilized, three hydrolytic enzymes were used: papain (3.5 m Anson μg/mg, Merck), chymotrypsin (3x cryst. bovine pancreas, salt free, Sigma), and lipoprotein lipase (LPL) (EC 3.1.1.3 from *P. fluorescens*). N-acetyl-L-tyrosine ethyl ester (ATEE), N-benzylarginine ethyl ester) (BAEE), p-nitrophenyl laurate (pNPL), and other chemicals were purchased from Nacalai tesque, Ltd., Japan.

2. Immobilization

Proteins were covalently fixed on the carrier surface. The spacers used are oligoglycine and oligonylon for PAM and ChB, respectively. The chemical structures are shown below.

PAM Carrier-CH_2-$(NHCH_2CO)_n$-OH NH_2-Enzyme   n = 0,2,3,4,6

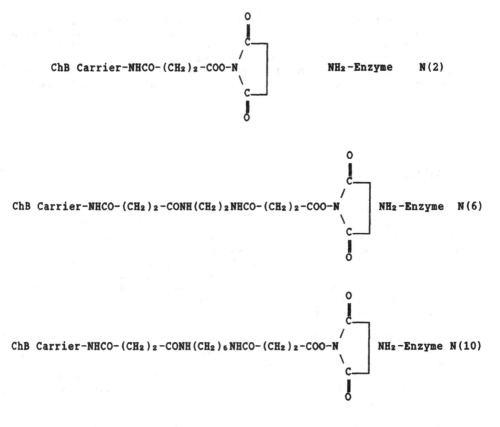

In the case of PAM, oligoglycine was first fixed to the carrier surface through Schiff's base formation, followed by reduction. After that, the enzymes were bound to the carboxyl end group of the spacer by using 1-ethyl-3-(3-dimethylaminopropyl)-carbodiimide, a water-soluble activator. The enzymes were fixed to the ChB through the iminosuccinic anhyride present at the end of the oligonylon spacers.

3. Activity Measures

Free and immobilized LPL were assayed using 0.01M p-nitrophenyl laurate (pNPL) in 0.05M PBS at pH 5.6. The amount of p-nitrophenyl produced was determined by measuring the absorbance of the solution or the supernatant at 400 nm with the Hitachi 101 spectrometer.

The hydrolytic activity of free papain and the immobilized papain was determined using N-benzyl-L-arginine ethyl ester (BAEE) as substrate. After predetermined periods of time, the enzymatic activity was calculated from the initial rate of BAEE hydrolysis by determining KOH consumed within the given period of time.

Free and immobilized chymotrypsin were assayed using 2mM of N-acetyl-L-tyrosine ethyl ester (ATEE) in 0.05M PBS at pH 8.0. The absorbance of the solution or the supernatant at 256 nm was plotted against the enzyme weight in the reaction mixture. The initial slope of the curve was used to evaluate the activity.

The casein hydrolysis was determined essentially according to Berg-meyer,[4] with minor modifications to overcome problems encountered with the insoluble conjugates. The absorbance of the solution or the superna-tant at 280 nm was plotted against the enzyme weight to evaluate the enzymatic activity. The hydrolytic activity of papain was determined in a similar manner, except that the assay medium contained 2mM EDTA and 5mM cystein.

The Michaelis constants, Km and Vm, of LPL immobilized on PAM were estimated employing pNPL solutions ranging from 1.25 to 12.5 mM.

4. Stability Measurements

The thermal stability of the immobilized enzymes was evaluated by measuring the residual activity of the enzyme exposed to various tempera-tures in 0.05M PBS of pH 7.4 for different periods of time. After heat-ing, the samples were quickly cooled and assayed for its enzymatic activ-ity at 37.0°C immediately or after storage at 4°C. The residual acti-vities were expressed as relative to the original assayed at 37.0°C with-out heating. The kinetics of thermal inactivation was investigated by determining the residual activity of the free and the immobilized enzymes after incubation at different temperatures.

The pH stability of the free and the immobilized enzymes was studied by incubating them in PBS including a substrate at 37°C and various pH regions for 20 min. The storage stability of the free and the immobilized enzymes was evaluated by placing them in 0.05M PBS of pH 7.4 at 25°C for various periods of time and the activity was assayed using the above mentioned techniques.

RESULTS AND DISCUSSION

1. Effect of Surface Concentration on the Activity

Figure 2 illustrates the effect of the surface concentration on the relative activity of the LPL immobilized onto the PAM microspheres with and without the oligoglycine spacer. The number of repeated glycine units is three. The different by saturated surface concentrations of immobiliz-ed enzymes were obtained by changing the initial concentration of enzymes in the reaction mixture for immobilization. It is clearly seen that the relative activity of the immobilized LPL without any spacer decreases gradually with the decreasing surface concentration of the immobilized LPL, whereas the relative activities of the immobilized LPL with the oligoglycine spacer are almost constant even at low surface concentra-tions and markedly higher than those without spacer. The effect of the surface concentration on the relative activity is shown in Figure 3 for papain immobilized onto ChB beads using oligonylon spacers at different enzyme concentrations. As can be seen, almost a similar result as given in Figure 2 is obtained for the papain immobilized with the oligonylon spacers.

The lower relative activities of the enzymes immobilized without spacer, compared to those of the enzymes immobilized with spacers, sug-gest that the enzymes undergo greater denaturation when immobilized with-out spacer than with spacers. This may be explained in terms of structur-al deformation of the immobilized enzyme molecules as illustrated in

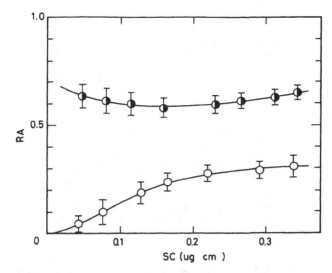

Figure 2. Effect of the surface concentration (SC) of the immobilized LPL on the relative activity (RA) (pNPL, pH 5.60, 37.0°C). (o) PAM-G(0)-LPL, (●) PAM-G(3)-LPL.

Figure 4. The covalently immobilized enzymes without spacer must undergo remarkable deformation in the lower surface concentration region, whereas the enzyme molecules immobilized with spacer may be protected from the heavy structural deformation even in the lower surface concentration region owing to the spacer chain. The increase in the relative activity of the enzyme immobilized without spacer with the increasing surface concentration may be ascribed to a reduced interaction with the substrate, because of high enzyme concentration during the immobilization reaction. In addition, the spacer on the carrier surface probably reduces the steric interference with the substrate binding process, especially toward high molecular weight substrates.

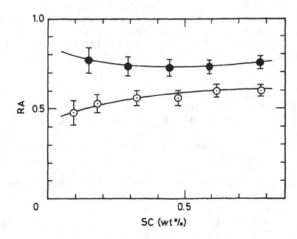

Figure 3. Effect of the surface concentration (SC) of the immobilized papain on the relative activity (RA) (BAEE, pH 8.0, 37.0°C). (●) ChB-N(2)-papain, (●) ChB-N(6)-papain.

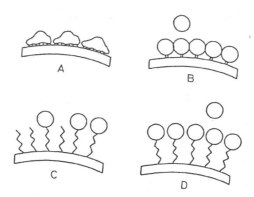

Figure 4. Schematic representation of the molecular state for
the enzymes immobilized on the surface of polymeric
carriers.

(A): sparse immobilization without spacer
(B): dense immobilization without spacer
(C): sparse immobilization with spacer
(D): dense immobilization with spacer

Table 1 summarizes the results of the effect of the spacer length on
the relative activity at almost the same surface concentration of the
immobilized enzymes. It can be seen that even without spacer the immobilized enzymes are still active in hydrolysis toward the low molecular
weight substrates, but practically inactive or less active toward casein,
a high molecular weight substrate. The low activity toward casein will be
explained in terms of the difficult approach of casein to the active site
of the enzymes because of steric hindrance caused by the enzyme immobilization and the large size of the macromolecular substrate. In addition,
it is apparent in Table 1 that the optimum spacer length exists for all
of the immobilized enzymes toward the respective low molecular weight
substrate. The highest activity was obtained with tri- or tetra-glycine
(n; 3-4) as the spacer for PAM microsphere series, as well as ChB-N(6)-
papain for ChB beads series. On the other hand, the enzymatic activity

Table 1. Effect of the length of the spacer (n) on the relative activity (RA) of the immobilized enzymes.

	PAM-G(n)-LPL		PAM-G(n)-papain			PAM-G(n)-chymotrypsin			ChB-N(n)-papain		
n	SC ($\mu g/cm^2$)	RA(%) pNPL	SC ($\mu g/cm^2$)	RA(%) BAEE	Cas	SC (μgcm^2)	RA(%) ATEE	Cas	SC (μgcm^2)	RA(%) BAEE	Cas
0	0.34	33	0.36	41	7	0.32	23	4			
2	0.36	50	0.38	45	15				0.78	60	25
3	0.35	69	0.35	54	25	0.34	38	13			
4	0.36	63	0.40	53	29	0.32	37	18			
6	0.33	55	0.29	48	31	0.30	31	23	0.80	79	37
10									0.77	75	40

toward the high molecular weight substrate monotonously increases as the spacer becomes longer at least in the length range examined here.

2. Determination of Michaelis Constant and Maximum Reaction Velocity

In order to examine whether or not this heterogeneous hydrolysis with the immobilized enzymes obeys the first-order kinetics with respect to the enzyme concentration, hydrolysis was carried out by varying the enzyme concentration over a wide range. Figure 5 shows the observed results on pNPL hydrolysis by the free and the immobilized LPL. Clearly, the slope of the curve is unity, indicating the first-order reaction with respect to the LPL concentration.

Lineweaver-Burk plotting was performed for the immobilized enzymes with different spacers. The Michaelis constant Km and the maximum reaction velocity Vm for the free and the immobilized LPL on PAM microspheres are tabulated in Table 2, together with those of the ChB-N(2)-papain series. The apparent Km of the immobilized LPL without spacer was higher than that of the immobilized LPL with spacer and the free one. This may be because the probability of formation of LPL-substrate complex will decrease for the LPL immobilized without spacer owing to the steric hindrance, compared with those of the enzymes with spacer and the free one. On the other hand, the Vm value of LPL immobilized without spacer gives the lowest value, suggesting that the relative activity of the directly immobilized LPL decreased in the course of the covalent fixation.

3. Thermal Stability of the Immobilized Enzymes

As is well known, the activity of immobilized enzymes, especially covalently bound systems, is more resistant against heat and denaturing

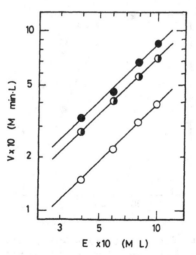

Figure 5. Effect of the LPL concentration (E_0) on the reaction velocity (V) (Hydrolysis: pNPL, 3.0×10^{-3} M, pH 5.60, 37.0°C). (●) free LPL, (◐) PAM-G(3)-LPL, (○) PAM-LPL.

Table 2. Michaelis parameters K_m and V_m at 37.0°C.					
Sample code	Substrate	pH	[E]$_0$ (M/L)	K_m (M/L)	V_m (M/min·L)
LPL	pNPL	5.60	4.0×10^{-7}	2.9×10^{-3}	3.3×10^{-5}
PAM-LPL	pNPL	5.60	4.0×10^{-7}	4.2×10^{-3}	1.5×10^{-5}
PAM-G(3)-LPL	pNPL	5.60	4.0×10^{-7}	3.5×10^{-3}	2.8×10^{-5}
papain	BAEE	8.0	6.0×10^{-7}	2.0×10^{-3}	9.4×10^{-5}
ChB-N(2)-pap	BAEE	8.0	6.0×10^{-7}	1.8×10^{-3}	5.5×10^{-5}

agents than that of the soluble form.[5] Figures 6 and 7 show the effect of temperature on the stability of the immobilized LPL onto PAM microspheres and papain onto ChB beads in PBS, respectively. Figure 6 indicates that the immobilized LPL, with and without spacer, are more stable than the free LPL in the range of high temperatures. The immobilized LPL treated at 55°C for 1 hr exhibits the activity 2 to 3 times higher than that of the free LPL. The higher stability of the immobilized LPL without spacer compared with ones with spacer is probably ascribed to the stabilization of the LPL molecules due to the multipoint attachment of the LPL molecule to the surface of the PAM microsphere when any spacer is not used, leading to reduction in molecular mobility which is greatly related to the enzyme stabilization.[6] A similar behavior is observed for the papain immobilized onto the ChB beads, The immobilized papain treated at 70°C for 1 hr have activities 3 to 4 times higher than that of the free one, as is seen in Figure 7. The papain immobilized onto the ChB-N(6) bead is slightly less stable than that onto the ChB-N(2). This result also suggests that the immobilization of papain with a shorter spacer stabilizes more markedly the papain molecule due to reduction in molecular mobility than with a longer spacer chain.

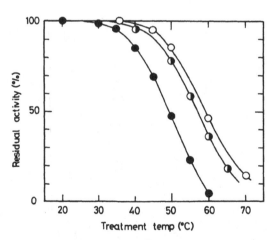

Figure 6. Effect of the heat treatment on the residual activity of the immobilized LPL (heat treatment: pH 5.60, 1 hr; hydrolysis: pNPL pH 5.60, 37.0°C). (●) free LPL, (◉) PAM-G(3)-LPL, (○) PAM-LPL.

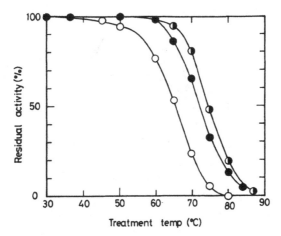

Figure 7. Effect of the heat treatment on the residual activity of the immobilized papain (heat treatment: pH 8.0, 1 hr; hydrolysis: BAEE pH 8.0, 37.0°C). (o) free papain, (●) ChB-N(6)-papain, (◐) ChB-N(2)-papain.

The kinetic curve of thermo-inactivation at 65°C for the LPL immobilized onto PAM microspheres is shown in Figure 8. It is interesting to note that the inactivity of the immobilized enzymes proceeds through a two-stage process characterized by the rate constants; $k_1 = 4.2 \times 10^{-2}$ min^{-1} and $k_2 = 1.3 \times 10^{-2}$ min^{-1}. The free LPL lost 95% of its initial activity by the heat treatment at 65°C for 20 min.

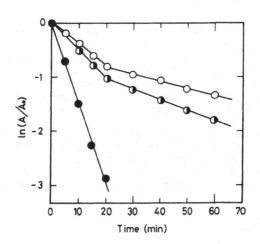

Figure 8. Activity change (A/A₀) of the immobilized LPL by heat treatment at 65°C (pNPL, pH 5.60, 37.0°C). (●) free LPL. (o) PAM-G(3)-LPL, (◐) PAM-LPL.

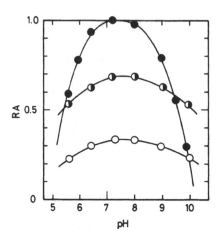

Figure 9   Effect of pH of the reaction medium on the relative
           activity (RA) of the immobilized LPL (pNPL,
           37.0°C). (●) free LPL, (◐) PAM-G(3)-LPL, (o) PAM-
           LPL.

4. Effect of pH on the Activity

The pH effect on the activity of the LPL immobilized onto PAM microsphere was studied for pNPL hydrolysis in PBS at 37°C over a wide pH
region. The results are presented in Figure 9, where it is seen that the
immobilized LPL has the same optimum at pH 7.0 as the free one, but the
pH range where the immobilized enzyme has high activities is considerably
widened, probably due to diffusional limitation of the immobilized enzyme
molecule.[7] Again, the directly immobilized LPL displays a greater stability against pH than the free and the immobilized enzymes with spacer.

5. Stability for Repeated Use

Figure 10 shows the residual activity of the immobilized enzymes when
they are used repeatedly. It can be seen that the activity is retained
without any significant loss for both PAM-G(n)-LPL and ChB-N(n)-papain,
irrespective of the spacer interposition, even if the batch reaction is
repeated 10 times. This high stability compared with that of the free
enzyme is in marked contrast with the rather poor stability of the enzyme
physically adsorbed on a polystyrene microsphere.[8]

The weight of the immobilized enzymes remaining after the last batch
was found to be practically the same as the original one for both cases,
with and without spacer, suggesting no leakage of the immobilized enzyme
during the repeated uses. This gives an evidence for the covalent fixation of the enzyme molecules onto the polymeric carrier surface.

6. Storage Stability

The immobilized papain could be stored in aqueous suspensions at 4°C

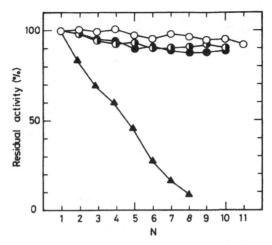

Figure 10. Effect of the number of repeated uses (N) on the residual activity of the immobilized enzymes (37.0°C). (o) PAM-LPL (pNPL, pH 5.60), (●) PAM-G(3)-LPL (pNPL, pH 5.60), (▲) polystyrene/LPL (pNPL, pH 5.60), (●) ChB-N(6)-papain (BAEE, pH 8.0).

for 6 months without a significant loss of activity, whereas the corresponding free papain lost more than 30% of the initial activity under the same conditions. The higher stability of the immobilized papain can be attributed to the prevention of autodigestion and the thermal denaturation owing to the fixation of papain molecules on the surface of ChB beads. However, it is often pointed out that lyophilization, by loss of enzymes directly from the water suspensions, is normally accompanied by loss of the enzymatic activity. Therefore, the enzymatic activity retained after lyophilization was determined for the immobilized and free enzymes. Very high residual activities were observed for the immobilized papain for BAEE hydrolysis; 90% for ChB-N(2)-papain, 84% for ChB-N(6)-papain, and only 70% for the free papain. There is a similarity among the thermal, the storage, the lyophilization stabilities. All of these findings can be explained in terms of the state of the covalent fixation between the polymeric carrier and the enzyme molecules. It is reported that hydrophilic carriers such as Sephadex, Sepharose, and polyacrylamide yield enzyme derivatives with high lyophilization and thermal stabilities.[9,10] The ChB bead belongs to the hydrophilic carrier.

To examine the enzymatic stability in a continuous reaction system under a rather drastic condition, the influence of the storage in PBS of pH 7.4 at 37°C was studied for the immobilized papain. The residual activity at BAEE hydrolysis is given in Figure 11. Apparently, the immobilized papain is much more stable than the free. Again, the immobilized papain with a shorter spacer shows a more stable activity than that with a longer spacer in spite of the initial lower activity.

Similar behaviors were observed for the LPL and chymotrypsin immobilized onto the PAM microspheres with and without spacer.

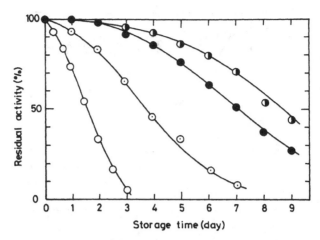

Figure 11. Effect of the storage time on the residual activity of the immobilized papain (storage: PBS, pH 8.0, 37.0°C; hydrolysis: casein, pH 8.0, 37.0°C). (o) free papain, (●) ChB/papain, (◐) ChB-N(10)-papain, (◑) ChB-N(2)-papain.

REFERENCES

1. E. K. Katchalski in: "*Insolubilized Enzymes*," M. Salmona, Eds., Plenum Press, New York, 1982, p. 12.
2. T. Hayashi and Y. Ikada, Biotech. Bioeng., **35**, 518 (1990).
3. T. Hayashi and Y. Ikada, Polymer Preprints, Japan, **38**, 2954 (1989).
4. H. Bergmeyer, Academic Press, New York and London, (1963).
5. R. Ulbrich, A. Schellenberger, and W. Damerau, Biotech. Bioeng., **28**, 511 (1986).
6. Y. Kulis and B. S. Kurtinaitene, Biokhim. Zh., **43**, 453 (1978).
7. P. F. Greenfield and R. L. Laurence, J. Food Sci., **40**, 906 (1975).
8. T. Hayashi and Y. Ikada, unpublished data.
9. L. Goldstein, Biochem. Biophys. Acta., **315**, 1 (1973).
10. D. Gebel, I. Z. Steinberg, and E. Katchalski, Biochem., **10**, 4661 (1971).

CONTRIBUTORS

David E. Albert,   301
Anatrace, Inc.,
1280 Dussel Drive,
Maumee, OH 43537

Lois Aldwin,   295
Cetus Corporation,
1400 53rd Street,
Emeryville, CA   94608

Ali Al-Hakim,   155
Division of Medicinal and Natural Products Chemistry,
College of Pharmacy,
University of Iowa,
Iowa City, IA   52242

Simon J. Archibald,   281
Colla-Tec, Inc.,
Plainsboro, NJ   08536

L. W. Barrett,   95
Materials Research Center,
Center for Polymer Science and Engineering,
Whitaker Laboratory #5, Lehigh University,
Bethlehem, PA 18015

Himangshu R. Bhattacharjee,   245
Allied-Signal, Inc.,
Biotechnology Department,
101 Columbia Road,
Morristown, NJ   07962-1057

C. Brucato,   53
University of Lowell,
Department of Chemistry,
Lowell, MA   01854

V. M. Cabalda,   119
School of Chemistry,
The University of Birmingham,
P. O. Box 363,
Birmingham B15 2TT, England

Charles E. Carraher, Jr.,   95, 111, 147
Department of Chemistry,
Florida Atlantic University,
Boca Raton, FL 33431

A. S. Chang,   275
Fibers and Polymers Laboratory,
Massachusetts Institute of Technology,
Cambridge, MA 02139

P. Dave,   53
University of Lowell,
Department of Plastics Engineering,
Lowell, MA 01854

Stoil Dirlikov,   79
Coatings Research Institute,
Eastern Michigan University,
Ypsilanti, MI 48197

Mary B. Douglas,   301
Anatrace, Inc.,
1280 Dussel Drive,
Maumee, OH 43537

Isabelle Frischinger,   79
Coatings Research Institute,
Eastern Michigan University,
Ypsilanti, MI 48197

Nobuaki Fukui,   181
Department of Agricultural Biochemistry and Biotechnology,
Tottori University,
Tottori 680 Japan

Charles G. Gebelein,   1
Department of Chemistry,
Youngstown State University,
Youngstown, OH 44555

Malay Ghosh,   63
Schering-Plough Corp.,
2000 Galloping Hill Road,
Kenilworth, NJ 07033

Steven A. Giannos,   69
University of Lowell,
Department of Chemistry,
Lowell, MA 01854

Ina Goldberg,   245
Allied-Signal, Inc.,
Biotechnology Department,
101 Columbia Road,
Morristown, NJ 07962-1057

Richard A. Gross,   53, 69
University of Lowell,
Department of Chemistry,
Lowell, MA   01854

R. D. Harris,   265
Laboratory of Molecular Biophysics,
School of Medicine,
The University of Alabama at Birmingham,
P. O. Box 300/University Station,
Birmingham, Alabama 35294

Toshio Hayashi,   321
Research Center for Biomedical Engineering,
Kyoto University,
Sakyo-ku, Kyoto 606, Japan

Donald Hilvert,   315
Departments of Chemistry and Molecular Biology,
Research Institute of Scripps Clinic,
10666 North Torrey Pines Road,
La Jolla, California 92037

Marcia A. Hintz,   301
Anatrace, Inc.,
1280 Dussel Drive,
Maumee, OH 43537

Shigehiro Hirano,   181
Department of Agricultural Biochemistry and Biotechnology,
Tottori University,
Tottori 680 Japan

Toshiro Iijima,   215
Dept. of Textiles,
Jissen Women's University,
Oosakaue, Hino 191, Japan.

Yoshito Ikada,   321
Research Center for Biomedical Engineering,
Kyoto University,
Sakyo-ku, Kyoto 606, Japan

Yoshiaki Inaki,   31
Department of Applied Fine Chemistry,
Faculty of Engineering,
Osaka University,
Suita, Osaka 565, Japan

M. Safiqul Islam,   79
Coatings Research Institute,
Eastern Michigan University,
Ypsilanti, MI 48197

Mamoru Iwata,   181
Department of Agricultural Biochemistry and Biotechnology,
Tottori University,
Tottori 680 Japan

J. Jaggard,   265
Laboratory of Molecular Biophysics,
School of Medicine,
The University of Alabama at Birmingham,
P. O. Box 300/University Station,
Birmingham, Alabama 35294

K. Jumel,   119
Chembiotech, Ltd.
Institute of Research and Development,
Vincent Drive,
Edgebaskon, Birmingham B15 2SQ, England

David Kaplan,   69
Science and Advanced Technology Directorate,
U. S. Army Natick Research,
Development and Engineering Center,
Natick, MA   01760-5020

J. F. Kennedy,   119
Research Laboratory for the Chemistry of
Bioactive Carbohydrates and Proteins,
School of Chemistry,
The University of Birmingham,
P. O. Box 363,
Birmingham B15 2TT, England

Melvin H. Keyes,   301
Anatrace, Inc.,
1280 Dussel Drive,
Maumee, OH 43537

Yasuo Kikuchi,   189
Marine Science Laboratory,
Faculty of Engineering,
Oita University,
Dannoharu, Oita 870-11, Japan

Kazukiyo Kobayashi,   167
Faculty of Agriculture,
Nagoya University,
Chikusa, Nagoya 464-01, Japan

Christian Krarup,   275, 281
Neurophysiology Laboratory,
Harvard Medical School,
Brigham and Women's Hospital,
Boston, MA   02115

Naoji Kubota,   189
Marine Science Laboratory,
Faculty of Engineering,
Oita University,
Dannoharu, Oita 870-11, Japan

T. J. Lepkowski,   79
Paint Research Associates,
430 W. Forest Avenue,
Ypsilanti, MI 48197

Hilton Levy,   11
Office of the Scientific Director,
NIAID,
Bethesda, Md.

Shu-Tung Li,   281
Colla-Tec, Inc.,
Plainsboro, NJ 08536

Robert J. Linhardt,   155
Division of Medicinal and Natural Products Chemistry,
College of Pharmacy,
University of Iowa,
Iowa City, IA 52242

Jian Liu,   155
Division of Medicinal and Natural Products Chemistry,
College of Pharmacy,
University of Iowa,
Iowa City, IA 52242

J. William Louda,   111
Department of Chemistry,
Florida Atlantic University,
Boca Raton, FL 33431

Roger D. Madison,   281
Colla-Tec, Inc.,
Plainsboro, NJ 08536

J. A. Manson,   95
Department of Chemical Engineering,
Department of Materials Science and Engineering,
Whitaker Laboratory #5, Lehigh University,
Bethlehem, PA 18015

R. H. Marchessault,   47
McGill University,
Chemistry Department,
3420 University St.,
Montreal, Canada H3A 2A7

Divakar Masilamani,   245
Allied-Signal, Inc.,
Biotechnology Department,
101 Columbia Road,
Morristown, NJ 07962-1057

Jean M. Mayer,   69
Science and Advanced Technology Directorate,
U. S. Army Natick Research,
Development and Engineering Center,
Natick, MA 01760-5020

S. P. McCarthy,   53
University of Lowell,
Department of Plastics Engineering,
Lowell, MA 01854

E. H. Melo,   119
School of Chemistry,
The University of Birmingham,
P. O. Box 363,
Birmingham B15 2TT, England

Eiko Mochizuki,   31
Department of Applied Fine Chemistry,
Faculty of Engineering,
Osaka University,
Suita, Osaka 565, Japan

C. J. Monasterios,   47
McGill University,
Chemistry Department,
3420 University St.,
Montreal, Canada H3A 2A7

V. J. Morris,   135
AFRC Institute of Food Research,
Norwich Laboratory,
Colney Lane, Norwich NR4 7UA, U.K.

Yoshinobu Naoshima,   147
Department of Biological Chemistry,
Okayama University of Science,
Ridai-cho, Okayama 700, Japan

Danute E. Nitecki,   295
Cetus Corporation,
1400 53rd Street,
Emeryville, CA 94608

T. V. Norregaard,   275
The Deaconess Hospital,
Boston, MA 02214

Mary A. Oleksiuk,   245
Allied-Signal, Inc.,
Biotechnology Department,
101 Columbia Road,
Morristown, NJ 07962-1057

T. Parker,   265
Laboratory of Molecular Biophysics,
School of Medicine,
The University of Alabama at Birmingham,
P. O. Box 300/University Station,
Birmingham, Alabama 35294

S. Perutz,   275
Fibers and Polymers Laboratory,
Massachusetts Institute of Technology,
Cambridge, MA 02139

Deborah A. Piascik,   245
Allied-Signal, Inc.,
Biotechnology Department,
101 Columbia Road,
Morristown, NJ 07962-1057

K. U. Prasad,   265
Laboratory of Molecular Biophysics,
School of Medicine,
The University of Alabama at Birmingham,
P. O. Box 300/University Station,
Birmingham, Alabama 35294

S. P. Qureshi,   95
Amoco Performance Products,
P. O. Box 409,
Bound Brook, NJ 08805.

Thomas H. Ridgway,   111
Department of Chemistry,
University of Cincinnati,
Cincinnati, Ohio 45221

Leszek M. Rzepecki,   229
College of Marine Studies,
University of Delaware,
Lewes, DE 19958

Andres Salazar,   11
Department of Neurology,
Armed Service University,
Bethesda, Md.

Anthony J. Salerno,   245
Allied-Signal, Inc.,
Biotechnology Department,
101 Columbia Road,
Morristown, NJ   07962-1057

Toshihiro Seo,   215
Department of Polymer Science,
Tokyo Institute of Technology,
Ookayama, Meguro-ku, Tokyo 152, Japan

Devang Shah,   69
University of Lowell,
Department of Chemistry,
Lowell, MA   01854

L. H. Sperling,   95
Materials Research Center,
Center for Polymer Science and Engineering,
Department of Chemical Engineering,
Department of Materials Science and Engineering,
Whitaker Laboratory #5, Lehigh University,
Bethlehem, PA 18015

Dorothy C. Sterling,   111
Department of Chemistry,
Florida Atlantic University,
Boca Raton, FL 33431

Kiichi Takemoto,   31
Department of Applied Fine Chemistry,
Faculty of Engineering,
Osaka University,
Suita, Osaka 565, Japan

339

Peter D. Unger,   245
Allied-Signal, Inc.,
Biotechnology Department,
101 Columbia Road,
Morristown, NJ 07962-1057

D. W. Urry,   265
Laboratory of Molecular Biophysics,
School of Medicine,
The University of Alabama at Birmingham,
P. O. Box 300/University Station,
Birmingham, Alabama 35294

Takehiko Wada,   31
Department of Applied Fine Chemistry,
Faculty of Engineering,
Osaka University,
Suita, Osaka 565, Japan

J. Herbert Waite,   229
College of Marine Studies,
University of Delaware,
Lewes, DE 19958

S. Wong,   53
University of Lowell,
Department of Chemistry,
Lowell, MA 01854

Zhen-Ping Wu,   315
Departments of Chemistry and Molecular Biology,
Research Institute of Scripps Clinic,
10666 North Torrey Pines Road,
La Jolla, California 92037

Ryuji Yamaguchi,   181
Department of Agricultural Biochemistry and Biotechnology,
Tottori University,
Tottori 680 Japan

I. V. Yannas,   275
Fibers and Polymers Laboratory,
Massachusetts Institute of Technology,
Cambridge, MA 02139

Christopher S. Youngen,   301
Anatrace, Inc.,
1280 Dussel Drive,
Maumee, OH 43537

N. T. Zervas,   275
Department of Neurosurgery,
Massachusetts General Hospital,
Boston, MA 02114

Printed in the United States
by Baker & Taylor Publisher Services